Embedded Software: The Works

Embedded Software: The Works

Second Edition

Colin Walls

AMSTERDAM ● BOSTON ● HEIDELBERG ● LONDON
NEW YORK ● OXFORD ● PARIS ● SAN DIEGO
SAN FRANCISCO ● SINGAPORE ● SYDNEY ● TOKYO

Newnes is an imprint of Elsevier

Newnes is an imprint of Elsevier
The Boulevard, Langford Lane, Kidlington, Oxford OX5 1GB, UK
225 Wyman Street, Waltham, MA 02451, USA

First edition 2006
Second edition 2012

Notice
No responsibility is assumed by the publisher for any injury and/or damage to persons or
property as a matter of products liability, negligence or otherwise, or from any use or operation
of any methods, products, instructions or ideas contained in the material herein. Because of rapid
advances in the medical sciences, in particular, independent verification of diagnoses and drug
dosages should be made

British Library Cataloguing-in-Publication Data
A catalogue record for this book is available from the British Library

Library of Congress Cataloging-in-Publication Data
A catalog record for this book is available from the Library of Congress

ISBN: 978-0-12-415822-1

For information on all Newnes publications
visit our web site at books.elsevier.com

Printed and bound in the US

12 13 14 15 16   10 9 8 7 6 5 4 3 2 1

Working together to grow
libraries in developing countries

www.elsevier.com | www.bookaid.org | www.sabre.org

ELSEVIER   BOOK AID International   Sabre Foundation

Dedication

To Blood Donors Everywhere

Your generosity saves lives every day.

Thank you.

Contents

Foreword

What Do You Expect—Perfection?

A few words from Jack Ganssle …

Why are so many firmware projects so late and so bug-ridden? A lot of theories abound about software's complexity and other contributing factors, but I believe the proximate cause is that coding is not something suited to *Homo Sapiens*. It requires a level of accuracy that is truly super-human. And most of us are not super-human.

Cavemen did not have to get every gazelle they hunted—just enough to keep from starving. Farmers never expect the entire bag of seeds to sprout; a certain wastage is implicit and accepted. Any merchant providing a service expects to delight most, but not all, customers.

The kid who brings home straight A grades thrills his parents. Yet we get an A for being 90% correct. Perfection isn't required. Most endeavors in life succeed if we score an A, if we miss our mark by 10% or less.

Except in software, 90% correct is an utter disaster, resulting in an unusable product. 99.9% correct means we're shipping junk. One-hundred K lines of code with 99.9% accuracy suggests some 100 lurking errors. That's not good enough. Software requires near perfection, which defies the nature of intrinsically error-prone people.

Software is also highly entropic. Anyone can write a perfect 100 line-of-code system, but, as the size soars, perfection, or near perfection, requires ever-increasing investments of energy. It's as if the bits are wandering cattle trying to bust free from the corral; coding cowboys work harder and harder to avoid strays as the size of the herd grows.

So what's the solution? Is there an answer? How good does firmware have to be? How good can it be? Is our search for perfection or near perfection an exercise in futility?

Complex systems are a new thing in this world. Many of us remember the early transistor radios which sported half a dozen active devices, max. Vacuum tube televisions, common in the 1970s, used 15–20 tubes, more or less equivalent to about the same number of transistors. The 1940s-era ENIAC computer required 18,000 tubes—so many that technicians wheeled shopping carts of spares through the room, constantly replacing those that burned out. Though that sounds like a lot of active elements, even the 25-year-old Z80 chip used a quarter of that many transistors, in a die smaller than just one of the hundreds of thousands of resistors in the ENIAC.

Now the Pentium IV, merely one component of a computer, has 45 million transistors. A big memory chip might require one-third of a billion. Intel predicts that later this decade, their processors will have a billion transistors. The very simplest of embedded systems, like an electronic greeting card, requires thousands of active elements.

Software has grown even faster, especially in embedded applications. In 1975, 10,000 lines of assembly code was considered huge. Given the development tools of the day—paper tape, cassettes for mass storage, and crude teletypes for consoles—working on projects of this size was very difficult. Today 10,000 lines of C, representing perhaps three to five times as much assembly, is a small program. A cell phone might contain 5 million lines of C or C++, astonishing considering the device's small form factor, miniscule power requirements, and breathtakingly short development times.

Another measure of software size is memory usage. The 256 byte [that's not a typo] EPROMs of 1975 meant even a measly 4k program used 16 devices. Clearly, even small embedded systems were quite pricey. Today? 128k of Flash is nothing, even for a tiny app. The switch from 8 to 16 bit processors, and then from 16 to 32 bitters, is driven more by addressing space requirements than raw horsepower.

In the late 1970s, Seagate introduced the first small Winchester hard disk, a 5 Mb 10-pound beauty that cost $1500. Five MB was more disk space than almost anyone needed. Now 20 Gb fits into a shirt pocket, is almost free, and fills in the blink of an eye.

So, our systems are growing rapidly in both size and complexity. Are we smart enough to build these huge applications correctly? It's hard to make even a simple application perfect; big ones will possibly never be faultless. As the software grows, it inevitably becomes more intertwined; a change in one area impacts other sections, often profoundly. Sometimes this is due to poor design; often, it's a necessary effect of system growth.

The hardware, too, is certainly a long way from perfect. Even mature processors usually come with an errata sheet, one that can rival the datasheet in size. The infamous Pentium divide bug was just one of many bugs. Even today, the Pentium 3's errata sheet [renamed "specification update"] contains 83 issues. Motorola documents nearly a hundred problems in the MPC555.

What is the current state of the reliability of embedded systems? No one knows. It's an area devoid of research. Yet a lot of raw data is available, some of which suggests we're not doing well.

The Mars Pathfinder mission succeeded beyond anyone's dreams, despite a significant error that crashed the software during the lander's descent. A priority inversion problem—noticed on Earth but attributed to a glitch and ignored—caused numerous crashes. A well-designed watchdog timer recovery strategy saved the mission. This was a very instructive failure as it shows the importance of adding external hardware and/or software to deal with unanticipated software errors.

It's clear that the hardware and software are growing in raw size as well as complexity. Pathfinder's priority inversion problem was unheard of in the early days of microprocessors, when the applications just didn't need an RTOS. Today most embedded systems have some sort of operating system, and so are at peril for these sorts of problems.

What are we as developers to do, to cope with the explosion of complexity? Clearly the first step is a lifelong devotion to learning new things. And learning old things. And even re-learning old things we've forgotten. To that end we simply must curl up in the evenings with books such as this, this compendium of the old and the new, from the essentials of C programming to UML and more. Colin is an old hand, who here shares his experience with plenty of practical "how to do this" advice.

We do learn well from experience. But that's the most expensive way to accumulate knowledge. Much better is to take the wisdom of a virtuoso, to, at least for the time it takes to read this book, apprentice oneself to a master.

You'll come away enriched, with new insights and abilities to solve more complex problems with a wider array of tools.

Perfection may elude us, but wise developers will spend their entire careers engaged in the search.

Preface to the First Edition

How This Book Came About

I wonder how many people formulate a plan for their lives and then follow through with it. Some do, I'm sure, but for most of us, even if we have ambitions and ideas for the future, we are often buffeted by the random events around us. And that is how it was for me. I work for Accelerated Technology, a division of Mentor Graphics, which is dedicated to providing embedded software design and development tools and operating systems. I have done a variety of jobs within the company over the years, throughout a number of acquisitions and name changes. Of late, I have been doing outbound marketing—traveling around Europe and North America getting excited about our products and the technology behind them. I often describe my job as "professional enthusiasm." It is a great job, and I work with great people. I enjoy it. But it is also very taxing, with a lot of time spent on the road (or in the air) or just waiting for something to happen. I tend to be away from home quite a lot, but I have always accepted that this price is one I have to pay for such a rewarding job. And then it all changed. …

In mid-summer 2004, my wife was suddenly diagnosed with acute myeloid leukemia and immediately admitted to hospital. This meant that my working life was suddenly put on hold. I needed to be home to support her and care for our two teenage daughters. Her treatment— aggressive chemotherapy—progressed through the rest of the year. There have been many ups and downs, but the outcome, at the time of writing at least, is that she is in remission, having monthly blood checks to ensure that this progress continues.

It is under these circumstances that you find out who your friends are—an old cliché but true nevertheless. In my case, I also found out that I was working for the right company. My management and colleagues could not have been more supportive, and I am indebted to all of them. (I'll name names later.) Clearly I could not continue with my usual job, for a while at least. I was put under no pressure, but I gave careful thought as to what I might do that was useful, but would be compatible with my new lifestyle. I had been harboring the idea of writing this book for a while; I suggested the idea to my management, who readily agreed that it would be a worthwhile project.

Where This Book Came From

It is nearly 20 years since I last wrote a book (*Programming Dedicated Microprocessors*, Macmillan Education, 1986). That book was about embedded software, even though the term was not in common use at that time. Writing that book was a realization of an ambition, and the couple of copies that I still have are among my most treasured possessions. Writing a book is a time-consuming activity. Back when I wrote my first book, my job was relatively undemanding, and we had no children. Finding the time then was not so difficult. I have long since wanted to write another book and have had various ideas for one, but spare time has been in short supply.

Nevertheless, I have still been writing over the years—lots of technical articles, conference papers, lessons for training classes, and presentations. Then I had a thought: why not gather a whole bunch of this writing together in a book? To simplify the process, I concentrated my search for material within Accelerated Technology because the company owns the copyrights. A few other pieces did eventually get donated, but that was later.

At Microtec Research, *NewBits*, a quarterly newsletter, evolved from a simple newsletter into a technical journal, and I was a regular contributor from the early 1990s. *NewBits* continued to be published after the acquisition of Microtec by Mentor Graphics, but the journal was eventually phased out. Recently, after revitalizing the embedded software team by acquiring Accelerated Technology, we decided to revive *NewBits*, and it has gone from strength to strength. Since I collected every issue of *NewBits*, I had all of the research material at my fingertips. I quickly determined that many of my articles and those by several other writers would be appropriate for inclusion in this book. Other sources of material were white papers and another Accelerated Technology newsletter called *Nucleus Reactor*.

I have endeavored to make as few alterations as possible to the articles. Some required almost no changes; others needed a little updating or the removal of product-specific information. Each article includes an introductory paragraph that describes the origins of the article and puts it into context.

What You Will Find Here

In selecting the material for this book, my goal was to ensure that every article was relevant to current embedded software development practice and technology. Of course, many pieces have a historical perspective, but that alone was not sufficient qualification for their inclusion. Quite a few articles were rejected because they covered a technology that did not catch on or has since been superseded by something else. Maybe those topics will appear when I write *A History of Embedded Software*, which I slate for publication on the fiftieth anniversary of the start of it all (in 2020—50 years after the Intel 4004 was announced).

In this book, you will find a selection of articles covering just about every facet of embedded software: its design, development, management, debugging procedures, licensing, and reuse.

Who This Book Is For

If you are interested in embedded software, this book includes something for you. The articles cover a wide range, and hence, it is of interest to newcomers to the field as well as old hands. If your background is in more conventional software, some pieces will give you a perspective on the "close to the hardware" stuff. If your background is in hardware design, you may gain some understanding of "the other side."

If you teach—either in a commercial or academic context—you may find some of the articles provide useful background material for your students. Some of the CD-ROM content is designed with just you in mind. See the section "What's on the CD-ROM?" following this preface.

How to Use This Book

There is no "right" way to use this book. I have attempted to sequence the articles so that reading the book from front to back would make sense. I have also done my best to categorize the articles across the chapters to help you find what might be of particular interest, with a few cross-references where I felt they might be helpful. I am frustrated by inadequate indexes in reference books, so I have tried to make the index in this book an effective means of finding what interests you.

Acknowledgments

It is difficult to write this section without it sounding like an acceptance speech for an Academy Award.

Before naming individuals, I want to make a more general acknowledgment. I have spoken to many colleagues within Accelerated Technology and in the wider world of Mentor Graphics about this book project. The response has been universal encouragement and enthusiasm. This helped.

When I started working with Elsevier, my editor was Carol Lewis, who started the project. In due course, the baton was passed to Tiffany Gasbarrini, who helped me see it through to conclusion. I enjoyed working with both of them, admired their professionalism, and appreciated their guidance.

I have always enjoyed reading the work of Jack Ganssle—his books and regular columns in *Embedded Systems Programming* magazine—who manages to convey useful technical information in a thought-provoking and, more often than not, humorous way. So, when I started planning this book, I sought his advice, which he enthusiastically provided. I was also delighted when he agreed to provide the Foreword. Thanks Jack.

I have a debt of gratitude to my management and colleagues, who have stood by me and gone that extra mile at a difficult time. Apart from (perhaps) maintaining some of my sanity, they were instrumental in making this book possible. To name a few: Neil Henderson, Robert Day, Michelle Hale, Gordon Cameron, Joakim Hedenstedt. There are many others. Thanks guys.

As I have mentioned, a great deal of the material in this book had its origins in the Microtec newsletter *NewBits*. So, I am indebted to all the folks who were involved in its publication over the years. Lucille Woo [née Ching] was the managing editor of *NewBits*, developing it from a simple newsletter into a solid technical journal, with graphics and design work by Gianfranco Paolozzi. At Microtec, various people looked after the journal over the years, including Eugene Castillo, Melanie Gill, and Rob van Blommestein. The "revival" of *NewBits* at Accelerated Technology was led by Charity Mason, who had a hard act to follow but rose to the challenge admirably.

I am allergic to one facet of business: anything with the word "legal" involved. I accept the necessity for all the procedures, but I am grateful that people like Jodi Charter in Mentor Graphics Procurement can take care of the contract processing for me.

I was pleased by the prompt and positive response I got from David Vornholt at Xilinx and Bob Garrett at Altera when I sought solid background material on FPGA processors.

Last, and by no means least, I would like to thank Archie's proud parents, Andrea and Barry Byford, for their permission to use his photographs in the Afterword.

Contributors

I wrote about half the words in this book. The rest were contributed cooperatively or unwittingly by these individuals: Zeeshan Altaf, Fakhir Ansari, Antonio Bigazzi, Sarah Bigazzi, Paul Carrol, Lily Chang, Robert Day, Michael Eager, Michael Fay, Jack Ganssle, Bob Garrett, Kevin George, Ken Greenberg, Donald Grimes, Larry Hardin, Neil Henderson, C.C. Hung, Meador Inge, Stephen Mellor, Glen Johnson, Pravat Lall, Tammy Leino, Nick Lethaby, Steven Lewis, Alasdar Mullarney, Stephen Olsen, Doug Phillips, Uriah Pollock, James Ready, John Schneider, Robin Smith, Dan Schiro, Richard Vlamynck, David Vornholt, Fu-Hwa Wang, John Wolfe.

I would like to thank every one of them for their contributions and apologize in advance if I managed to forget attributing anyone.

It is easy to think of a book as just a bunch of words between two covers. But you will find that many articles here, as in most technical publications, include illustrations that help reinforce the message. I am happy with writing words, but I am no artist. So I am grateful for the help I received from Chase Matthews and Dan Testen, who provided the necessary artistic talent.

A Good Cause

I was delighted when my management generously agreed that all proceeds from this book could be donated to a charitable organization of my choice. Naturally, I thought about my circumstances and decided to support a group that produced tangible results. I chose the LINC Fund (Leukaemia and Intensive Chemotherapy—www.lincfund.co.uk), which is based at the center where my wife was treated.

LINC is a charitable organization dedicated to the support and well-being of patients with leukemia and related conditions. Money collected by LINC is used to purchase equipment and aid research at Cheltenham General Hospital Oncology Unit. Since the onset of these conditions can be very sudden, with patients embarking on lengthy treatment programs with little or no notice, some people find themselves in financial difficulties. The LINC Fund also provides help to such families at a time when they need it most.

On their behalf, thank you for your contribution, which I know will be spent wisely.

Contact Me

If you have comments or questions about this book, or indeed anything to do with embedded software, I would be very pleased to hear from you. Email is the best way to reach me:

```
colin_walls@mentor.com
```

If you wish to reach other contributors, please email me, and I will endeavor to put you in contact with them.

To see the latest information about this book—updates, errata, downloads—please visit www.EmbeddedSoftwareWorks.com.

Colin Walls, 2005

Preface to the Second Edition

Since the first edition of this book was published, the world has moved. Many things have changed, but others have remained constant. The world of embedded software has progressed, with new processors and technologies becoming common, resulting in changes to the world of the embedded software engineer. With this in mind, the folks at Elsevier approached me with a view to updating the book for a new edition and I was delighted to accept.

This book has a selection of new articles in many chapters and two new chapters that look at open source software and multicore. Of course, I removed a few articles that seem less relevant.

I have left the old Preface unchanged, but I need to augment the list of contributors to whom I am endebted: Vlad Buzov, Dan Driscoll, Christopher Hallinan, Waqar Humayan, Geoff Kendall, Russell Klein, John Lehmann, Shabtay Matalon, Mark Mitchell.

The other updated information is about my personal circumstances. Although my wife received superb medical care, without which she would only have lived a few weeks, she died in June 2006. The treatment gave her an extra 2 years, at least some of which was quality time. My daughters and I have got on with our lives and I am still at Mentor Graphics and still very much involved with embedded software.

I was pleased that, again, my employers have agreed that all proceeds from this edition can be donated to LINC. I continue to be confident that the money will be spent wisely.

Colin Walls, 2012

What's on the Website?

To provide as much value as possible, we decided to create a website to provide
supplementary material—http://www.elsevierdirect.com/companions/9780124158221.
Here is a guide to what you can find there.

Code Fragments

Many of the articles include listings of C or C++ code. These are all provided on the website so that you can use them without needless retyping. They have all been carefully checked, so they should build and run as expected. These are provided as plain text files (with the extension .txt) that pertain to relevant articles. You can simply copy and paste the code as you want. There are 23 files:

Chapter 1: Embedded Software
Migrating Your Software to a New Processor Architecture

Chapter 2: Design and Development
Embedded Software and UML

Chapter 3: Programming
Programming for Exotic Memories
Self-Testing in Embedded Systems
A Command-Line Interpreter
Traffic Lights: An Embedded Software Application

Chapter 4: C Language
Interrupt Functions and ANSI Keywords
Optimization for RISC Architectures
Bit by Bit
Programming Floating-Point Applications
Looking at C—A Different Perspective
Structure Layout—Become an Expert

Chapter 5: C++
Why Convert from C to C++?
Clearing the Path to C++

C++ Templates—Benefits and Pitfalls
Exception Handling in C++
Looking at Code Size and Performance with C++
Write-Only Ports in C++
Using Nonvolatile RAM with C++

Chapter 6: Real Time
Programming for Interrupts

Chapter 7: Real-Time Operating Systems
A Debugging Solution for a Custom Real-Time Operating System
Introduction to RTOS Driver Development

Chapter 8: Networking
Who Needs a Web Server?

Training Materials

The subjects of many of the articles lend themselves to training classes or seminars. Indeed, many of my articles originated as pieces for training classes or seminars. Where possible, I have created a slide set to accompany an article (but obviously I was unable to do so for every article).

These files are in pairs. One is a Microsoft PowerPoint (.PPT) file. To use this file effectively, you need a license for PowerPoint. You can then not only present the slides, but modify and customize them. The other file is an Adobe PDF file. To view this file, you need to obtain the free Adobe Reader software. If you simply open the PDF, it will display full screen (there's no need to press CTRL/L), and you can make the presentation (press Esc to finish). Adobe Reader is an effective presentation tool but does not facilitate making changes to the PDF files.

All the slide sets are designed to be coherent sequences, following the theme of the relevant article, but you may also find individual slides are useful for illustrating points in other contexts. There are over 400 slides, in 69 pairs of files:

Chapter 1: Embedded Software
What Makes an Embedded Application Tick?
Memory in Embedded Systems
Memory Architectures
How Software Influences Hardware Design
Migrating Your Software to a New Processor Architecture
Embedded Software for Transportation Applications
How to Choose a CPU for Your System on Chip Design
An Introduction to USB Software
Toward USB 3.0

Roadmap to Embedded Software Development

In the time since embedded systems first emerged, the role of software has changed, along with the tools used and the overall approach to its development.

In the early days, the software was just an afterthought—something that the designer quickly put together when the hardware was complete. It was written in assembly language and implemented by someone who understood the hardware intimately.

Over time, as systems became a little more complex, there was an increasing trend to use a software specialist to write the code for embedded systems. Typically, these guys were quite knowledgeable about electronics and very comfortable working close to the hardware.

The last 10–15 years has seen the emergence and rapid growth of embedded software development teams in response to the explosion in the size and complexity of the software. Such teams are not homogeneous—they have not grown simply to add more manpower to get the code cranked. An embedded software team is composed of numerous specialists—perhaps experts in networking, user interface design, the application of the device itself or traditional embedded software skills, for example. Many engineers are likely to have little understanding of the special nature of embedded programming, as they have been drawn from a desktop programming context.

This book is intended to provide introductory guidance to many kinds of readers, including engineers who have a background primarily in hardware design. However, developers, who are very expert at software on the desktop, who need to understand embedded, deserve particular attention, as this transition is becoming increasingly common. With this in mind, there are two key questions to be answered:

1. How does embedded software really differ from desktop applications?
2. What tools and components are utilized to build embedded software?

Embedded Versus Desktop Software

There are numerous factors that render embedded software different from its desktop equivalent. Not recognizing one or more of these may be a significant factor in bad embedded software design.

Memory Size

Programmers of desktop systems develop code with no regard to its memory footprint. Memory is cheap and most desktop computers are designed to be fitted with a copious amount. If the computer becomes low on memory, desktop operating systems, like Windows, use hard drive space to augment it—swapping data on and off of disk as needed.

Embedded systems always have limited memory. It may not be the tiny amount found on devices in the past, but the memory size will be bounded—adding extra is rarely an option. Excess memory is rarely provided because cost and power consumption demands mean that minimizing the size of memory is desirable. This has implications on the choice of programming language (bad implementations of C++ can be problematic), the way in which development tools are utilized (e.g., compiler optimization) and the selection of operating system.

Articles that appertain to memory issues may be found in Chapters 1, 3, and 4.

CPU Power

Desktop computers routinely use state-of-the-art processors. Many users have the equivalent of a 1990s supercomputer on which they do a little light word processing and Web surfing. But, the CPUs are quite cheap and power consumption and heat dissipation have historically been ignored.

An embedded system, on the other hand, is likely to have a CPU that is only just powerful enough for the required function, as cost and power consumption requirements demand conservative design. This has implications for code and operating system efficiency.

Code Optimization

All modern compilers perform code optimization. For desktop applications, the priority is speed—users are interested in how fast the software can do its job. If that speed comes at the expense of the program size—i.e., it is bigger—that is not regarded as a problem.

Embedded developers need to produce optimized code too, but the priority of speed over size is not so clear-cut. A key feature of an embedded compiler is controllability. Developers need to be able to fine tune the optimization to their precise needs. This topic is addressed by articles in Chapters 4 and 5.

Operating System

A desktop computer has an operating system, which is almost always Windows, Linux, or Mac OS X. The developer is unlikely to have needed to select the OS and there are very large pools of expertise with using all three options.

An embedded system may have an OS. This might even be a special variant of Windows or Linux, but it might equally be one of the numerous real-time operating systems (RTOSes) on the market. The developers may have chosen to implement a proprietary, in-house developed OS. For certain systems there may be no OS at all, which is termed running on "bare metal."

RTOSes are addressed extensively in Chapter 7; Chapters 8 and 9 include further coverage of operating system issues.

Real-Time Behavior

A real-time system is not necessarily fast, but it is predictable—deterministic is the usual term. Desktop systems are rarely real time. They may be fast, but there is no critical requirement for software to respond at a specific time. Many embedded systems are real time, with some control systems having very stringent timing constraints. This requirement has a bearing on choice of OS and on program design. Chapter 6 includes articles that look at issues around real-time system development.

Development Paradigm

Software for desktop computers is normally developed on the same machine on which it will be executed. This has always been the way with computers as software development is performed using software tools.

Embedded systems are different because they are very unlikely to have the resources needed to develop software. Normally, code is developed on a desktop computer (the "host") and later executed on the embedded system itself (the "target"). This has implications in the procurement of tools, which must be specific for the type of target. Also, the process is different, because, although an edit/compile/debug cycle is still followed, the execution phase is more complex, as the code must either be transferred to a target or run under some kind of simulation environment.

The selection of development tools is addressed by articles in Chapters 2 and 9.

Execution Paradigm

Programs of desktop computers are normally user initiated. The operating system is told what to run and this software runs until it either completes its job or is terminated by the user. Programs are read off of disk into memory ready for execution.

Most embedded systems execute specific software from start-up and that software simply continues execution until the device is powered down. Sometimes software is just executed straight out of read-only memory (ROM), but it might be copied from persistent memory (like flash) into RAM for execution.

Every Embedded System Is Different

Maybe the biggest difference between desktop and embedded systems is philosophical. To the first approximation, all desktop systems are the same—or, at least, there are only a few variations. It is possible to write some software and distribute it to a great many people with a very high confidence that they will all be able to execute it with no problems.

On the other hand, every embedded system is different. These differences may be technical—different CPU, memory architecture, peripherals, application priorities, and operating systems. But they are also commercial variations. The business model for the development, production, and sale of cell phone handsets is, for example, quite different from MRI scanners, even though they are both embedded systems.

It is this variability that colors many aspects of embedded software development and makes for interesting challenges. Embedded software engineers may need to learn about multiple CPU architectures, different operating systems, and a selection of development tools. This is all in addition to mastering the key programming skills around real-time systems and working close to the hardware.

Embedded Software Tools and Components

As with desktop software development, embedded software engineers need to make use of tools to build and test their software and are likely to incorporate reusable software components into their design.

Development Tools

Superficially, the embedded code development process looks very similar to that on the desktop: code modules are compiled, linked together along with libraries of reusable components, and executed under the control of a debugger. But this similarity is more apparent than real, as, at every step, there are marked differences in the tools and the way that they are used.

Embedded compilers are actually cross compilers—they run on one machine (the host) and generate code for another (the target). A compiler is likely to need to be set up to generate code for a specific target CPU, as it is common for a whole family to be supported. Fine grain control over optimization is a must. Most embedded compilers generate assembly language output.

An assembler is needed to translate compiler output to machine code, but it is also possible that small parts of an embedded application will be handwritten in assembly.

An embedded linker is a critical tool. Apart from combining multiple object modules together with the extraction of library routines, the linker's job is to locate code and data correctly in memory.

The memory map requirements for an embedded system may be very precise and quite complex, so this tool must have the flexibility to fulfill those requirements.

An embedded debugger is not a single item, but normally a family of tools, which facilitate different execution environments. Code may be run purely on the host computer. This can be achieved by either running the code natively or using an instruction set simulator. Alternatively the debugger might connect to a target—either a real prototype or an evaluation board with a similar architecture—to execute the code there. The connection to the target is commonly JTAG, but other options, like Ethernet, may be available.

Apart from these tools, which will appear somewhat familiar to the desktop programmer, the embedded developer may have other, more specialized tools. Typical options include profilers that show how real time is being utilized and power analyzers that track how device power consumption relates to code execution.

Software Components

In the earliest days of embedded programming, the developer probably wrote every last byte of code that was executed by the device. As with desktop programming, embedded software development has increasingly become an exercise in connecting up components.

The simplest and most obvious reusable software components are libraries. Some libraries are provided with compilers, which include explicitly called standard functions (like **printf**(), for example) and implicit functions, calls to which are made by generated code. The libraries supplied with embedded compilers should have all the right characteristics for this type of application: they should be reentrant and ROMable. But this should be verified.

Other libraries are available, particularly for C++ , which provide the developer with reusable, commonly needed functionality. An example is the Standard Template Library. Such libraries should be used with great care, as they may not be in a suitable form to use for an embedded application.

On a desktop computer, the operating system is given. Embedded systems may or may not use an OS. There is a wide choice of RTOSes available, along with other options like embedded Linux and Windows CE.

A traditionally challenging part of embedded development is interfacing with hardware. A benefit of using an OS is that this code is localized to device drivers. A commercial OS or Linux will have a very wide range of available device drivers for standard hardware available for reuse.

It is increasingly common for embedded systems to be connected. So, networking support in some form is needed and this is often available from OS suppliers. Networking may simply be TCP/IP, which may include an alphabet soup of other protocols. There is also bus connectivity like CAN and I2C, along with other technologies like SNMP, ZigBee, Bluetooth, WiFi, and USB. With all networking software, it is essential to ensure that it is fully validated to ensure standards compliance and guaranteed interoperability with other systems.

Although hard drives are rarely employed in embedded devices, a need for organized data storage is not uncommon. This is normally provided using flash memory, which may be organized to behave like a disk drive. Again, support for files systems is typically available from OS vendors. For embedded applications, data storage support has particular requirements, notably resilience to power failure and support for multithreading.

The plummeting cost of large, color LCD displays, often with touch sensitivity, has resulted in their being incorporated in many embedded designs. The software challenge in supporting sophisticated graphics and/or user interface requirements is not inconsiderable. A number of commercial packages are available that provide for these needs in a form that is optimal for embedded applications.

Although not yet commonplace, it is becoming possible to source a large part of the code for a number of common embedded applications. Even if not ready to use "out of the box," such packages are a framework to which the developer can add value.

Embedded Software

Chapter Outline

This first collection of articles either set the scene, providing a broad view of what embedded software is all about, or address a specific area that is not really encompassed by another chapter.

1.1 What Makes an Embedded Application Tick?

This is very much a "setting the scene" article, based upon one that I wrote for NewBits in late 2003 and a talk that I have delivered at numerous seminars. It introduces many embedded software issues and concepts that are covered in more detail elsewhere in this book.

CW

Embedded systems are everywhere. You cannot get away from them. In the average American household, there are around 40 microprocessors, not counting PCs (which contribute another 5–10 each) or cars (which typically contain a few dozen). And these numbers are predicted to rise by a couple of orders of magnitude over the next decade or two. It is rather ironic that most people outside of the electronics business have no idea what "embedded" actually means.

Marketing people are fond of segmenting markets. The theory is that such segmentation analysis will yield better products by fulfilling the requirements of each segment in a specific way. For embedded, we end up with segments like telecom, mil/aero, process control, consumer, and automotive. Increasingly though, devices come along that do not fit this model.

For example, is a cell phone with a camera a telecom or consumer product? Who cares? An interesting area of consideration is the commonality of such applications. The major comment that we can make about them all is the amount of software in each device is growing out of all recognition. In this article, we will take a look at the inner workings of such software. The application we will use as an example is from the consumer segment—a digital camera— which is a good choice because whether or not you work on consumer devices, you will have some familiarity with their function and operation.

1.1.1 Development Challenges

Consumer applications are characterized by tight time-to-market constraints and extreme cost sensitivity. This leads to some interesting challenges in software development.

Multiple Processors

Embedded system designs that include more than one processor are increasingly common— market research suggests that, before very long, multicore designs will be the norm. A digital camera typically has two CPUs: one deals with image processing and the other looks after the general operation of the camera. The biggest challenge with multiple processors is debugging. The code on each individual device may be debugged—the tools and techniques are well understood. The challenge arises with interactions between the two processors. There is a clear need for debugging technology that addresses the issue of debugging the system—that is, multicore debugging.

Limited Memory

Embedded systems almost always have limited memory. Although the amount of memory may not be small, it typically cannot be added on demand. For a consumer application, a combination of cost and power consumption considerations may result in the quantity of memory also being restricted. Traditionally, embedded software engineers have developed skills in programming in an environment with limited memory availability. Nowadays, resorting to assembly language is rarely a convenient option. A thorough understanding of the efficient use of C and the effects and limitations of optimization are crucial.

If C++ is used (which may be an excellent language choice), the developers need to fully appreciate how the language is implemented. Otherwise, memory and real-time overheads can build up and not really become apparent until too late in the project, when a redesign of the software is not an option. Careful selection of C++ tools, with an emphasis on embedded support, is essential.

User Interface

The user interface (UI) on any device is critically important. Its quality can have a very direct influence on the success of a product. With a consumer product, the influence is overwhelming. If users find that the interface is "clunky" and awkward, their perception of not just the particular device, but also the entire brand will be affected. When it is time to upgrade, the consumer will look elsewhere.

So, getting it right is not optional. But getting it right is easier to say than do. For the most part, the UI is not implemented in hardware. The functions of the various controls on a digital camera, for example, are defined by the software. And there may be many controls, even on a basic model. So, in an ideal world, the development sequence would be:

1. Design the hardware
2. Make the prototypes
3. Implement the software (UI)
4. Try the device with the UI and refine and/or reimplement as necessary.

But we do not live in an ideal world. …

In the real world, the complexity of the software and the time-to-market constraints demand that software is largely completed long before hardware is available. Indeed, much of the work typically needs to be done even before the hardware design is finished. An approach to this dilemma is to use prototyping technology. With modern simulation technology, you can run your code, together with any real-time operating system (RTOS) on your development computer (typically Windows or Linux), and link it to a graphical representation of the UI. This enables developers to interact with the software as if they were holding the device in their hand. This capability makes checking out all the subtle UI interactions a breeze.

1.1.2 Reusable Software

Ask long-serving embedded software engineers what initially attracted them to this field of work and you will get various answers. Commonly though, the idea of being able to *create* something was the appeal. Compared with programming a conventional computer, constrained by the operating system and lots of other software, programming an embedded system seemed like working in an environment where the developer could be in total control. (The author, for one, admits to a megalomaniac streak.)

But things have changed. Applications are now sufficiently large and complex that it is usual for a team of software engineers to be involved. The size of the application means that an individual could never complete the work in time; the complexity means that few engineers would have the broad skill set. With increasingly short times to market, there is a great incentive to reuse existing code, whether from within the company or licensed from outside.

The reuse of designs—of intellectual property in general—is common and well accepted in the hardware design world. For desktop software, it is now the common implementation strategy. Embedded software engineers tend to be conservative and are not early adopters of new ideas, but this tendency needs to change.

Software Components

It is increasingly understood that code reuse is essential. The arguments for licensing software components are compelling, but a review of the possibilities is worthwhile.

We will now take a look at some of the key components that may be licensed and consider the key issues.

1.1.3 Real-Time Operating System

The treatment of an RTOS as a software component is not new; there are around 200 such products on the market. The differentiation is sometimes clear, but in other cases, it is more subtle. Much may be learned from the selection criteria for an RTOS.

RTOS Selection Factors

Detailed market research has revealed some clear trends in the factors that drive purchasing decisions for RTOS products.

> **Hard real time:** "Real time" does not necessarily mean "fast"; it means "fast enough." A real-time system is, above all, predictable and deterministic.
> **Royalty free:** The idea of licensing some software, and then paying each time you ship something, may be unattractive. For larger volumes, in particular, a royalty-free model is ideal. A flexible business model, recognizing that all embedded systems are different, is the requirement. Of course, open source is another possibility.
> **Support:** A modern RTOS is a highly sophisticated product. The availability of high-quality technical support is not optional.
> **Tools:** An RTOS vendor may refer you elsewhere for tools or may simply resell some other company's products. This practice will not yield the level of tool/RTOS integration required for efficient system development. A choice of tools is, on the other hand, very attractive.
> **Ease of use:** As a selection factor, ease of use makes an RTOS attractive. In reality, programming a real-time system is not easy; it is a highly skilled endeavor. The RTOS vendor can help by supplying readable, commented source code, carefully integrating system components together, and paying close attention to the "out-of-box" experience.
> **Networking:** With approximately one-third of all embedded systems being "connected," networking is a common requirement. More on this topic later.
> **Broad CPU support:** The support, by a given RTOS architecture, of a wide range of microprocessors is a compelling benefit. Not only does this support yield more portable code, but also the engineers' skills may be readily leveraged. Reduced learning curves are attractive when time to market is tight.

RTOS Standards

There is increasing interest in industry-wide RTOS standards, such as OSEK, POSIX (Portable Operating System Interface), and μiTRON. This subject is wide ranging, rather beyond the scope of this article and worthy of one devoted exclusively to it.

OSEK: The short name for the increasingly popular OSEK/VDX standard, OSEK is widely applied in automotive and similar applications.

μiTRON: The majority of embedded designs in Japan use the μiTRON architecture. This API may be implemented as a wrapper on top of a proprietary RTOS, thus deriving benefit from the range of middleware and CPU support.

POSIX: This standard UNIX API is understood by many programmers worldwide. The API may be implemented as a wrapper on top of a proprietary RTOS.

1.1.4 File System

A digital camera will, of course, include some kind of storage medium to retain the photographs. Many embedded systems include some persistent storage, which may be magnetic or optical disk media or nonvolatile memory (such as flash). In any case, the best approach is standards based, such as an MS-DOS-compatible file system, which would maximize the interoperability possibilities with a variety of computer systems.

There is increasing demand for a fault-tolerant flash file system with wear-leveling support to protect against power failure and maximize the usable lifetime of media.

1.1.5 USB

There is a seemingly inviolate rule in the high-tech world: the easier something is to use, the more complex it is "under the hood."

Take PCs for example. MS-DOS was very simple to understand; read a few hundred pages of documentation and you could figure out everything the OS was up to. Whatever its critics may say, Windows is easier to use, but you will find it hard (no, impossible) to locate anyone who understands everything about its internals; it is incredibly complex.

USB fits this model. Only a recollection of a few years' experience in the pre-USB world can make you appreciate how good USB really is. Adding a new peripheral device to a PC could not be simpler. The electronics behind USB is not particularly complex; the really smart part is the software. Developing a USB stack, either for the host computer or for a peripheral device, is a major undertaking. The work has been done for host computers—USB is fully supported on Windows and other operating systems. It makes little sense developing a stack yourself for a USB-enabled device. Many off-the-shelf USB packages are available.

1.1.6 Graphics

The LCD panel on the back of a camera has two functions: it is a graphical output device and part of the UI. Each of these functions needs to be considered separately.

As a graphical output device, an LCD is quite straightforward to program. Just setting RGB values in memory locations results in pixels being lit in appropriate colors. However, on top of this underlying simplicity, the higher-level functionality of drawing lines and shapes, creating fills, and displaying text and images can increase complexity very rapidly. A graphic functions library is required.

To develop a GUI, facilities are required to draw screen elements (buttons, icons, menus, etc.) and handle input from pointing devices. An additional library, on top of the basic graphics functions, is required.

Over the last few years, user expectations of UIs have been raised substantially. With the low cost and ready availability of touch-sensitive LCD panels, highly interactive UIs are common. From the developer's point of view, this is a big challenge, as such displays are complex to program from scratch. Fortunately, a number of commercial UI development packages are available that make the rapid creation of a sophisticated UI quite straightforward.

1.1.7 Networking

An increasing number of embedded systems are connected either to the Internet or to other devices or networks. This may not sound applicable to our example of a digital camera, but Bluetooth connectivity is quite common and even Wi-Fi-enabled cameras have been demonstrated.

A basic TCP/IP stack may be straightforward to implement, but adding all the additional applications and protocols is quite another matter. Some key issues are worthy of further consideration.

IPv6
IP is the fundamental protocol of the Internet, and the currently used variant is v4. The latest version is v6. (Nobody seems to know what happened to v5.) To utilize IPv6 requires new software because the protocol is different in quite fundamental ways. IPv6 addresses a number of issues with IPv4. The two most noteworthy are security (which is an add-on to IPv4 but is specified in IPv6) and address space. IPv6 addresses are much longer and are designed to cover requirements far into the future (see Figure 1.1).

If you are making Internet-connected devices, do you need to worry about IPv6 yet?

The answer is yes. Until recently, IPv6 has only been essential for military applications and European and Asian markets, but, as the last blocks of IPv4 addresses have been allocated, the move to IPv6 has become urgent worldwide. You also need to consider support for dual stacks and IPv6/IPv4 tunneling.

Who Needs a Web Server?
The obvious answer to this question is "someone who runs a web site," but, in the embedded context, there is another angle.

Standard format:

    3ffe:2900:0102:0001:0000:0000:0000:0002

Leading zeros removed:

    3ffe:2900:102:1:0:0:0:2

Double colon notation:

    3ffe:2900:102:1::2

Figure 1.1
IPv6 addresses

Imagine that you have an embedded system and you would like to connect to it from a PC to view the application status and/or adjust the system parameters. This PC may be local, or it could be remotely located, connected by a modem, or even over the Internet; the PC may also be permanently linked or just attached when required.

What work would you need to do to achieve this?

The following tasks are above and beyond the implementation of the application code:

- Define/select a communications protocol between the system and the PC.
- Write data access code, which interfaces to the application code in the system and drives the communications protocol.
- Write a Windows program to display/accept data and communicate using the specified protocol.

Additionally, there is the longer-term burden of needing to distribute the Windows software along with the embedded system and update this code every time the application is changed.

An alternative approach is to install web server software in the target system.

The result is:

- The protocol is defined: HTTP.
- You still need to write data access code, but it is simpler; of course, some web pages (HTML) are also needed, but this is straightforward.
- On the PC you just need a standard web browser.

The additional benefits are that there are no distribution/maintenance issues with the Windows software (everything is on the target), and the host computer can be anything (it need not be a Windows PC). A handheld device like a smartphone or a tablet is an obvious possibility.

The obvious counter to this suggestion is size: web servers are large pieces of software. An embedded web server may have a memory footprint as small as 20 K, which is very modest, even when storage space for the HTML files is added.

SNMP

SNMP (Simple Network Management Protocol) is a popular remote access protocol, which is employed in many types of embedded devices. The current version of the specification is v3.

SNMP offers a very similar functionality to a web server in many ways. How might you select between them?

If you are in an industry that uses SNMP routinely, the choice is made for you. If you need secure communications (because, for example, you are communicating with your system over the Internet), SNMP has this capability intrinsically, whereas a web server requires a secure sockets layer (SSL). On the other hand, if you do not need the security, you will have an unwanted memory and protocol overhead with SNMP.

A web server has the advantage that the host computer software is essentially free, whereas SNMP browsers cost money. The display on an SNMP browser is also somewhat fixed; with a web server, you design the HTML pages and control the format entirely.

1.1.8 Conclusion

As we have seen, the development of a modern embedded system, such as a digital camera, presents many daunting challenges. With a combination of the right skills and tools and a software-component-based development strategy, success is attainable. But what will be next?

What Is So Simple About SNMP?

SNMP, which stands for "Simple Network Management Protocol," can be described in many ways (e.g., versatile, standardized, secure, and widely used), but simplicity is not one of its attributes. So, why the name?

The answer is that it is not a *simple* network management protocol; it is a *simple network* management protocol. It is designed for the management of equipment on "simple" networks—LANs or maybe even point-to-point links. So the name does make sense.

1.2 Memory in Embedded Systems

I am often asked to explain what embedded systems actually are. This tends to lead to a supplementary question about how they differ from "normal" computers, from a software point of view. The nature and treatment of memory is a fine example of how the two worlds often differ. This is as true today as it was when I wrote an article about the topic in NewBits in the early 1990s, upon which this article is based. The concepts have not changed, but some of the numbers have. I observe that, in the original text, I described 4 M as the normal RAM size for a PC!

CW

1.2.1 Memory

When people discuss the advances in microelectronics over the last few years, chances are they are raving about the speed of the latest RISC chip or just how fast a Pentium can go if you keep it cold enough. However, a much quieter revolution has been taking place in the same context: memory has been growing.

Consider some of the implications of this revolution. The PC is just over 20 years old; the original model had just 16 K of memory, whereas 2 G is now considered an average amount. The mainframe I used 20-odd years ago had rather less memory than my phone does now. Where is it all going to end?

I suppose the theoretical limit of memory density would be 1 bit per atom. I have even read that this limit is being considered as a practical proposition. With that kind of memory capacity, how big of an address bus do you need? Thirty-two bits is certainly not enough. How about 64? Maybe we should go straight to a 128-bit bus to give us room to move. Will that last for long? The answer is easy this time: we will never need an address bus as big as 128 bits because 2^{128} is well beyond our current (or any likely future) capacity.

What Is Memory?

That really should be an easy question to answer, but you would get a different response from different people.

A hardware engineer would respond:

Memory is a chip in which you can keep bits of data. There are really two kinds: ROM and RAM. These, in turn, come in two varieties each. There is masked programmed ROM and programmable devices, which you can program yourself. RAM may be static, which is easy to use but has less capacity; dynamic is denser but needs support circuits.

A typical software engineer would respond:

Memory is where you run your program. The code and data are read off of the disk into memory, and the program is executed. You do not need to worry too much about the size, as virtual memory is effectively unlimited.

An embedded systems programmer would respond:

Memory comes in two varieties: ROM, where you keep code and constants, and RAM, where you keep the variable data (but which contains garbage on startup).

A C compiler designer would say:

There are lots of kinds of memory: there is some for code, variable data, literals, string constants, initialized statics, uninitialized statics, stack, heap, some is really I/O devices, and so forth.

Four differing answers to the same question! These responses do not contradict one another, but they do reflect the differing viewpoints.

Memory in Embedded Systems

These differing views of the nature of memory are quite likely to come sharply into focus on an embedded system project. The hardware designer puts memory on the board, the C compiler designer provides the software development tools, and software engineers often end up doing the programming. An experienced embedded systems programmer has learned to reconcile the differences, understands what the hardware engineer has provided, and understands how to use the development tools to make the program fit into that environment.

1.2.2 Implementation Challenges

What problems need to be overcome when implementing software for an embedded system?

ROMable Code

The first and most obvious challenge when implementing software for an embedded system is to arrange for code to be stored in ROM and variable data to be assigned RAM space.

This radically differs from a "normal" computer, where code and data are simply loaded into (read/write) memory as a unit. Data may, therefore, be mixed up with the code, and its values may be set up at compile time. Furthermore, although not a sensible practice, programs may be made self-modifying since the code is stored in read/write memory. A true cross-compiler generates code that is ROMable (i.e., will execute correctly when stored in ROM). In addition, a linker intended for such applications facilitates the independent positioning of code into ROM and data into RAM.

Many modern systems work in a slightly different way. On start-up, the code is copied out of ROM (typically flash) into RAM and then executed from there. This is largely because RAM access speed is much higher. However, the requirement for code to be clearly separated from (variable) data remains.

Program Sections

To accommodate the need to treat various types of memory differently, the concept of a program "section" evolved. The idea is that memory could be divided into a number of named units called sections. While coding in assembly, the programmer could specify the sections where code, data, constants, etc., are placed.

The actual allocation of memory addresses to sections takes place at link time. The linker is provided with start addresses for each section or the start address where a sequence of sections, in a specified order, may be placed. Contributions from a number of object modules may be made to a given section. Normally, these are concatenated and placed at the address specified.

For an embedded system, at the simplest level, just two program sections are needed: one for code and constants (ROM) and one for data (RAM). In each module, the programmer takes care to indicate the appropriate section for each part of the program. At link time, all the code and constants are gathered together and placed in ROM, and all the data is placed in RAM.

Static Variables

In C, any variable that is not automatic (i.e., on the stack or held in a register) is stored statically. A static variable has memory allocated for it at compile time. When such a variable is declared, there is the option of providing an initial value. If no value is specified, the variable is set to zero.

Clearly there is a potential problem with the use of static variables in an embedded system. The values set up at compile time would be lost, as only code (and constant data) is blown into ROM. There are three possible solutions:

1. Do not use initialized static variables. Use explicit assignments to set up their values and never assume an unassigned variable contains zero. Although this approach is possible, it is inconvenient.
2. Map the initialized statics into ROM. This means that although they do have the required initial value, they cannot be changed at all later. While this may sound restricting, it is often useful to have look-up tables of variables (more likely structures), which are actually treated as constants.
3. Map the variables into RAM, if the compiler package in use permits, with their initial values being mapped into ROM. It is a simple matter to copy one area of (ROM) memory to another (RAM) at startup, before `main()` is called.

1.2.3 When All Goes Wrong

When developing software for an embedded microprocessor, the reconciliation of the various views of memory may not be easy. If the compiler simply divides the code and data into two sections, destined for ROM and RAM, respectively, the possibilities for resolving the initialized statics problem are limited. Furthermore, the compiler may treat literal text strings as data, which would be inconvenient because they need to reside in ROM with the code.

1.2.4 When All Goes Right

A true embedded toolkit can make dealing with memory in an embedded system very straightforward. The use of named program sections is employed to the full. This support begins with the compiler, which generates a number of sections. A typical list is as follows:

- `code`—The program code
- `zerovars`—Uninitialized static
- `initvars`—Initialized statics
- `const`—Variables declared constant
- `strings`—Constant text strings
- `literals`—Compiler-generated literals
- `tags`—Compiler-generated tags

The names of the sections may be changeable and will vary from one compiler to another. Using the assembler, any assembly code may contribute to these sections or create others.

The final stage is the use of the linker, which enables each section to be placed in the appropriate part of the memory map. In addition, there may be a command that permits initialized data to be set up from values held in ROM at startup, with minimal effort.

So the story has a happy ending and the problems of diverse memory in an embedded system may be solved by the use of the right tools.

1.3 Memory Architectures

The perception of what a memory address really is can be challenging for many engineers, who are unaccustomed to thinking in such "low-level" terms. I wrote a piece for NewBits in 1992 that investigated this topic, and it was the basis for this article.

CW

I remember when the term "architecture" always seemed to refer to buildings—their style and design. I guess this viewpoint was initiated, or at least reinforced, by the letters "RIBA" (Royal Institute of British Architects) on my father's business card. Nowadays, everything seems to have an architecture. Chip architecture is a good example, where it has spawned the term "silicon real estate," which carries the analog with the construction industry a little bit further. In this article I will give some consideration to memory architecture, as implemented by microprocessors used in embedded systems.

1.3.1 The Options

In general, memory architectures fall broadly into five categories:

- Flat single space
- Segmented
- Bank switched
- Multiple space
- Virtual.

Figures 1.2–1.5 illustrate diagrammatically the first four schemes. Every microprocessor may employ one or more of these possibilities. Each chip offers a different combination, and each option may be more or less appropriate for a specific application. Cache memory, while not strictly a memory architecture, is also discussed.

1.3.2 Flat Single-Space Memory

Flat memory is conceptually the easiest architecture to appreciate. Each memory location has an address, and each address refers to a single memory location. The maximum size of addressable memory has a limit, which is most likely to be defined by the word size of the chip. Examples of chips applying this scheme are the Freescale Coldfire and the Zilog Z80.

Typically, addresses start at zero and go up to a maximum value. Sometimes, particularly with embedded systems, the sequence of addresses may be discontinuous. As long as the programmer understands the architecture and has the right development tools, this discontinuity is not a problem.

Most programming languages, like C, assume a flat memory. No special memory-handling facilities need be introduced into the language to fully utilize flat memory. The only possible problems are the use of address zero, which represents a null pointer in C, or high addresses, which may be interpreted as negative values if care is not exercised.

Linkers designed for embedded applications that support microprocessors with flat memory architectures normally accommodate discontinuous memory space by supporting scatter loading of program and data sections. The flat address memory architecture is shown in Figure 1.2.

1.3.3 Segmented Memory

A perceived drawback of flat memory is the limitation in size, determined by the word length. A 16-bit CPU could only have 64 K of memory and a 32-bit architecture may be considered overkill for many applications. The most common solution is to use segmented memory (see Figure 1.3). Examples of chips applying this scheme are the Intel 8086 and the Hitachi H8/500.

The idea of segmented memory addressing is fairly simple. Addresses are divided into two parts: a segment number and an offset. Offsets (usually 16 bits) are used most of the time, where the additional high-order bits are held in one or more special segment registers and assumed for all operations. To address some memory over a longer range, the segment registers must be reloaded with a new value. Typically, there are individual segment registers for code, data, and stack.

The use of segmented memory necessitates the introduction of the concepts of "near" and "far" code and data. A near object may be accessed using the current segment register

0

MAX

Figure 1.2
Flat address memory architecture

settings, which is fast; a far object requires a change to the relevant register, which is slower. Since segmented memory is not immediately accommodated by high-level languages, near and far (or _near and _far) keywords must be introduced. With these keywords, you can specify the addressing mode used to access a code or data item. Using a "memory model," you can specify default modes. For example, a "large" model would access all objects as "far," and a "small" model would use "near" for everything. With segmented memory, the size of individual objects (e.g., modules or data arrays) is generally limited to the range addressable without changing the segment register (typically 64 K).

Compilers for chips with segmented memory typically implement a wide range of memory models and the far and near keywords.

1.3.4 Bank-Switched Memory

Another approach to extending the addressing range of a CPU is bank switching. This technique is a little more complex than segmented memory, but it has the advantage that such memory may be implemented with a processor that does not itself support extended memory.

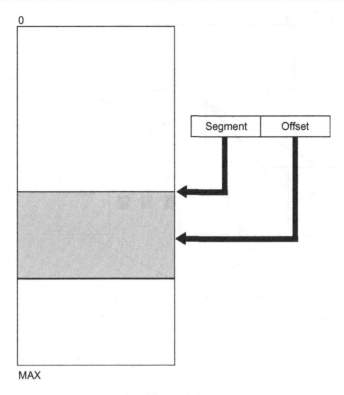

Figure 1.3
Segmented memory architecture

A bank-switched memory scheme comprises two parts: a range of memory addresses, which represent a "window" into a larger memory space, and a control register, which facilitates the moving of this window (see Figure 1.4). Accessing the bank-switched memory area requires the control register settings to be verified and adjusted, if necessary, before the required location is accessed within the window.

Little, if anything, can be done with a C compiler to accommodate bank-switched memory. However, a linker may provide an "overlay" scheme to enable a number of data items to exist apparently at the same address. Better still, the linker could implement the concept of "logical views"—that is, groups of modules within which a certain setting of the control register(s) may be held constant. Transfers (jumps or calls) between logical views are performed by code that performs the necessary bank switch.

1.3.5 Multiple-Space Memory

Yet another approach to increasing the memory-addressing range of a limited number of addresses is the use of multiple memory spaces (see Figure 1.5).

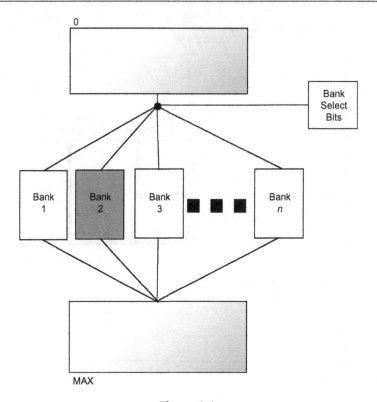

Figure 1.4
Bank-switched memory architecture

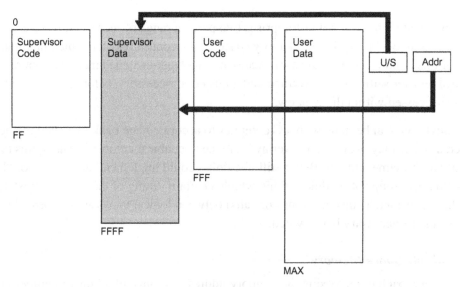

Figure 1.5
Multiple-space memory architecture

A Freescale Coldfire, for example, may optionally have four address spaces: user code and data and supervisor code and data. Since the functions of these spaces are essentially noninterchangeable, no particular provision is required within a high-level language.

The Intel 8051 family employs multiple address spaces, which map to different types of memory: on-chip and off-chip ROM and RAM, etc. Since such memory may not be addressed unambiguously (as a given address may correspond to several actual memory locations, any one of which may be validly addressed), special data type modifiers need to be added to high-level languages, like C, to qualify variables appropriately.

1.3.6 Virtual Memory

The virtual memory scheme has a long history but has remained appropriate to modern computer systems, where memory demands continually escalate. Since it is very rarely employed in embedded systems, little space will be devoted to its consideration here. Virtual memory gives you the impression that memory is of indefinite size. This impression is achieved by automatically swapping areas of memory on and off of the hard disk using specialized hardware facilities. A delay occurs whenever a memory location that is currently swapped out must be read in. The delay reduces the usefulness of virtual memory for real-time systems.

1.3.7 Cache Memory

Although not strictly a memory architecture by the definition of those described previously, memory caches are becoming a common feature of many modern, high-performance microprocessors. A full discussion of memory cache design and implementation would fill an entire article or more by itself. Bearing in mind that the presence of a cache may generally be ignored by a programmer, only a brief comment is appropriate here.

A memory cache is a special area of high-speed memory within or adjacent to the CPU. The cache controller reads from regular memory into the cache when a memory location is accessed. Subsequent accesses to the same or nearby locations can then be performed from the cache, without accessing the slower main memory. The performance of a small code loop may often be significantly increased when the entire code can reside in the cache.

By careful sequencing of instructions, the best performance may be extracted from a cache facility. Recent compilers for chips with such an architecture often take advantage of this technique.

1.3.8 Memory Management Units

Many 32-bit microprocessors, which are used for embedded applications, include a memory management unit (MMU). It is either built-in or may be included as an option. An MMU provides a means to protect memory from corruption. Typically it is used with a real-time operating system (RTOS), where each task and the OS itself can be protected from malfunctioning code in another task. This may be done by simply blocking (or write-protecting) memory areas or through the implementation of a "process model," where each task apparently has a complete private address space starting at zero. From the embedded applications programmer's point of view, very little action needs to be taken to accommodate the use of an MMU. It is largely an issue for the RTOS.

1.3.9 Conclusions

Understanding the memory architecture of a chip is essential to determine its appropriateness for a specific application. For very large or very small applications, flat memory is usually best. For small programs with a lot of data, bank-switched memory may be particularly suitable.

1.4 How Software Influences Hardware Design

> *There is always a tension between hardware and software specialists. Even today, when the boundaries are becoming increasingly indistinct, the labels persist. I have always felt that I am on a crusade to bring the two sides together. This is what I had in mind when I wrote a piece for NewBits in 1992, which was the basis for this article.*
>
> *CW*

At the time of writing, the Festive Season still seems mercifully far off. I do not like to think about Christmas at all until a couple of weeks before. You may well ask: what has this got to do with embedded systems design? The answer lies in an old English Christmas custom, which has an interesting parallel in the electronics world that I will discuss further. In the nineteenth century, December 26 was designated "Boxing Day"; the day upon which tradesmen received their "Christmas Boxes" (i.e., bonus money or another gift). This seems slightly illogical, as I am sure they would rather have had the extra money before Christmas than after, but I digress. The tradition on Boxing Day was for masters and servants to change places for the day; the same practice in the armed forces resulted in officers and men swapping roles. The objective, in both cases, was to enable each "side" to see the other's point of view more clearly. Could the same idea be applied in the embedded system development lab?

1.4.1 Who Designs the Hardware?

In a small company, or even a small development department of a larger operation, there may well be a limited number of engineers (or even just one) who design embedded systems in their entirety. Imagine the answer to your question about the staff deployment: "Fred over there does our embedded stuff." The definition of "stuff" would cover both hardware and software. This situation is fine for the development of small systems. Fred probably understands all the aspects of the application and can optimize the interaction of hardware and software accordingly, assuming that he is sufficiently expert in both disciplines to evaluate the trade-offs properly.

In a larger lab, there is much more likely to be a sharper differentiation between hardware and software, with just a small number of "gurus" who "have a foot in both camps."

Decisions made at the start of an embedded systems project have implications that last through its design and often into its production. Skimping on hardware to reduce cost or power, for example, could cost you extra programming time at the end. Splurging on hardware by overestimating your needs, however, burdens your design with extra cost on each unit shipped. Designers usually have hardware or software expertise, but not both.

1.4.2 Software Leading Hardware

Typically, the design of an embedded system starts with the hardware. Only when that has been (irreversibly) designed is software taken into consideration. The efficiency of the implementation would be ultimately enhanced if software were considered at a much earlier stage. It would be ridiculous to actually suggest that the software engineers should do the hardware guys' work, but their early involvement would be useful.

If nothing else, it would minimize the "finger pointing" during the integration phase, when each "side" blames the other for any problems that arise.

1.4.3 Software/Hardware Trade-Offs

The consideration of software requirements and capabilities has a real influence at many points during the hardware definition.

Processor Selection
A factor in processor selection is the availability of advanced software development tools. Although low-power chips or cores are specified for many applications, the saving of a few microwatts of CPU power consumption is often used up by the additional memory needed to store inefficient code.

Memory Size and Mix
A decision on the exact amount of memory and the mix of ROM (flash, etc.) and RAM should be made as late as possible in the design cycle, since this is often hard to predict. Cramming code and data into too small a memory area is a problem, and having too much memory has price and power implications. Sometimes memory sites can be designed to take either RAM or ROM, which means extra traces for the Write Enable signal. This strategy allows the memory type mix be determined late in the design cycle and offers the ability to replace most of the ROM with RAM for debugging purposes (see the section, "Debug Hardware," that follows).

Peripheral Implementation

The inclusion of other peripheral devices in the design should be carefully considered. Timers and serial I/O, for example, can be implemented in software when necessary. There is a greater load on software when each bit of serial communication has to be processed instead of having hardware handle a whole byte. However, if CPU processing power is available and hardware cost is a concern, having the software do the work is often a reasonable solution. After all, you pay for the software development just once, but an extra chip adds cost to every unit shipped. This trade-off must be made carefully, because the considerations of a low-volume application (where software development cost is dominant) differ from those for a product shipping in large volume (where unit cost is the issue).

1.4.4 Debug Hardware

Adding hardware to help debug software does not occur to many hardware designers. It is a matter worthy of further thought.

In-Circuit Emulators

Historically in-circuit emulators (ICEs) were the instrument of choice for embedded software development. They provided a totally unintrusive way to debug code at full speed on a real target. But as processor complexity increased and, more importantly, clock speeds escalated, ICEs became more and more expensive and their availability declined. Now they tend to be available for only low-end devices.

If an ICE is available and the cost is not a problem, then this may offer the very best debugging tool. However, in reality, a full team of software developers rarely can be equipped with ICEs for software debugging. These instruments are ideal for addressing complex problems resulting from the close interaction of the software and the hardware, and if they are in short supply, should be reserved for this kind of job.

Monitor Debuggers

A good solution, in many cases, is to make use of a monitor debugger. This requires some provision in the target hardware: ROM must be replaced temporarily by RAM and an additional serial port or other I/O device must be available at debug time. Neither of these requirements presents a problem in most circumstances if the need is considered early in the design phase. The result is the ability to debug the code at full speed on the target hardware. This does not have all the facilities of an ICE, but quite enough for most purposes. In particular, a monitor-based debugger can be the basis of a run-mode debug facility with a real-time operating system. Typically, a debug monitor communicates with the host via Ethernet.

JTAG Support

Most modern microprocessors provide on-chip support for software debugging. Most devices use a JTAG interface for this purpose.

To use JTAG for software debug, an adaptor is needed between the target and the host machine to handle the special synchronous protocol and manage the interface to a high-level language debugger. Connecting this adapter to the target board requires an appropriate connector, but the cost and space to add it are negligible. However, this must be considered during the design phase.

1.4.5 Self-Test Support

Most embedded systems have a degree of built-in self-testing. Sometimes self-testing is executed only at power-up to check for any chip failures that may have occurred. In other systems, the self-test code may be run as a background task when no other processing is required.

The programming techniques employed for self-testing really deserve an article to themselves; so, for the moment, we will concentrate on the provisions that can be incorporated into the hardware design.

I/O Circuits

The self-test facilities that may be incorporated into I/O circuits are device-specific. A typical example is a serial line (RS232) interface. To confirm that the serial interface chip is transmitting and receiving correctly, a "loopback" can be implemented. This capability results in each transmitted character being sent straight to the receiver. The self-test code first sends a sequence of characters and then verifies that they are being received correctly. The loopback channel in the I/O circuit must be controllable by the software.

On-Board Switches

An on-board switch (or, better still, a small switch bank), jumpers, or press buttons can provide test commands to the software. With a four-way switch bank, for example, one switch could activate the self-test mode, and the other three could select one of eight possible baud rates for a serial line.

Status Displays

A display could be as complex as a line of LCD characters upon which messages can be written, or as simple as a single LED, which can be turned on or off by the software. A surprising amount of information can be conveyed by such a single LED. It can be in one of three simple states: on, off, or flashing. A good approach is to use flashing to indicate the software is OK, because this requires action to maintain the flashes and is, therefore, fail-safe. Of course, the instructions to turn on and off the LED should ideally be embedded in the main software loop. The fixed on and off states can then be used to indicate two possible failure conditions. Just like with lighthouses, the way the LED flashes can indicate a status. There are two variations of flashing that can be employed: bursts of flashes separated by a pause (where the number of flashes in the burst can indicate a particular status) or variable duty cycles (ratio of on and off times) can convey different states. Each of these has its attractions, but the duty cycles method is probably a little easier to implement and understand.

1.4.6 Conclusions

Addressing software early, while the hardware design is still fluid, is the way to avoid software/hardware mismatches in an embedded systems project. A serial port or other I/O device on the target hardware, for example, lets a monitor-based debugger work at full speed on the final hardware design to debug optimized code. Since debugging hardware may be absent from the production boards or may have other uses in the application, there may be no additional cost at all.

Making the right decisions at the start of an embedded systems project is important because skimping or splurging on hardware costs time and money.

1.5 Migrating Your Software to a New Processor Architecture

In the mid-1990s, Motorola (now Freescale) dominated the high-end embedded processor market, but things were changing fast. It was with this change in mind that, in 1996, I wrote a piece on migrating between processors for NewBits, which was the basis for this article. Another article in this book extends on the concepts: "On The Move—Migrating from One RTOS to Another" (Chapter 7) expands on issues only touched upon here.

CW

A few years ago, the basic decision involved in developing a high-end embedded system—"which processor should I use?"—was fairly easy to answer. It could, more often than not, be translated into "which 68 K family device should I choose?" Things have changed. Freescale offers a number of architectures for high-end applications. At the same time, a plethora of other 32-bit devices compete for attention, all with strong differentiating factors that can make them an attractive choice.

Having chosen the processor for the next project, the next issue to address is the development of new expertise in programming the device and porting existing code. There are three major topics to cover:

- Avoiding target-specific code
- Real-time operating system (RTOS) issues
- Open standards and how they aid target processor migration.

1.5.1 Target Specifics

The primary aspects of an embedded application, which have a degree of target specificity, are the code, data, and real-time structure.

Code
Code written in C or C++ will itself tend to be portable. The only real concern is the precise dialect of the language that has been used. If the change of processor necessitates a change of

compiler supplier, there may be some issues. In particular, extensions to the language (e.g., interrupts) may be implemented differently or may include a proprietary set of keywords.

A nonlinear change is likely in the runtime performance of the code. Assuming, for example, that the new chip is supposed to deliver a 2× performance increase, this will not be the case with all language constructs; some will be even faster than this, others will benefit less from the higher processor performance. Some architectures lend themselves more readily to a particular program structure or language feature.

Assembler code will inevitably require a rewrite. Even the migration from 68 K to ColdFire requires a very careful review of the code to ensure that it is confined to the instruction subset supported by these devices. This is an ideal time to review whether some of the assembler code can be replaced by a (portable) high-level language implementation.

Data and Variables

Since the definition of the storage allocation for data types in C/C++ is, by definition, target-specific, a change in this area may be anticipated. For example, an int may be allocated 32 bits instead of 16 bits. This is quite likely to affect the performance and functionality of the code as well as the RAM requirements of the application, which may increase or decrease unexpectedly. In practice, the storage allocation schemes for most 32-bit target compilers are essentially identical, so little or no difficulty should arise.

To completely eliminate problems, some engineers implement a defensive strategy, which renders their code more portable. This approach is very simple: use `typedef` statements to implement some bit-size-specific data types in unsigned and signed variants: `U8`, `U16`, `U32`, `S8`, `S16`, and `S32`. Porting the code to a new device and/or compiler is simply a matter of making a slight edit to a header file.

Structure layout and alignment may change, as well as bit field allocation. This problem is unlikely to be encountered, because it is bad practice to design code that relies on these factors.

According to the language definitions, `enum` variables should be allocated the same space as `int`. However, many compilers optimize the space, allocating 1, 2, or 4 bytes, depending upon the range of values required. Again, this could result in an unexpected change in the performance or memory requirements for an application.

Native compilers, used for the development of desktop applications, are solely focused on the generation of fast code, at the expense of compactness if necessary. For embedded systems software development, such an assumption about a user's requirements would be inappropriate. For a program to achieve maximum speed, data must be laid out in memory such that it can be accessed efficiently. This necessitates waste; extra bytes are added for "padding" to facilitate the correct alignment. Cross-compilers generally have a facility to optionally pack data—i.e., they eliminate the padding and generate any extra code that is required to access the data. This may be implemented differently for a different processor, leading to code incompatibility and an increase or decrease in memory requirements.

Data and Function Parameters

The conventional method for function parameter passing is to push each value onto the stack, from right to left. Any code that relies on this mechanism will not be portable. Here is an example:

```
fun(int n)
{
    int *p, x;
    p = &n;
    x = *++p;
    ...
}
```

High-performance devices tend to have an RISC architecture, with a large set of registers. These may be more efficiently employed for parameter passing, which would break this code.

For straightforward C or C++ code, the details of parameter passing are unimportant. Parameter details are important when assembler code is involved or when libraries of code built with a different compiler are to be used.

Data and Register Usage

A modern optimizing compiler will make effective use of CPU registers for storing automatic variables in C and C++ programs. High-performance RISC devices, with much larger register sets, will benefit even more from this optimization. Also, a "smart" optimizer will perform "register coloring," which makes even more efficient use of the available registers. This technique is even a benefit to performance when the microprocessor itself performs register renaming. In this example, i and j are likely to be allocated to the same register:

```
color()
{
    int i, j;
    for (i=0; i<4; i++)
        ...

    ...
    for (j=0; j<4; j++)
        ...
}
```

Although the use of registers to hold variables is transparent to the C or C++ programmer, except possibly at debug time, it may affect the runtime performance and stack requirements unexpectedly. Also, the use of the register keyword, which advises the compiler that a particular variable is heavily used, may be more effective than usual.

Data and Endianness

Memory is normally just a series of byte-size storage locations. The way these are grouped together to form 16- or 32-bit words is arbitrary. There are broadly two approaches: the most significant byte of the word can be allocated the lowest address (which is "big-endian") or it can have the highest address ("little-endian"). These two approaches are illustrated in Figure 1.6.

For most code, the "endianness" (or "endianity") is irrelevant. Certain program constructs are sensitive, and they may lead to nonportable code. They should be avoided.

There can be a problem when a big-endian processor exchanges data with a little-endian machine. It is necessary for one or another format to be established as the standard for such an interchange and any necessary byte-swapping is performed in software.

Many modern microprocessors, like the PowerPC, are flexible and may be operated in either big- or little-endian mode. Most compilers will generate code for either model, but care may be required because libraries may be available in only one of the two formats.

Real-Time Concerns

Moving to a higher-performance microprocessor will obviously increase the speed of the application. As mentioned previously, this performance increase may be nonlinear for various reasons. This is rarely a source of difficulty. Of more significant concern are constructs that characterize the real-time structure of the program. The most significant examples are interrupts.

The interrupt mechanism on various chips is different. Also, it is likely that the design of the hardware around the CPU will differ in a new design. The net result is that a rewrite of the interrupt code is very likely to be necessary.

To maximize the possibilities for portability, interrupt service routines should be written in C or C++ (assuming that the compiler supports this option). The interrupt vector can most likely be coded in high level too; an array of pointers to functions is usually sufficient.

1.5.2 RTOS Issues

For higher-end embedded systems, the use of a real-time operating system (RTOS) is common. Historically, for many designs, an in-house kernel has been utilized for reasons

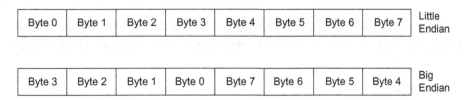

Figure 1.6
Endianness and byte allocation

of ready availability and perceived cost advantages. This situation complicates issues when migration to an alternate processor architecture is planned. An RTOS often has a high assembler code content, requiring a rewrite, and specific architectural features of the device are frequently exploited, which may necessitate a total redesign.

As designs become more complex, the need for an RTOS becomes even more apparent. Also, time-to-market pressures are always increasing. As a result, three other requirements emerge:

- More comprehensive tool support is essential. A debug technology, which is capable of tackling the multitasking environment, and the special problems that can result, is no longer optional.
- To speed application development, it is very beneficial to use an RTOS that has been designed for a specific environment: telecoms, handheld equipment, automotive, etc.
- The requirements do not end at the RTOS itself. More complex applications, associated with short times to market, demand that, wherever possible, standard application code is brought in, rather than developed in-house. This "intellectual property" may be as simple as hardware drivers or may be more sophisticated middleware or application-enabling technology, such as TCP/IP, SNMP, or Wi-Fi.

These factors point to a requirement to employ a commercial RTOS product, where its availability for the new architecture may be assured and the selection of support products verified. In the longer term, when the next change in processor architecture is necessitated, the vendor is most likely to be motivated to provide support in good time.

1.5.3 Processor Migration and Open Standards

Although changing microprocessor architecture is part of the problem, many of the headaches associated with migration to a new device come from the necessity to adopt different programming techniques and new tools. The way a compiler makes use of an architecture can strongly affect the way an engineer works. In most cases, compiler design is at the discretion of the tool developer. The result is that a change of processor may require the use of very different tools. Also, an embedded systems developer is likely to find very little interoperability between different vendors' tools.

An open standard governing the application program's interface to the chip architecture was defined with the launch of the PowerPC: the Embedded Application Binary Interface (EABI).

EABI—Introduction

EABI is based upon the UNIX SVR4 desktop Application Binary Interface for PowerPC, optimized for memory usage, while retaining runtime performance. Compliance to EABI facilitates a number of kinds of tool interoperability:

- Generally any compiler may be used with any debugger.
- Code from multiple compilers may be mixed.
- Third-party binary libraries may be used.

The key matters addressed by the standard are:

- Function-calling conventions
- Register usage
- Data types and alignments
- File/debug formats.

Although this article reviews some aspects of EABI, a detailed discussion of the standard is beyond its scope.

EABI and Register Usage

EABI dictates the way the machine registers are applied by a compiler. Three categories of register are defined:

- **Dedicated registers** are used for one specific purpose at all times; for example, gpr1 is the stack pointer.
- **Volatile registers** may be corrupted through a function call; the calling function must preserve them if their values are required later.
- **Nonvolatile registers** are preserved through a function call; the called function must preserve them if their use is required.

The following table illustrates the allocation of registers and condition register (CR) fields:

Register	Type	Application
grp0	Volatile	Language-specific use
grp1	Dedicated	Stack pointer
grp2	Dedicated	Read-only data area anchor
grp31/Ngrp4	Volatile	Parameter passing/return values
grp51/Ngrp10	Volatile	Parameter passing
grp111/Ngrp12	Volatile	
grp13	Dedicated	Small data area anchor
grp141/Ngrp31	Nonvolatile	
fpr0	Volatile	Language-specific use
fpr1	Volatile	Parameter passing/return values
fpr21/Nfpr8	Volatile	Parameter passing
fpr91/Nfpr13	Volatile	
fpr141/Nfpr31	Nonvolatile	
Fields CR21/NCR4	Nonvolatile	
Other CR fields	Volatile	

EABI and Stack Frames

The C function requirement for local storage is often beyond what can be accommodated in registers. For this purpose, EABI describes the format of stack frames. In Figure 1.7, the format of a complete stack frame is illustrated.

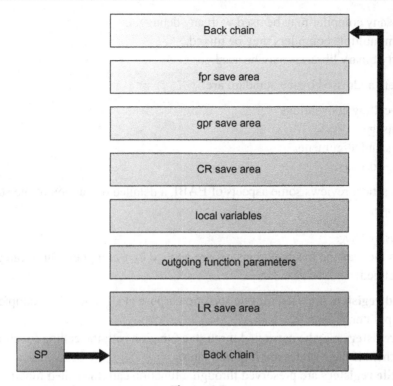

Figure 1.7
EABI stack-frame format

In practice, it may not be necessary to employ all the components of the frame. A minimal stack frame may consist of just 8 bytes. A true leaf function (i.e., a function that does not call any others) may have no need for its own stack frame at all.

EABI and Data Types
Data types, which correspond to those in common high-level languages, are outlined by EABI: byte, half word (2 bytes), word (4 bytes), double word (8 bytes), and quad word (16 bytes). Alignment rules for data structures are also specified, which ensures that the use of memory is consistent from one compiler to another.

The EABI definition encompasses ABIs for both big-endian and little-endian operations. However, these two ABIs are not interoperable.

EABI and File/Debug Formats
To permit tool interoperability, compatibility of file formats is essential. EABI specifies two formats:

- For object files, *Executable and Linking Format* (ELF) is used. This format is well specified and widely understood.

- For debug information, *Debug With Arbitrary Record Format* (DWARF) is used. This format has been extended by different vendors in different ways, which unfortunately reduces interoperability.

1.5.4 Conclusions

Migrating to a new processor is an inevitable challenge faced by embedded developers. With careful planning and a good understanding of the issues, the problems may be minimized. Furthermore, a forward-looking strategy can result in code and development procedures that are ready for porting again, when the need arises.

1.6 Embedded Software for Transportation Applications

Interest in automotive electronics and, hence, embedded software has escalated drastically in recent years, as such systems have become more extensive. Other forms of transportation make similar demands upon software, which is why embedded software is discussed in more general terms in this article, based upon a piece I did for NewBits in early 2003.

CW

The extent and complexity of electronics in all modes of transportation (automobiles, trains, airplanes) has been steadily increasing in recent years, and no end is in sight for this trend. Although, in essence, such systems have much the same characteristics as other embedded, real-time systems, a number of specific tools, technologies, and techniques have been developed. This article reviews the topic, covering programming techniques, with an emphasis on the appropriate use of C and real-time issues, including an introduction to OSEK/VDX.

1.6.1 Introduction

Consideration of the nature of embedded software in automotive and transportation applications quickly reveals a key difference from other types of systems: failure really matters. Many such systems, perhaps not all, have a safety-critical element, and this factor necessitates an appraisal of the programming tools and techniques. We will initially consider the approach to writing the code, then move on to consider real-time issues and how they relate to the use of a real-time operating system (RTOS).

1.6.2 Transportation System Characteristics

It is possible to identify some common characteristics of embedded systems used in transportation and automotive applications. It is very common for such systems to be distributed—multiple microprocessors or microcontrollers are deployed, with appropriate interprocessor communication facilities. For example, interaction in a car may occur between the engine management, antilock braking, and driver information systems. The devices employed may be of widely differing architectures, depending upon the functionality demanded of them; they could be 8-, 16-, or 32-bit parts and may be standalone processors or highly integrated microcontrollers. This diversity has significant impact upon programming techniques and RTOS selection. Both of these issues will be considered further.

Almost all transportation-oriented embedded systems can be described as "real-time," a common definition of which is as follows: "A real-time system is one in which the correctness of the computations depends not only upon the logical correctness of the computation, but also upon the time at which the result is produced. If the timing constraints of the system are not met, system failure is said to have occurred." In other words, a real-time system may not be fast (although it often will be)—its key characteristics are predictable response and reliability. This definition also implies that system failure is not unacceptable, but it should be anticipated and its behavior well defined. These parameters place constraints on the approach to the system design and limit the selection of a real-time operating system, which will be addressed in more detail in the sections that follow.

1.6.3 Programming Issues

Many matters must be addressed before planning the development of software for a safety-critical system. The first thing to consider is the choice of programming language.

It is widely accepted that a high-level programming language, as opposed to an assembly language, not only enables the programmer's time to be used more efficiently, but will most likely result in more accurate, reliable, and maintainable code. The choice is likely to be limited to C, C++, and Ada, at best. Most other languages either are unavailable for the type of processor used in such an application or would be unsuitable for other reasons. As we have seen, a given project may require programming multiple devices of widely differing architectures.

All embedded systems tend to have limited resources (particularly memory and CPU power), but, with many transportation and automotive applications, this limit is even more acute, as cost issues are often paramount. Processors are selected that are just powerful enough and sufficient memory (but no more) is provided. The result of such limitations tends to eliminate C++ and Ada because their demands on resources tend to be too great.

The result is that almost all such systems are programmed in C. This is a concern because C is not really 100% suitable for such applications. It can generate compact, fast code, which is good. It may also be written in a clear maintainable style, if the right guidelines are followed. However, being a very powerful language means that it is possible to write "dangerous" code, which could produce unpredictable results. Also, the definition of the language standard contains ambiguities, which can also lead to unexpected behavior. These two factors indicate the need for some very firm guidelines if C is to be used successfully.

It is useful to use a "style guide" to ensure that code is written in a consistent manner within an organization. There are many publications that address this topic. The need for guidance in writing safer, more secure code is much more than a question of a style and methodology. Using the complete C language is very inadvisable, and the use of a carefully selected subset is the best approach. Books on writing "safer" C are available (see the "Further Reading" section that follows), but the clearest and most concise guidance is provided by the Motor Industry Software Reliability Association (MISRA), who have specified "MISRA C," a well-defined C subset, avoiding all the major pitfalls. MISRA C is really just a set of rules; a certain number of which are "required," and the remainder are "advisory." There are accommodations for "deviations," but the rule set is well suited to real programming situations, based upon actual experience in developing safety-critical systems. Applying MISRA C does not put any particular constraints upon the choice of programming tools. Any suitable C compiler which supports ISO/ANSI C and includes accommodation for embedded systems requirements is acceptable. Various tools are also available to analyze MISRA C compliance.

1.6.4 Real-Time Operating System Factors

An understanding of the characteristics of embedded systems for transportation applications yields some definitive requirements for an RTOS. It is also worthwhile to consider available options.

RTOS Requirements

We can highlight key features required for an RTOS used for this type of application:

- **Statically defined data structures:** Most RTOSes on the market offer the capability to create objects dynamically. This functionality is rarely required in practice and provides a potential opportunity for failure and unexpected behavior. An RTOS with objects defined at build time would be more suitable.
- **Multiple processor architecture support:** Since transportation systems tend to be distributed, with many different microprocessor architectures employed, it is useful to consider an RTOS that is supported on all (or as many as possible) of the devices. This simplifies the selection process and staff deployment.
- **Interprocessor communications:** An RTOS that incorporates a well-defined interprocessor communication mechanism is attractive for distributed systems.
- **Certifiability:** Systems used in safety-critical applications commonly require certification by relevant authorities. Using an RTOS which is known to be certifiable reduces risk. It is not possible to have an RTOS that is itself "certified."
- **Standard API:** If the RTOS supports a standardized application program interface, the learning curve is shortened and staff deployment is simplified.

OSEK/VDX

The OSEK/VDX standard was developed by a group of European automobile manufacturers and is finding broad acceptance across the industry worldwide. The standard addresses most of the issues previously identified but goes further, including standards for:

- The operating system (OS)
- Communications (COM)
- Network management (NM)
- OS implementation language (OIL).

The OSEK/VDX standard has been implemented for a wide variety of chips by various RTOS vendors. The application of this standard permits coding to a standard API and methodology, which results in portability of both code and expertise. Details may be found at www.osek-vdx.org.

1.6.5 *Conclusions*

The successful development of embedded systems for transportation applications depends upon many factors. Understanding the architecture and unique demands of such systems is a start. Care with the code-development process is essential, and MISRA C is a useful guideline for the use of this language. Real-time issues are also important, and the OSEK/VDX standard for RTOSes directly addresses the requirements of this kind of application.

Examples of MISRA C Rules

Advisory Rules

Rule 66: "Only expressions concerned with loop control should appear within a for statement." Good style, which leads to more readable code.

Rule 86: "If a function returns error information, then that error information should be tested." Good practice, but checking input parameters is more effective.

Rule 93: "Function should be used in preference to a function-like macro." Macros have no type-checking and are, hence, less secure. Many compilers will inline small functions anyway to improve speed.

Required Rules

Rule 21: "Identifiers in an inner scope shall not use the same name as an identifier in an outer scope, and therefore hide that identifier." Good practice for clear code.

Rule 71: "Functions shall always have prototype declarations and the prototype shall be visible at both the function definition and call." A common-sense rule that should be followed everywhere.

Rule 76: "Functions with no parameters shall be declared with parameter type void." Although the syntax may be a little odd, this avoids ambiguity. It lines up with a rule that disallows variable numbers and/or types of arguments (Rule 69).

1.7 How to Choose a CPU for Your System on Chip Design

The choice of CPU for any embedded design is interesting, but often surprisingly arbitrary.
Stephen Olsen wrote a paper considering the matter in the context of an SoC design, where
many of the factors are particularly critical, but mostly turn out to be just as applicable to
conventional, board-based designs. This article is based upon that paper.

CW

There are several factors to consider when choosing a CPU for your next system on chip (SoC) design. If you consider that the CPU is to the SoC what an engine is to an automobile, you would not put a Volkswagen engine into a Hummer and expect it to perform. Similarly, a Ferrari engine would also be unsuitable in such a vehicle. Although it may deliver similar horsepower to the Hummer engine, it would fail due to a lack of torque. Simple assessments of "horsepower" are just as misleading in CPU selection as they are in the automobile world. There is an optimal solution for the desired functionality. The same holds true for the CPU choice in an SoC. Many times the CPU is chosen based purely on the system architect's knowledge of, and past experience with, a particular device. The decision of which CPU to use should also consider the overall system metrics: complexity of overall design, design reuse, protection, performance, power, size, cost, tools, and middleware availability.

1.7.1 Design Complexity

The design's complexity is critical to the choice of CPU. For example, if the design calls for a single-state machine to be executed with interrupts from a small set of peripherals, then you may be better off with a small CPU and/or microcontroller such as the 8051 or the Z80. Many systems may fit this category initially. An example might be a pager. The memory footprint is small, the signal is slow, and battery consumption is required to be extremely low.

The algorithms and their interaction will dictate the complexity. They may or may not also dictate the need for an RTOS. Typically, as the application complexity increases, the need for a greater bit-width processor increases.

1.7.2 Design Reuse

Designs are continuing to be reused and are growing in complexity; that pager designed in 2000 may have to be upgraded to play MP3s in 2005; now it needs to support a touch-screen display. The 8-bit CPU is not enough to keep up with the task at hand. How many interfaces a design contains is a good indicator of the amount of processor power required. In our pager example, initially there were two main interfaces: the UI and the radio link. For the newer design, which adds an MP3 player, we will need to add a memory interface for storing and transferring the data, and an audio interface for playing the data. Now the system complexity is greatly increased from its initial conception, and if we have taken a forward-looking approach to the design, we can reuse much of this earlier work.

Make sure that you have room for growth. Today your 8-bit design may be good for the MP3 player, but when the design gets reused and placed in a set-top box application, which has a much higher bandwidth peripheral set, you may need to reengineer the complete solution to migrate to an ARM-, MIPS-, or PowerPC-based architecture to deal with the new constraints.

1.7.3 Memory Architecture and Protection

The system may need to protect itself from outside attack or even from itself. This causes us to look at CPUs that include (or can include) memory management units (MMUs) to address this issue. Virtual memory will allow trusted programs access to the entire system, and untrusted ones to access just the memory they have been allocated. A 3G cell phone is a prime example of the need for protection. No longer can you use a CPU that lacks an MMU, since a rogue program will crash your phone. Although an MMU does not eliminate the possibility of system crashes, it reduces the likelihood of hard-to-resolve system failures.

Three main CPU architectures center around 8-, 16-, and 32-bit data registers with 16-, 24-, and 32-bit address buses. The main difference between these CPUs is how much information one particular register can hold and how much it can address directly:

- 8-bit data/16-bit address = (0–256), with 64 K address space
- 16-bit data/24-bit address = (0–65536), with 16 M address space
- 32-bit data/32-bit address = (0–4 billion), with 4 G address space.

Why would an embedded system ever need to access 4 G of address space? The answer is simple: as the system is asked to perform more complex tasks, the size and complexity of the code it runs increases. In the early days, CPM on a Z80 utilized a process of banking memory and page swapping in order to run more complex programs on an 8-bit machine. Space of 64 K was not sufficient, and a solution was to make the system more complex by overlaying memory and pages to get more out of the CPU.

It seems like a 24-bit address bus would be adequate for many designs. A couple of factors drive us to a 32-bit address space: protection and pointers. For protection, the CPU with virtual memory can use the entire address range to divide up the physical memory into separate virtual spaces, thus providing protection from bad pointers. And the ability for any register to become a pointer to memory without the need for indexing simplifies the software.

1.7.4 CPU Performance

The performance of the overall system will be greatly impacted by the selection of CPU. Specifically, features like cache, MMU, pipelining, branch prediction, and superscalar architecture all affect the speed of a system. Depending on the needs of the SoC, these features may be necessary to achieve system performance.

1.7.5 Power Consumption

The end use of the SoC will determine how much power your design can consume. If your design is battery operated, the CPU will need to be as power conscious as possible.

For instance, some CPUs have the ability to sleep, doze, or snooze. These modes allow the CPU, when idle, to suspend operation and consume less power by shutting down various parts of the CPU. Different CPUs perform the same task with different results.

1.7.6 Costs

The cost of the CPU can be measured in several ways. First the intellectual property (IP) cost, which is the cost to acquire the IP for your SoC and any derivative products. Then there is the system integration cost. Which tools are available for design and implementation of your SoC? Finally, is the CPU variant silicon proven, and is it available on the bus architecture that your SoC is utilizing?

1.7.7 Software Issues

The availability of an RTOS and middleware may dictate your choice as well. For instance, in designing a PDA, you may want the middleware that is available for Linux, but the choice of a virtual operating system will dictate that you migrate away from small non-MMU CPUs.

Is there a graphics system or a file system necessary in the design? If so, then the choice of RTOS will dictate the type of CPU that is needed as well. Many RTOS vendors target specific families, leaving others untouched. Most 8-bit CPUs have simple schedulers that are adequate for small designs that utilize little outsourced code. They are not likely to be adequate for designs that consume any quantity of outsourced code. The outsourcing of the solution will strongly influence the RTOS choice, which, in turn, dictates what types of CPU are possible.

The tools necessary to do the design: are they available for the standard ANSI C/C++ compiler that you may use? How will you debug your design, either in the hardware/software cosimulation environment or on the SoC after it exists? Does a JTAG port exist, and is the CPU using this channel for debug, or is a dedicated serial port necessary? The choice of a higher-level language like C++ or code generated from a design in UML may also dictate the need for a higher bus width and clock frequency to deal with the code size and complexity.

1.7.8 Multicore SoCs

The SoC may be better off if partitioned into several processor subsystems that communicate via a loosely connected FIFO or serial channel. Many designs incorporate a DSP (digital signal processor) and an RISC CPU to share the workload and simplify the design of each processor domain. But this further complicates the CPU choice, which may now be multiplied several times over.

1.7.9 Conclusions

Modern SoC design has presented new challenges for the system architect. No longer is the choice of CPU trivial. Utilizing metrics such as the complexity of overall design, design reuse, protection, performance, power, size, cost, tools, and middleware availability can simplify the decision.

1.8 An Introduction to USB Software

The success of USB on the desktop is undeniable and has led to a surge of interest in its incorporation into embedded devices. This interest, in turn, resulted in numerous questions about just how USB worked and how it might be implemented. In response, C.C. Hung and I wrote a tutorial in NewBits in the summer of 2004. This article is based on that tutorial.

CW

However you look at it, USB is a good thing.

If you have worked with PCs for a few years, you will remember how it used to be. You would buy a new peripheral device, spend lots of time and effort ripping apart your PC to install the card, and then start worrying about software. It took forever. By the time you'd finished, any excitement you had about this cool new device had long since evaporated.

USB changed all that. Nowadays, you just plug a standard USB cable into the back of the computer and into the device and switch on. Sometimes the computer will take a moment or two to figure out drivers and so forth, but in no time, you are up and running.

USB highlights an interesting phenomenon in the high-tech world: the simpler something is to use on the outside, the more horribly complex it is on the inside! We will take a look at how USB works, and you will see what this means.

1.8.1 What Is USB?

USB was designed as an alternative to replace the plethora of different serial and parallel interfaces used to connect peripherals to PCs. It is a single standard, which minimizes the number of different interfaces, cables, and connectors. The whole input/output system is simplified, and it offers the potential for real plug and play.

There is also plenty of capacity. A USB system may support up to 127 devices. A wide range of performance options are available to support a selection of requirements. The original USB specification allowed 1.5 or 12 Mb/s; the latest spec increases this by up to 40×.

The structure of a USB system is hierarchical. It is termed a "tiered star" topology. A USB host includes the "root hub," which forms the nexus for all device connections. The host controls the bus, and there may only be a single host in any USB system (see Figure 1.8). Hubs have a single upstream connection but may have many downstream. Hubs increase the

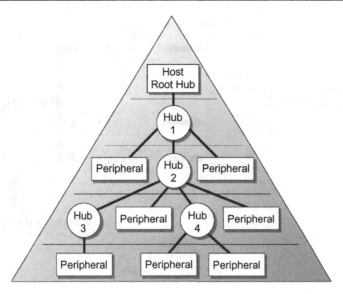

Figure 1.8
USB topology

logical and physical fan out of the network. "Peripherals"—the devices controlled by the host or hub—are also referred to as "functions" in the USB world.

1.8.2 A USB Peripheral

The typical architecture of a USB peripheral device is shown in Figure 1.9. The USB peripheral controller is interfaced to the controlling microprocessor or microcontroller. This, in turn, is interfaced to the rest of the electronics of the peripheral device.

Communication to and from the host or hub is transmitted serially in the form of frames. All communications are sent and received via addressable buffers called "endpoints."

Transfers take place through logical channels called "virtual pipes," which connect the peripheral's endpoints with the host. Each connection always has a control pipe (endpoint zero) and one or more data pipes (endpoints 1, 2, etc.), which may be configured as "IN" or "OUT" endpoints. IN denotes device to host communication; OUT is host to device.

1.8.3 USB Communications

When establishing communication with the host, each endpoint returns a descriptor. This is a data structure that tells the host about the peripheral's configuration, protocol, message transfer type, packet size, data transfer interval, and so on. USB is a pure master/slave architecture. The host initiates and controls all communication. The host sends a control packet to the peripheral device for "enumeration"—the process in which the peripheral describes itself to the host. The host chooses which peripherals are allowed to connect.

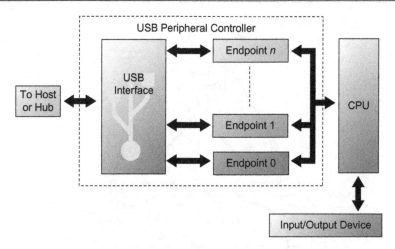

Figure 1.9
A USB peripheral

Four types of data transfer are supported:

- **Control transfers** exchange configuration, setup, and command information between the host and device. This is required for every connection and is associated with endpoint zero.
- **Bulk transfers** move large amounts of data, when timely delivery is not critical. There is no guaranteed throughput. Typical applications include printers and scanners.
- **Interrupt transfers** are confusingly named, as they have nothing to do with interrupts in the CPU-diverting sense. They are used to poll devices to ascertain whether they require service. Mice and keyboards are typically handled this way.
- **Isochronous transfers** handle streaming data and offer a guaranteed throughput. They are typically used for audio and video applications.

1.8.4 USB Software

From a software perspective, the implementation on host and peripheral—or function or device, as you prefer—is quite symmetrical and comprises a series of layers, as shown in Figure 1.10.

As with most communications protocols, USB can most easily be described as a series of discrete layers:

- At the top is the **application**. On the host, this is the program that is utilizing the device. In the function, this is the embedded software that controls the device.
- **Middleware** enables the easy integration of the application with the underlying components.

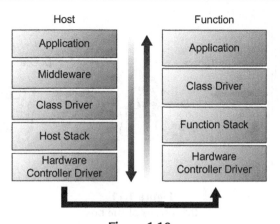

Figure 1.10
USB system architecture: software layers

- **Class drivers** implement capabilities such as storage, printing, audio, imaging, and human interface. They provide the real *characterization* of the device. They may be implemented according to the USB standard specifications, or they may be vendor-specific. This will depend upon the kind of device being supported. Here's some useful advice: if you are building a USB device, making it look like a standard peripheral—complying with a standard class driver—will make life much easier, obviating the need to develop host (Windows) drivers.
- The **Stack** layer processes the USB protocols and manages all data transactions for devices on the bus.
- The **Controller driver** layer manages USB device power, enumeration, and other low-level control functions.

1.8.5 USB and Embedded Systems

USB is generally thought of as being a means to connect devices to a PC. In this case, the PC is the USB host and the devices or peripherals are USB functions. But we are interested in embedded systems, which requires a different kind of perspective.

The obvious application of USB in an embedded system is when you are building a device—a peripheral that may connect to a PC. This is a USB *function* and requires software support for that end of the bus. Examples of USB functions include hard drives, printers, audio devices, medical equipment, cameras, keyboards, mice, … and the list just goes on. You name it nowadays, and it has a USB interface. There are times when the USB *host* need not be a PC and is an embedded system of some kind. Possibilities here include set-top boxes, point-of-sale devices, and healthcare monitoring systems. Of course, the interface board that goes in a PC could well be an embedded system in its own right. These are USB hosts and need software support for that end of the bus, which, as we have seen, has distinct differences from the function end.

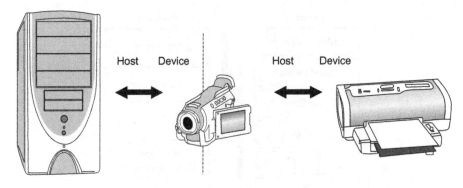

Figure 1.11
Both sides of the bus

It is possible to identify examples of embedded systems that may act as both USB host and function, at different times. The example shown in Figure 1.11 is a camera, which may be a function, when a PC is the host, for uploading pictures. It can be the USB host when it connects directly to a printer to get a hard copy of images. In this context, the camera is actually supporting two separate USB systems. It needs software support for both ends of the bus.

Recognizing this kind of situation, an extension to the USB 2.0 specification was defined. This extension is called "USB On-the-Go" or "OTG." This capability enables a device to behave as a limited USB host and connect to other USB functions. The primary limitation is the inability to support hub functionality. It is ideal for portable devices, such as our example. For this to work, both the devices must be OTG-compatible. Although greeted with much enthusiasm, OTG has, to date, not been widely adopted.

1.8.6 Conclusions

Although familiar to almost any user of a modern computer, the underlying functionality of USB is not widely understood. It is well defined, but complex. Its successful application requires absolute adherence to published standards, which may most readily be achieved by licensing a commercial embedded USB stack. The standards are evolving rapidly, as higher-speed interfaces become standard; USB 3.0 is the next step.

1.9 Toward USB 3.0

USB continues to be a hot topic and the arrival of USB 3.0 seemed to pique interest further. I found that Web seminars covering USB seemed to attract record audiences, for example. With this in mind, my colleague Waqar Humayan wrote a paper that reviewed the state of USB technology and introduced USB 3.0. I have adapted that paper to draft this article, which gives a flavor of what USB 3.0 is all about.

CW

1.9.1 Introduction

USB 3.0 or "SuperSpeed" USB is the next evolutionary step toward wired USB. Released in November 2008, USB 3.0 takes theoretical bandwidth of USB to around 5 Gb/s, which is more than 10 times faster than its predecessor USB 2.0 (Hi-Speed USB). This significant increase in speed has made USB 3.0 an extremely lucrative option for bandwidth-hungry applications like storage and multimedia.

Although, much like previous USB standards, USB 3.0 is backward compatible, it is not merely an extension of USB 2.0, rather, it's a completely new bus specification from the hardware point of view. Still, it's the beauty of design that old class drivers and applications will want to migrate over to USB 3.0 either with no modifications or with minor changes.

1.9.2 Bus Architecture

USB 3.0 is dual bus architecture, in which both USB 3.0 and USB 2.0 signals run in parallel to each other. USB hosts enumerate devices on the highest supported signaling rate. Besides being a dual bus architecture, USB 3.0 shares the same architectural components with USB 2.0, i.e., host, hubs, and devices. USB 3.0 is also a tiered star topology and maintains the same four transfer types: control, bulk, interrupt, and isochronous. Operations of these transfer types in USB 3.0 are very similar from a software perspective, but are entirely different from the protocol point of view.

1.9.3 Cables and Connectors

USB 3.0 cable carries both USB 2.0 and USB 3.0 signals. USB 3.0 is a dual, simplex bus which supports simultaneous bidirectional flows. USB 3.0 cable carries two differential pairs for USB 3.0 signaling (SSTX+, SSTX−, and SSRX+, SSRX−) and one differential pair for USB 2.0 signaling (DP, DM). In order to house an additional four USB 3.0 signals, USB 3.0 connectors and receptacles have also been modified.

1.9.4 Packet Routing

Another major difference in USB 3.0 is packet routing. In USB 2.0 and previous standards, packets were broadcast on the bus, each device received all packets and rejected the packets that were not meant for it. In USB 3.0 packets are explicitly routed. This leads to the provision of suspending a link instead of a whole tree under a port.

1.9.5 Bidirectional Protocol Flow

Up until USB 2.0, USB only supported unidirectional data flow and the direction of bus was swapped after negotiation between host and function. As there are separate Rx and Tx channels in USB 3.0, simultaneous bidirectional flows are now possible. Looking at a bulk IN endpoint in USB 2.0 as an example, the host has no way of determining if the device wants to send the data, therefore it keeps on polling the bulk IN endpoint. If a device has no data to send on this endpoint, then it simply sends a negative acknowledgment handshake (NAK) unless data is available on this endpoint. Benefiting from the bidirectional bus in USB 3.0, the host initially asks for data on a bulk IN endpoint, if the device has no data to send then it simply sends a not ready handshake (NRDY) and the host will no longer poll this endpoint. Once data is available on this endpoint, it will send an endpoint ready handshake (ERDY) to the host. This protocol sequence saves bus bandwidth and power by allowing the host to stay idle if the device has no data to send. In order to improve performance of the system, USB 3.0 also uses bursts. Bursts allow an exchange of multiple packets and the receiver can send the status for all packets out of order.

1.9.6 Bulk Streaming

Another significant change in USB 3.0, with respect to achieving better throughput, is "bulk streaming." Bulk streaming is when a standard USB bulk endpoint supports a single stream of data between host and device via a host memory buffer and device endpoint. New USB SuperSpeed provides protocol level support for multi-stream model and uses stream pipe communications mode.

Bulk streaming allows command queuing and out of order completion of these commands. Typical application of this scenario is USB mass storage devices. Streaming allows the host to queue in multiple commands to the device and also allows a device to return data out of order to optimize performance.

In order to better understand this concept, let's take a look at a USB mass storage device. USB mass storage Bulk Only Transport (BOT) protocol uses one bulk IN and one bulk OUT endpoint. It involves sending a 31 byte Command Block Wrapper (CBW) on bulk OUT endpoint. As a result, data will be exchanged between host and device. If the host wants to read data then the device will send data on bulk IN endpoint otherwise if the host wants to write data, it will send data on bulk OUT endpoint. After the data is exchanged, the device sends a 13 byte Command Status Wrapper (CSW) reporting the result of this command. Observing this protocol keenly reveals that the host has to wait for completion of CBW before moving to the data stage and has to wait for completion of the data stage before scheduling the CSW.

Now let's look at the same BOT protocol using bulk streaming. The host can send multiple CBWs through one bulk OUT endpoint and data transfers are scheduled through other bulk IN and bulk OUT endpoints. CSWs are scheduled through one other bulk IN endpoint. Each command has stream ID associated with it to identify the transfer. Multiple commands and data transfers can be scheduled and the device and host hardware handle the stream switching on their own. As the host and device do not need to wait for the completion of the previous stage, the host can send back-to-back commands and the device completes these commands as they appear on bus and queue the status of each command.

1.9.7 USB 3.0 Power Management

Another important design goal of USB 3.0 is aggressive power management. Despite port level and device level suspension in USB 2.0, USB 3.0 introduces the following multi-level power management attributes:

- **Link:** USB 3.0 allows suspension of a particular link contrary to only port level suspend in USB 2.0.
- **Device:** Device level suspends are the same as in USB 2.0.
- **Function:** If there is a composite device, then USB 3.0 allows suspending only one function of this device. An example would be a composite device having a mass storage and LAN functions. If the host system has nothing to do on the LAN, and it wants to turn it off while still being able to access the storage contents, it can thus suspend only the LAN function while storage remains active. Since most of the USB devices are bus powered, putting a portion of device to sleep would allow considerable power savings in the device.

1.9.8 USB 3.0 Hubs

USB 3.0 hubs are special devices in the sense that they operate at both speeds while all other SuperSpeed devices can only operate in one mode at a given time, either SuperSpeed or non-SuperSpeed. Once connected with a SuperSpeed host, USB 3.0 hubs enumerate as two distinct devices, one as SuperSpeed and the other as non-SuperSpeed. A SuperSpeed device connected on a downstream port of a USB 3.0 hub connects to a host through the SuperSpeed part of hub, while a non-SuperSpeed device connects through the non-SuperSpeed part of the USB 3.0 hub.

1.9.9 xHCI—New Host Controller Interface

With the release of SuperSpeed USB specifications, USB-IF has also released a new host controller interface called **eXtensible Host Controller Interface** (xHCI). Apart from handling SuperSpeed devices, xHCI alone is capable of handling non-SuperSpeed devices. Today many PCs and laptops are entering the market with native support for xHCI controllers. In addition, xHCI PCIe adapter cards are also becoming available.

1.9.10 Future Applications for USB

Most PC and embedded vendors are providing USB as the only option for external device connectivity. Almost every kind of device we use in our daily lives, such as a mouse, keyboard, mobile phones, and printers, is using USB connectivity. It's anticipated that with SuperSpeed capability, USB will gain momentum in bandwidth-hungry applications. The following areas are where USB will excel in future.

Storage

So far the storage community is the most excited about USB 3.0. Earlier USB mass storage transport protocol, the "bulk only transport" (BOT) protocol, was sequential in nature. A sequential protocol can't take maximum benefit of a bidirectional bus and undermines SuperSpeed capability of USB 3.0. USB-IF's device working group for USB mass storage devices has deprecated the BOT protocol and adopted a new transport protocol standard called **USB Attached SCSI Protocol** (UASP). UASP is fully compatible with the T10 committee's SCSI architecture model standard and capable of utilizing maximum bandwidth of SuperSpeed USB.

Although external SATA (eSATA) matches the SuperSpeed USB bandwidth, the storage community seems more tilted toward USB. All of the leading vendors from the storage industry have worked aggressively in formalizing the UASP and related standards. There are two obvious reasons for this. First, all PCs and laptops have USB host ports, only a few support of eSATA. Second, adoption of UASP has eradicated the limitations introduced from the mass storage BOT standard.

At the time of this writing, a number of SuperSpeed USB storage devices have already hit the market and many more are expected. Most of these devices are still using the BOT protocol because operating systems lack native support of UASP. This trend will change once more operating systems become SuperSpeed and UASP aware.

Multimedia

Multimedia devices are the next obvious application area which can take benefit of SuperSpeed USB. Recently, USB-IF announced a new class specification for USB audio and video (USB AV). The use case of such a device can be found in a monitor with built-in speakers, microphone, and web camera.

USB Device Authentication

Current USB standards lack a device authentication mechanism. USB device authentication can add significant value for the end-user. This will help prevent unauthorized use of content on removable devices. Both the USB device and host will exchange public keys to authorize each other. The device will only be enumerated if the authorization was completed successfully.

1.9.11 Conclusions

Recent studies have shown that more than 10 billion USB units are in use all over the world—with more than 3 billion new USB units added every year. USB is the connectivity standard that has shown tremendous growth in recent times and has surpassed competing technologies. With a wide adoption from key players and entire ecosystems, and continuous improvement in terms of speed, power management, and user experience USB has dominated in the wired connectivity space. There's no question that USB 3.0 will maintain this position into the future as well.

Further Reading

[1]   Guidelines for the use of the C Language in vehicle based software, the original version of MISRA C. Publication details are available at www.misra.org.uk.
[2]   L. Hatton, Safer C, McGraw-Hill, New York, 1995.
[3]   J. Lemieux, Programming in the OSEK/VDX Environment, CMP Books, Lawrence, KS, 2001.

Design and Development

Chapter Outline

Finding a logical flow for the chapters in this book proved challenging. I more or less gave up and simply grouped articles that seemed to belong together. Design and development methodologies and tools seem to fit alongside one another. But where should they all go in a sequence? After all, you do design first, but design methodologies are seen as "high level," so shouldn't the "low-level" stuff come before it? See what I mean?

What did prove interesting is that there are two "buzzwords" that are evoking a lot of interest in the embedded world at the beginning of the twenty-first century: UML and Eclipse. Both are addressed in articles in this chapter.

2.1 Emerging Technology for Embedded Systems Software Development

Crystal ball gazing is a hazardous occupation. No matter how well you know a technical subject, new developments will arise that you were unable to foresee. I wrote an "agenda setting" piece for NewBits in the late 1990s, and while reviewing it for use in this book, I was surprised at how much had "gone according to plan." This article for the 1990s needed surprisingly little adaptation to be developed into an article for the twenty-first century. The beginnings of multicore design were foreseen, which gets a whole chapter later in this book. The only key new technology is the UML, which is addressed in more detail in an article in this chapter.

CW

It is easy to think of embedded systems development as state of the art and leading edge. However, since microprocessors were first introduced in the early 1970s and the business has been developing over 30 years—more than a quarter of a century—it is now a mature

Embedded Software: The Works. DOI: 10.1016/B978-0-12-415822-1.00002-7

technology. By "mature," I do not mean "stagnant" or "boring." Embedded systems software development is far from boring. It is hard to identify any other business that is more dynamic, fast moving, and forward looking.

That maturity may be used to real advantage. After 30 years of growing, it is possible to identify a number of clear trends in the evolution of embedded systems development as a whole. Those trends point to the emergence of key technologies, upon which we may confidently focus to address the challenges ahead.

In this article, we endeavor to identify some of those trends and single out the technologies that they drive, resulting in an agenda for our attention over the coming months and years.

2.1.1 Microprocessor Device Technology

The earliest microprocessors were 4- and 8-bit devices. As fabrication techniques became more sophisticated, integrated 8-bit microcontrollers began to appear and the first 16-bit microprocessors came into use. Once again, silicon technology moved on, and 16-bit microcontrollers were introduced and widely applied, as demand grew for more sophisticated embedded systems. Devices with 32-bit architecture gradually took hold in higher-end applications, and these too were complemented by highly integrated microcontrollers. The first 32-bit devices were all CISC architecture, but increasingly RISC chips are providing even higher performance.

It would be easy to interpret this "potted history" of the embedded microprocessor, as illustrated in Figure 2.1, as a description of a timeline: 8-bit micros were yesterday; 32-bit RISC is today. However, this is not the case. As the more powerful devices have become available and found application, they have not, for the most part, replaced the earlier parts but have augmented the range of options available to the designer. An embedded systems designer has a wider choice of microprocessors than ever before and must make a choice based upon functionality, specification, support, availability, and price.

This increasingly wide range of devices has a number of possible impacts on the software designer. Obviously, suitable programming tools must be available to support this array of processors; it is preferable that the tools are consistent from one device to another. More importantly, the necessity of migrating both code and programming expertise from one device to another is becoming commonplace. This need not present major problems. By careful code design, use of off-the-shelf components, and adherence to recognized standards, porting may be quite straightforward.

2.1.2 System Architecture

As microprocessors have evolved, the architecture of the systems in which they are used has progressed as well. The earliest systems were comprised of the CPU and a selection of logic devices. More highly integrated devices reduced the chip count, and

Figure 2.1
Microprocessor technology

higher-performance devices presented many design challenges to the hardware developer. From the software engineer's point of view, nothing really changed. For many years, the same debugging techniques could be employed as the system became more complex: in-circuit emulation, on-chip debug, ROM monitors, and instruction set simulation. This situation began to change.

As embedded systems become more powerful, with ever-increasing levels of demanded functionality, many designers are taking a fresh look at their use of microprocessors and microcontrollers. In many cases, instead of following the obvious path of simply incorporating more powerful devices, an alternate choice is made: the application of multiple processors. This choice may be driven simply by a desire to distribute the processing power (which would be typical in a telephone switch, for example). Alternatively, one or more additional processors may be added to provide specific functionality (e.g., a DSP—digital signal processor—in a mobile phone) just extra computing power.

One of the biggest challenges faced by software developers when confronted with a multiprocessor system is debugging. It is, of course, possible to simply run one debugger for each device. However, that is not really addressing the problem. What is needed is the means to debug the system; the functioning of each processor and the interaction between them needs to be debugged. The requirement is for a debug technology that supports multiple processors in a single debug session, even when a variety of architectures are represented.

2.1.3 Design Composition

In the earliest days of embedded systems, all of the development—both hardware and software design—was typically undertaken by a single engineer. The software element represented a small part of the entire effort: perhaps 5–10%. As illustrated in Figure 2.2, over time, the proportion of the engineering time dedicated to software development increased substantially. By the mid-1980s, this work was done by software specialists and comprised more like 50% of the development effort.

In the last few years, although hardware design has become more complex, the amount of software has grown drastically, now often being 70–80% of the total design effort. The result is that teams of software engineers are involved, and new challenges arise. Among these is the availability of hardware to facilitate software testing. Since more software needs to be developed (in a shorter time), an environment for testing is required sooner. Various solutions are available, including native code execution prototyping environments, instruction set simulation, and the use of standard, low-cost, off-the-shelf evaluation boards. In addition,

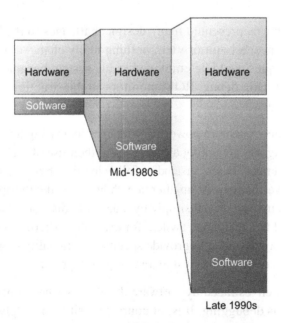

Figure 2.2
Design composition

low-cost host–target connection technologies are becoming common, typically using a JTAG interface.

This climate represents an ideal opportunity for hardware and software teams to work together. By using codesign and, in particular, coverification techniques, software engineers can test on "real" hardware sooner, and the hardware designers are able to prove their designs earlier, with less prototyping cycles.

2.1.4 Software Content

The proportion of development time dedicated to software has been increasing. Meanwhile, under pressure from worldwide trade and truly global competition, time to market has been decreasing. This has radically influenced the design strategy. The earliest designs were quite simple, being comprised solely of in-house designed applications code. As systems became more complex, a multithreading model was widely adopted for software development, and many developers opted for standard, commercial real-time operating system (RTOS) products.

As shown in Figure 2.3, the proportion of bought-in software, or "intellectual property" (a term borrowed from the hardware design world), has steadily increased, as further standards are adopted.

This trend has a number of implications for the software developer. The integration of standard software components—with the applications code and with one another—is a matter of concern. Debugging in a multithreading context is another issue. The business decision associated with the selection of intellectual property is particularly complex; future (e.g., migration to different processors) as well as immediate requirements must be taken into consideration.

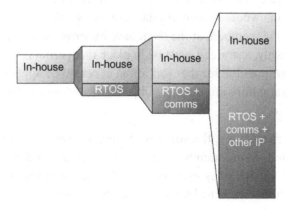

Figure 2.3
Software content

2.1.5 Programming Languages

For the first 4- and 8-bit microprocessors, there was no choice of programming language. Assembler was the only option. Since the applications were relatively simple, this was not a big problem.

As 16-bit technology became viable, the need for a practical high-level language became apparent, and several options emerged. Pascal and C were both in use on the desktop, and these languages were adapted for embedded systems. Intel developed PL/M specifically for this kind of application. Forth was also very popular for certain types of systems. Over time, with the increasing use of 32-bit technology, the two languages that persisted were C and Ada. The latter is prevalent in defense-oriented systems.

It has been known for some years that C++ would start to replace C for embedded software development. Now, between one-quarter and one-third of embedded systems code is written in C++. What was not anticipated a few years ago was the emergence of new languages and approaches, which are set to play a strong role in applications development in the future. The Java language was developed specifically for embedded applications and has found a niche where runtime reconfigurability is demanded. The Unified Modeling Language (UML) has become the most popular choice for a higher-level design methodology but has yet to be adopted universally.

2.1.6 Software Team Size and Distribution

As discussed earlier, the initial embedded system designs were one-man efforts. In due course, specialization resulted in engineers being dedicated to software development. The next step was the establishment of embedded software development teams. Managing software development is challenging in any context; embedded systems development is no exception and brings its own nuances. Using conventional programming techniques—procedural languages like C and assembler—most members of the team need to have a thorough knowledge of the whole system. As the team grows, this becomes less and less feasible. Typically, specific members of the team have expertise in particular areas. To manage the team effectively, a strategy must be in place that permits the encapsulation of their expertise. It must be possible for the work of an expert to be applied by the nonspecialist in a safe, secure, and straightforward manner. Object-oriented programming techniques find application in this context.

With many very large companies, the software teams are not simply growing; they are becoming distributed. Some members of the team are located at one site, while others are elsewhere. The sites may even be in different countries. This arrangement is common in Europe, where (spoken) language may be a concern. Elsewhere, time zones may be an issue (or an advantage, as a distributed team can work around the clock).

This is increasingly the case as emerging technology centers (e.g., in India) are widely utilized. The need for reusable software components becomes even more apparent in this context.

2.1.7 UML and Modeling

The UML has become a key design methodology in recent years, which goes hand in hand with increasing embedded software team size. There are broadly two ways to use a design tool: either as a guide to writing the actual code or as a means of generating the code directly. Code generation is controversial for embedded software, as it may be argued, quite validly, that every system is different and has very specific needs in this respect. This is where xtUML (executable and translatable UML) is attractive because it enables the application and architecture to be clearly separated. This follows the same philosophy as object-oriented programming—leveraging expertise through tools and technology.

2.1.8 Key Technologies

All of these trends, which have become established over 30 years of embedded systems development, point to some key technologies:

- **Microprocessor technology:** Leading to a proliferation in devices which involved the consideration of migration issues and writing portable code. That, in turn, drives a requirement for **compatible tools** and **RTOS products** across microprocessor families.
- **System architecture:** Progressing so that multiprocessor embedded systems are becoming commonplace. This drives a requirement for a **debug technology** that addresses these needs.
- **Design composition:** Changing, with a much greater part of the design effort being expended on software. This drives a requirement for **instruction set simulator technology** and **host-based prototyping**, and the application of **on-chip debug facilities** and **hardware/software coverification**.
- **Software content:** Moving from entirely in-house design to the wide use of intellectual property. This drives a requirement for **standards-based RTOS technology** and appropriate **debug technology**.
- **Programming language:** Narrowing choices somewhat. Although a strong requirement for **C tools** still prevails, compatible **C++ products** are in strong demand.
- **Software team size and composition:** Changing from one engineer (or less) to the employment of large, evenly distributed, teams. This drives a requirement for tools to support **object-oriented programming** and **RTOS technology with a familiar or standard API**. There is also an increasing demand for **modeling and design tools**.

2.1.9 Conclusions

Tracking all the emerging technologies, which are driven by the ongoing trends in embedded systems development, is no easy task. Taking any one in isolation is also fruitless because of the many interrelationships. For example, multithreading and multiprocessor debugging go hand in hand; standards-based RTOS technology is a real boon to processor migration; using a design methodology that flows naturally toward an implementation makes complete sense.

2.2 Making Development Tool Choices

Developing embedded systems software is a complex matter, and the tools are necessarily complex. Although today the focus is heavily on integrated development environments, the selection of the tools to be used in that context is just as important as it was when I wrote a detailed review of the topic for NewBits in the mid-1990s. That piece is the basis for this article. Many of the topics highlighted here are expanded upon in other articles in this book, notably in Chapters 4, 5, and 7.

CW

This article is a review of available tools and techniques for program development in embedded systems, and it also discusses the implications of the availability of development tools on selection of a target microprocessor and real-time operating system. This article addresses the following questions regarding the selection process:

- What build tools will be needed?
- What features should be sought?
- What about the debugging parameters and options?
- What about tool integration?

2.2.1 The Development Tool Chain

A useful way to view software development tools for this purpose is from the perspective of a tool chain, where each component forms a tight link to the next: from a high-level language, through the assembler, linker, and so on, to one or more debugger variants (see Figure 2.4).

There are two distinct parts to the tool chain:

1. The body of the chain consists of the tools that take source code and generate an executable: the *build tools*.
2. The base of the chain includes the tools used to execute and verify the executable program: the *debug tools*.

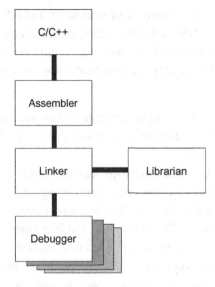

Figure 2.4
The development tool chain

Almost without exception, the use of the build tools (options, controls, formats, etc.) should be quite unaffected by the proposed execution environment and the variant of debug tool employed. For example, it should not be necessary to build using a special library in order to use a particular debug tool. The clear requirement is to test exactly the same code at all stages of the development process.

The options, with respect to the execution environment, offered by debug tools are numerous. These options will be reviewed in turn, but let us first consider the build tools.

2.2.2 Compiler Features

Most commonly, software for embedded systems is written in C. However, C++ is increasing in popularity, as object-oriented design becomes the norm. The main parameters governing the choice of cross-compiler are similar to those applied to native products, but other factors must be taken into consideration, such as:

- **Programming language accepted**
  The primary requirement in a compiler is that it accept the programming language in use. For both C and C++, full compliance with the ANSI specifications is essential.
- **Libraries provided**
  An ANSI-compliant C compiler need not, according to the specification, include a full runtime library. In reality, such a library is very useful, and its absence would

hinder efficient program development. Unfortunately, a number of the standard library functions, as specified by ANSI, are intrinsically nonreentrant, which may be a problem for some embedded system designs. Because particular demands may be placed on library code by an embedded system, access to the library source code is particularly desirable.

A common reason for using C++ is to facilitate code reuse and to be able to employ standard class libraries. It is, therefore, reasonable to expect a compiler to be supplied with such a library as standard.

- **Build tools that support an entire microprocessor family**
  Typically, an engineer selects a cross-compiler to support development for a specific target microprocessor that will be used for the current project. It is quite likely that future projects will use a different device, but it is commonly another member of the same family. With this in mind, choose build tools that support an entire microprocessor family. This support should, of course, go beyond the generation of code for the "baseline" device and should generate appropriate instructions for the specific variant in use.
- **Manufacturer support**
  Beyond the technical requirements of the build tools, it is at least as important to look at the "pedigree" of the build tools: consider the reputation of the company who produces them, their technical support facilities, and the size of the current user base.

2.2.3 Extensions for Embedded Systems

A cross-compiler is intrinsically a more complex tool than its native equivalent. This primarily comes about because very few assumptions about the target environment may be made by the compiler developer. To maintain the appropriate level of flexibility, the compiler manufacturer must implement a number of special features.

In particular, embedded systems almost always have complex memory configurations. The simplest have read-only memory (ROM) for code and constant data and random access (read/write) memory (RAM) for variable data. To accommodate this, the minimum required of the compiler is the generation of ROMable code, with the data clearly separated from the code. In most systems, a greater degree of control is needed, and being limited to this simple memory model would be a serious restriction.

A further implication of the memory structure of embedded systems is a clash with a language construct in C. In C, a static variable may be given an initial value. This was intended to avoid the necessity for initialization code for variables whose location in memory could be predicted at compile time. The intention was that such variables would be preset to their starting value in the executable file (memory image) on disk and loaded into memory with the program. For an embedded system, where the program is already in ROM, this mechanism does not work. This situation has three possible outcomes:

1. Static variables cannot be initialized.
2. Initialized statics can only be used as constants because they must be stored in ROM.
3. The build tools must readily accommodate the copying of data from ROM to RAM at start-up.

Since the C and C++ languages permit direct access to specific memory addresses, these languages are often useful for embedded system development, particularly for code that is closely associated with the hardware. Naturally, the compiler should not restrict this capability in any way.

As for assembler code, even though nowadays a high-level language is almost always chosen for software development, it is inevitable that, at some time, the programmer will write some assembler code. The use of assembler code may be necessary to permit the programmer to extract the last ounce of performance from the target chip, but, more likely, the programmer uses assembler code to access some microprocessor facility that does not map into C (e.g., enabling or disabling of interrupts). In the interests of efficiency and code portability, the use of assembler should be minimized, and the facilities for its development should be as flexible as possible. The ability to write a complete assembler module should be augmented by the means to include one or more lines of low-level code among the C language.

Impact of Real-Time Systems

The majority of embedded microprocessors are employed in real-time systems, and this puts further demands on the build tools. A real-time system tends to include interrupt service routines (ISRs); it should be possible to code these in C or C++ by adding an additional keyword `interrupt` declaring that a specific function is an ISR. This performs the necessary context saving and restoring and the return from interrupt sequence. The interrupt vector may usually be defined in C using an array of pointers to functions.

Furthermore, in a real-time system, it is common for code to be shared between the mainline program and ISRs or between tasks in a multithreading system. For code to be shared, it must be reentrant. C and C++ intrinsically permit reentrant code to be written, because data may be localized to the instance of the program (since it is stored on the stack or in a register). Although the programmer may compromise this capability (by using static storage, for example), the build tools need to support reentrancy. As mentioned previously, some ANSI-standard library functions are intrinsically nonreentrant, and some care is required in their use.

Since memory may be in short supply in an embedded system, cross-compilers generally offer a high level of control over its usage. A good example is a language extension supporting the `packed` keyword, which permits arrays and structures to be stored more efficiently. Of course, more memory-efficient storage may result in an access time increase, but such trade-offs are typical challenges of real-time system design.

One of the enhancements added to the C language during the ANSI standardization was the `volatile` keyword. Applying this qualifier to a variable declaration advises the compiler that the value of the variable may change in an unexpected manner. The compiler is thus prevented from optimizing access to that variable. Such a situation may arise if two tasks share a variable or if the variable represents a control or data register of an I/O device. It should be noted that the `volatile` keyword is not included in the standard C++ language. Many compilers do, however, include it as a language extension. The usefulness of a C++ cross-compiler without this extension is dubious.

2.2.4 Optimizations

All modern compilers make use of optimization techniques to generate good-quality code. These optimization techniques can result in software that rivals handcrafted assembler code for size and efficiency. Optimizations can be local to a function or global across a whole module.

Many optimizations may be applied in a generalized way, regardless of the target. More interesting are those that take specific advantage of the architectural characteristics of a specific microprocessor. These include instruction scheduling, function inlining, and `switch` statement tuning.

Instruction scheduling is a mechanism by which instructions are presented to the microprocessor in a sequence that ensures optimal usage of the CPU. This technique is a common requirement for getting the best out of RISC architectures. However, CISC devices can benefit from such a treatment.

The inlining of functions is the procedure whereby the actual code of a (small) function is included instead of a call. This is useful to maximize execution speed of the compiled code. Some compilers require specific functions to be nominated for this treatment, but automatic selection by the compiler is preferable. The optimization can yield very dramatic improvements in runtime performance.

In C, `switch` statements lend themselves to optimal code generation. Depending upon the values and sequence of the `case` constants, quite different code-generation techniques may be appropriate. Explicit tests, lookup tables, or indexed jump tables are all possible, depending upon the number and contiguousness of the `case` constants. It may even be efficient to generate a table with dummy entries if the constants are not quite contiguous. Since the compiler can "rewrite" the code each time it is run, efficiency rather than future flexibility can be the sole priority. This example gives a compiler a distinct advantage over a human assembler code writer.

Manufacturers of development tools for embedded systems have very limited knowledge of the architecture of individual configurations—every system is unique. As a result, fine control over the optimization process is essential. At a minimum, there should be a user-specified bias toward either execution time or memory usage.

2.2.5 Build Tools: Key Issues Recapped

In selecting build tools for embedded system software development, consider these two key issues:

1. Do the tools provide extensive accommodation for the special needs of embedded system development?
2. Does the compiler perform a high standard of optimization, with extensive user control of the process?

2.2.6 Debugging

Having designed and written a program and succeeded in getting it compiled and built, the programmer's next challenge is to verify the program's operation during execution. This is a challenge for any programming activity and never more than when working on an embedded system. In this context, many external influences on the debugging process and stringent requirements dictate the selection of tools.

Debugger Features

Some debugger features are desirable or vital in any context; others are specific to embedded systems work. We will concentrate on the latter.

A key capability of a debugger is the ability to debug fully optimized code. Although this sounds quite straightforward, it is not a facility offered by all debuggers. Often, there is a straight choice: ship optimized code or fully debugged code. It is common to select a microprocessor on the basis of its performance and to rely upon the compiler to deliver this performance. This is particularly true of high-performance RISC devices. It is unacceptable to be limited by available debugging technology.

In reality, debugging fully optimized code may be challenging for the programmer. The results of some optimizations (e.g., code motion and register coloring) can make it difficult to follow the execution process. It is, therefore, common to perform initial debugging with optimization "reigned in." However, for the final stages of testing, the debugger should not preclude the use of maximum optimization.

Programmers write software in a high-level language (usually C or C++) primarily in the interests of efficiency. The debugger should pursue this philosophy fully and operate entirely in high-level terms. Code execution should be viewed on a statement-by-statement basis; line-by-line is not good enough. Data should be accessible in appropriate terms. Expression evaluation, structure expansion, and the following of pointers should all be straightforward. On the other hand, low-level access to code and data should also be available, when required.

C++ presents additional requirements: function names should always be shown "unmangled" and constructors and destructors should be visible, for example.

Since the suppliers of tools for embedded systems development cannot predict exactly what a given embedded system is like, they are unable to predict the precise functional requirements of the debugger. In the same way as with the build tools, this problem may be circumvented by providing enough flexibility to the user. For a debugger, this flexibility is manifest in the availability of a powerful scripting language. This might permit I/O device modeling, test automation, and code patching, for example.

The user interface of a debugger is of primary concern, because its design can directly affect the user's productivity. It must be powerful, consistent, and intuitive, which is particularly important when debuggers are to be used in a variety of execution environments. It is clear that if a single debugger family can fulfill all the differing requirements, hours of operator training time can be saved.

Development Phases
Before considering how code debugging can be performed, it is useful to review the total development cycle of the embedded system. This process can be divided into five phases, as illustrated in Figure 2.5. At each phase, work on the software may progress, but the scope for progress and the techniques employed change as the system development continues.

Figure 2.5
Development phases

In phase 1, although the system hardware is undefined, initial work developing algorithms and trying ideas can proceed. At this stage, it is wise to train the engineers who are going to use the debugger in the use of the software development tools.

At phase 2, since the hardware configuration is known, the engineer performs detailed software design and a large part of the implementation.

In phase 3, although hardware is available, a software engineer may often wish that it was not because the hardware will probably be unstable. However, the engineer can now begin the software/hardware integration.

In phase 4, the availability of stable hardware, maybe in multiple units, permits the engineer to complete final integration and testing.

Some development projects can be completed entirely within the factory, without requiring phase 5. Commonly, however, on-site installation requires final tuning of the software. At some later time, enhancements to the system may necessitate on-site work on the software.

Each development phase calls for a particular type of debugger, as described in the sections that follow.

Native Debugger

At first sight, a native debugger (i.e., one running on the host computer, executing code in that environment) seems inappropriate for embedded systems development. However, there are two contexts in which such a tool may be useful.

During phase 1 of the project, with no clear idea of the target hardware configuration, a native debugger can provide a useful environment in which to develop ideas and formulate algorithms, particularly for sections of code that are not time critical. This idea can be extended further if a host-based prototyping environment is available. This permits a significant amount of development to proceed on parts of the application that interact with the hardware.

If a native debugger is available, one that has the same (or very similar) user interface to debuggers being used at later stages of the project, the native debugger can offer an ideal training ground, since even if the target hardware is available for training purposes, it may not be the safest place to "play around." The worst that can happen with a native debugger is to crash the computer. The consequences of some embedded systems going out of control may be more dire.

Debugger with Simulator

The simulation of the target chip, instruction by instruction, on the host computer provides a very useful environment for software testing at almost any phase of the project. In particular, at phase 2, when the hardware is known but unavailable, a simulator will make rapid progress possible.

A simulator allows very detailed debugging to be performed. Although not running at anything like full speed, the simulator keeps track of execution time and permits accurate timings to be taken. This means the engineer can fine-tune critical code sections early in the development cycle. Since the simulator can effectively add functionality to the microprocessor it is simulating, the execution of the code may be monitored in great detail without any intrusion at all. This facilitates 100% performance analysis and code coverage, which is not possible using other techniques.

Of course, a simulator limited to the simulation of just the core CPU would be of limited utility. The simulator must also address the interrupt and I/O systems.

Debugger with Monitor

Once stable and fully functional hardware is available, debugging on target hardware should be more straightforward. This is generally facilitated by use of a monitor debugger where the target hardware is connected to the host computer by a communications link (serial line, Ethernet, etc.) and the target runs a small (<10 K) monitor program that provides a debug environment to the debugger itself, which runs on the host. The result is a low-cost, highly functional debugging solution that enables code to be run at full speed on the target with very little overhead.

For a monitor debugger to be viable, the monitor itself must be highly configurable. Standard boards (evaluation boards, for example) should be supported "out of the box." Tools and services must be available to facilitate the rapid accommodation of custom hardware.

Although the use of a monitor debugger is most common during phase 4 of a project, it can also be used in phase 5. If the target monitor is included in the shipped software (after all, its memory overhead is likely to be very small), on-site debugging may be possible using just a laptop computer running the debugger.

JTAG Debugger

As the speed and complexity of microprocessors increase, the likely cost (and lack of feasibility) of in-circuit emulators (ICEs) increases. As a result, semiconductor manufacturers are increasingly adding debug facilities to the silicon itself. This may vary from the provision of hardware breakpoints (address/data comparators), which should be supported by a monitor debugger, to a special "debug mode" that requires specific debugger support.

Commonly such a debug mode uses a JTAG connection to provide on-chip debug (OCD). Assertion of OCD mode stops the processor and enables a debugger to read and write information to and from the machine registers and memory. To utilize OCD, an appropriate connector must be included on the target board, but this low-cost connector does not represent a significant overhead. Between the host computer and the target board, an OCD adapter is required. Like a monitor, a debugger with OCD (also termed "hardware assist") provides some of the functionality provided in the past by ICE at a much lower cost. Unlike a monitor, such a technique does not require an additional debug communications port(s) or code on the target.

Debugger with RTOS

As embedded applications become more complex, the use of a real-time operating system (RTOS) is increasingly common. Debugging such a system has its own challenges, and they dictate specific requirements in a debugger. Two particular areas of functionality are required in an RTOS debugger:

- Code debugging must be "task aware." Setting a breakpoint on a line of code should result in a break only when the code is being executed by the task being debugged. Code shared between tasks is very common, so this requirement can easily arise. Similarly, data belonging to a specific task instance must be accessible to the engineer.
- Information about the multithreading environment (system data) is required: task status, queues, inter-task communications, and so on.

It is clearly desirable that both these requirements are addressed in the same debug tool.

If an in-house designed RTOS is used, particular debug challenges arise.

RTOS awareness may be implemented using all of the previously mentioned debug technologies. In particular, OCD and monitor debuggers are most commonly adapted. It is, however, quite possible to enhance simulators or even native debug environments to be RTOS aware.

2.2.7 Debug Tools: Key Issues Recapped

In selecting debug tools for embedded systems software development, there are two key issues:

1. Does the debugger permit the use of fully optimized code?
2. Do the tools provide support for a wide selection of execution environments used in various phases of the development?

2.2.8 Standards and Development Tool Integration

When selecting development tools, attention to standards is essential. For build and debug tools, it is worth investigating the tools that colleagues and associates are using. Industry standards are likely to enjoy long-term support and "grow" with the target chip. Apart from the development tools themselves, integration with standard version management systems is increasingly a requirement with larger project teams. Similarly, clear links to design techniques must be sought.

Beyond industry standards, attention should be paid to the adherence to "real" standards; i.e., those set by international standards bodies. An obvious starting point is the programming language itself. Although the use of pure ANSI C/CC++ is desirable, in reality, a few specific language extensions are essential to make the language useful and efficient for embedded systems development. Such extensions are provided by suppliers of appropriate compilers (i.e., compilers specifically designed for working with embedded systems), and their use is, of course, very reasonable. A good example of an essential extension to the C language is the keyword `interrupt`, which enables a C function to be declared an interrupt service routine. Then the compiler can take care of the necessary context saving and restoring. However, some nonessential language extensions, provided by a few suppliers, should be avoided to aid code portability between compilers. Similarly, the use of a standard object module format (OMF) for relocatable and absolute binary files may remove the necessity of using build and debug tools from a single source.

In broad terms, choosing tools developed with open interfaces ensures interoperability with other products now and in the future.

2.2.9 Implications of Selections

Although selection of the software development tools is important in itself, it is one of a number of such selections that must be made during the development of an embedded system. Other selections include the target microprocessor, the development host computer, and the RTOS. It is important to appreciate the interaction between these various selection processes, some of which may be less obvious than others.

Target Chips

Many reasons can be cited for the selection of a particular microprocessor:

- It has the right range of features.
- The price was right.
- Low power consumption.
- It is fast.
- I have used it before.
- A colleague is using it.
- I liked the salesman.

These reasons are all valid, and a combination of them may be justification for selecting a device. However, another criterion should also be applied:

- A good range of software development tools is available.

Purchasing something from a single, unique source rarely is an acceptable decision. Why should it be the case with software tools? If a microprocessor is supported by a very limited range of tools—perhaps from a single vendor—its use should be called into question.

Host Computers

The choice of development platform is largely driven by the local culture. It is likely to be a PC running Windows or some flavor of Linux. Software tools vendors offering support on an incredibly wide selection of hosts may be guilty of redefining the word "support." Often, on the less-popular platforms, the product versions on offer are extremely old and have not been maintained.

RTOS

The choice of an RTOS (along with the decision to use one or not) is influenced by a number of factors. This topic is worthy of an article by itself; however, the availability of development tools is a significant factor, which I address here.

An RTOS with a suitably open architecture makes the most sense. It should accept the output generated by a range of build tools. Suitable debugging tools must also be available.

An option, which is considered under some circumstances, is the use of an in-house developed RTOS. This often represents the worst case in terms of tool availability and compatibility.

2.2.10 Conclusions

The selection of development tools for embedded systems software is not an easy task, with many vendors offering partial or even complete selections of products. A good appreciation of the possibilities and a checklist of questions to pose to vendors are key prerequisites.

2.3 Eclipse—Bringing Embedded Tools Together

In 2004 interest in Eclipse for embedded development applications snowballed. Late in the year Sarah Bigazzi wrote a piece for NewBits discussing the technology and its application, which I have adapted for this article.

CW

2.3.1 Introduction

Development tools are widely known to be key to the success of microprocessors. Although powerful embedded tools have been developed over the last two decades, little progress has been made in integrating multivendor tools on multiple hosts. Without good integration, communication between tools is restricted, and the full potential of the tools is untapped.

Proprietary IDEs (integrated development environments) limit integration and prevent use of best-in-class or preferred tools. This inflexibility frustrates developers and curbs productivity. *De facto* proprietary standards partially address this problem but are restricted to a single host. Thus, embedded developers have long wished for a host agnostic open IDE that they can enhance with their own or third-party tools.

The Eclipse platform, an open host-independent, industry-standard base, makes this possible.

On the desktop, the Eclipse platform is already noted for its excellence and is used in numerous business applications. The benefits seen on the desktop—a common tool interface and integration platform—can be brought to the embedded world.

In this article I introduce Eclipse and describe how it can be adapted and enhanced to make an extensible embedded platform without compromising key Eclipse concepts—standard interface and plugability.

2.3.2 Eclipse Platform Philosophy

During the Internet boom days, the availability of tools mushroomed for the various Internet business applications. Since these tools were built by diverse organizations, most of them had their own GUI paradigms and rarely worked well with each other. It became apparent that a standard IDE and framework were required. To address this need, IBM started the Eclipse project to build a well-designed tool integration platform so that independently built tools could be part of a single environment. The result was the Eclipse platform.

Originally, IBM released the Eclipse platform into Eclipse Open Source, and later, on February 2, 2004, the Eclipse Foundation reorganized into a not-for-profit corporation. "Eclipse became an independent body that would drive the platform's evolution to benefit the providers of software development offerings and end users. All technology and source code provided to this fast-growing ecosystem will remain openly available and royalty-free."

Unlike other open source organizations, the Eclipse Foundation is driven by business needs; hence, it is also known as the "directed" open source organization.

A major goal for Eclipse is to provide a well-planned and secure platform for commercial tool vendors. In addition, the Eclipse Foundation constantly works to remove hurdles in licensing the platform for commercial use. Contributed code is thoroughly scrubbed before it is committed; to ensure ease of licensing, there are plans to replace the existing CPL (Common Public License), which is already much simpler than the GPL (General Public License), with a more relaxed EPL (Eclipse Public License).

2.3.3 Platform

The Eclipse design focuses on a new paradigm—an open platform to integrate tools. In the old paradigm, individual tools are integrated, one at a time, either into an IDE or with another tool. This is a workable patch for a small set of proprietary tools but fails to scale in the larger multivendor context.

To address scalability, the Eclipse platform uses the innovative plug-in architecture. The platform, developed from the ground up, comprises well-defined GUI and framework

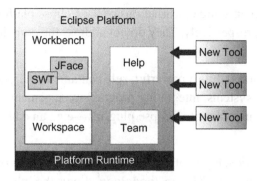

Figure 2.6
Eclipse platform architecture

mechanisms that provide a standard interface, facilitate integration, and are extensible. Tool developers, who no longer have to worry about GUI and framework issues, can concentrate on their tool-specific advancements—e.g., multicore debug.

Extension points, extensions, and plug-ins form the underlying mechanisms of the plug-in architecture. Plug-ins are the smallest functional entities. Eclipse plug-ins from any source can be plugged into the platform for a single integrated environment. Except for the platform runtime, Eclipse itself is implemented as a set of plug-ins as shown in Figure 2.6.

The core of Eclipse is its user interface made up of the Workbench, JFace, and the Standard Widget Toolkit (SWT). The combination of these plug-ins is known as the Rich Client Platform (RCP).

- **SWT and JFace:** SWT and JFace take care of the windowing system in an OS-independent way allowing portability across hosts.
- **Workbench:** The Workbench is the Eclipse UI. It is a collection of editors, views, perspectives, and dialogs provided as a common base for tools to use and extend.
- **Workspace:** Resources—projects, folders, and files—reside in the Eclipse Workspace where you can navigate them at will. The manipulation of resources, of course, provokes automatic incremental builds.
- **Team:** The Team plug-in takes care of source control. CVS (Concurrent Versions System) is the default, but other source control systems, including ClearCase, Source Integrity, and Visual SourceSafe, can be plugged in.
- **Help:** The Help plug-in does what the name implies. Integrated tools can extend Help for tool-specific needs.

2.3.4 How Eclipse Gets Embedded

As many in the Eclipse community have said, "the Eclipse platform by itself is an IDE for everything and nothing in particular." CDT (C/C++ Development Tools) and JDT (Java

Development Tools) are open source incarnations of the Eclipse platform for the desktop C++ and Java developer, respectively. They do not address the complexities of embedded development.

An Eclipse-based embedded IDE is a powerful, self-contained environment for building and debugging embedded systems, integrating program management and debug tools, and allowing users to drop in their favorite Eclipse plug-ins—e.g., an editor or source control system.

It is important that such a development strictly adheres to Eclipse principles and is itself implemented as a set of plug-ins. This methodology allows the inheriting of today's key features in Eclipse but also future ones as they become available. For example, platform runtime changes made in Eclipse 3.0 would be automatically reflected in the embedded IDE.

The key embedded development technologies that need to be made available as Eclipse plug-ins include:

- **Build tools:** To accommodate the great variability between embedded systems, compilers, assemblers, linkers, and so on, build tools tend to be significantly more complex than their native development counterparts. The return—in terms of improved usability—of their incorporation into an IDE is very significant.
- **Debug:** Software engineers spend more time debugging than anything else, and embedded programmers tend to use a variety of debug tools, which may accommodate different execution environments, RTOS awareness, or multicore debug capabilities.
- **Target connection:** An embedded development environment generally consists of a host computer with one or more target devices connected to it. These targets may be local, or they may be located remotely and reached via a network; they may be "real" targets (i.e., actual boards) or "virtual" (i.e., provided by some kind of simulation or emulation facility). Selecting and configuring the connections to multiple targets is another complex matter that may be readily simplified in this way.
- **Simulation:** Availability (or, rather, nonavailability) of hardware is an increasingly difficult challenge for embedded software developers, as the beginning stages of software development are brought forward to an earlier point in the project cycle. A number of simulation tools may be employed—native execution, instruction-set simulation, and hardware/software coverification are all options. These also need to be brought into the IDE.
- **Profiling:** Ensuring that an embedded application functions correctly is the first priority, but, since resources are always scarce, profiling tools are employed and need to be contained within the IDE. These tools analyze how resources—time and memory primarily—are used by the application (and any associated RTOS).

2.3.5 Conclusions

The need for an IDE for embedded software is apparent. The use of Eclipse as the basis for such an environment is clearly a very flexible approach, which is gaining ground across the embedded software development industry and will yield benefits for both suppliers and users alike.

2.4 A Development System That Crosses RTOS Boundaries

At the end of 2004, Robert Day wrote a paper that looked at the use of Eclipse for embedded software development from a different perspective. That paper was the basis for this article.

CW

In the embedded software world, the choice of real-time operating system has typically dictated the choice of development tools. In the 1980s and 1990s, close relationships were established between RTOS vendors and tool providers.

At the end of the 1990s, industry consolidation resulted in major RTOS vendors and device providers owning their own tools technology. With the increase in software in every new embedded design, the reuse of software is becoming critical, making it very unattractive to keep switching embedded software tools.

This leaves us in an interesting tools conundrum. Which tools do you choose if you are using a given RTOS and device combination? And how can you protect your investment in tools (and code generated using those tools) if you switch your RTOS and/or devices? This situation is further exacerbated if more than one RTOS and/or more than one device are used in the same design—a common occurrence in the System-on-Chip (SoC) era.

2.4.1 Are Standards the Solution?

Standards are often an effective way of dealing with code reusability. The widespread use of ANSI C as a development language across embedded systems gives developers a chance for reuse. However, there are no real standards for tools and RTOSes. Therefore, each design has to use a specific compiler, which generates code for a specific device, to work with a specific RTOS. The RTOS and the tools have a proprietary API from compiler flags, through debug GUI, through IDE interface. This means that each project involves a substantial amount of relearning and retooling.

What is really needed is an environment that is standard across all embedded systems tools. This would allow the tools and RTOS vendors to plug and play with each other, and it would give users the benefit of a single tool to manage their projects and code. The environment would also provide standard interfaces and interoperability across the multiple tools and operating systems that are available today. This same environment could also support embedded projects that use a proprietary RTOS, or no RTOS at all.

One of the issues associated with creating such an environment touches on embedded systems politics. If either a device company or an RTOS company created this environment, it would become very difficult for their competitors to embrace it as their standard because it would leave them somewhat at the mercy of their competitor. So having an environment developed and maintained by an agnostic third party is very appealing.

Looking to the desktop or enterprise world is an interesting exercise because it inherits the benefits of a huge developer network, and the ability to use desktop productivity tools (such as standard editors and version management systems). The problem with this approach is that the desktop world is very host-specific (e.g., Microsoft Visual Studio). In addition, if embedded features are required, the desktop tools providers are not readily open to develop those features because they would have a relatively small base of embedded customers.

2.4.2 The Eclipse Solution

The good news for the embedded world is that a solution exists. While it may not be well known in the embedded world yet, Eclipse is set to revolutionize the embedded software developer's environment. It is as much a culture as it is a product, and it has been embraced by major embedded solution providers as their standard tools environment.

Eclipse is an open platform for tool integration built by an open community of tool providers. To quote the Eclipse web site: "The Eclipse Platform is an open IDE for anything, and for nothing in particular." It was developed by IBM and first released in 2001. In 2004, it was spun out of IBM into a nonprofit corporation called the Eclipse Foundation.

The technology is an open framework that is available in source code. The framework is written in Java and is highly portable across host environments. To date, it is available under Windows, Linux, Solaris, HP-UX, Mac OS, and IBM AIX.

2.4.3 Eclipse Plug-Ins

A key factor that makes Eclipse useful is the notion of plug-ins. Each tools provider can build their tools according to a certain set of rules and APIs that allow them to plug-in to the Eclipse framework. In the embedded world, this enables embedded tools providers to build true embedded products that will simply plug into the IDE, and it allows them to focus on their core competencies without worrying about developing IDEs.

If the Eclipse plug-in rules are adhered to, the embedded tools will track the latest versions of the Eclipse framework, as well as plug into other Eclipse-based environments.

2.4.4 Eclipse Licensing

Another key factor that makes Eclipse useful is the licensing model under which the Eclipse framework is provided. The Common Public License (CPL) provides royalty-free source code and world distribution rights and allows tools developers to offer the Eclipse framework and their plug-in products without putting their own intellectual property (IP) back into the community.

This model makes for a very viable business by allowing an open source framework to be a common vehicle, with a common look and feel. The open source framework can also be maintained by an RTOS- and device-agnostic organization and contain detailed embedded functionality provided by the embedded companies.

2.4.5 Eclipse User Advantages

Many of the embedded RTOS companies are now providing an Eclipse-based solution. This means that the embedded user will have a common platform for the common development functions such as project management, editing, file navigation, and build management. The user will also have a common interface to compilation and debugging tools.

If the RTOS providers have implemented their plug-ins correctly, then this one environment can host multiple operating systems without having to change environments. Each RTOS vendor can provide tools that will have a common Eclipse look and feel, but with specific features that help build and debug applications using that RTOS.

An example of where this is particularly beneficial is in companies with a large product portfolio that serve the same market but necessitate different user requirements—such as cell phone companies.

Cell Phone Example

Smart phones use a desktop-like operating system such as iOS or Android, whereas middle- and low-end phones may use more of a true embedded RTOS. Much of the software IP will be the same, so having it under a single project manager is very beneficial. The devices are also likely to be the same, so the compilation system can be consistent. Only the RTOS environment and high-end applications will change, which Eclipse can facilitate.

When considering a multicore architecture, the cell phone example also applies because most cell phones use at least one standard processor core and at least one DSP. Eclipse can host the different compilers, debuggers, and RTOS choices for each processor in one environment.

2.4.6 Perspectives

Eclipse, as do engineers, operates using the notion of perspectives. When building, engineers use a build perspective, and when debugging, they use a debug perspective, and so forth.

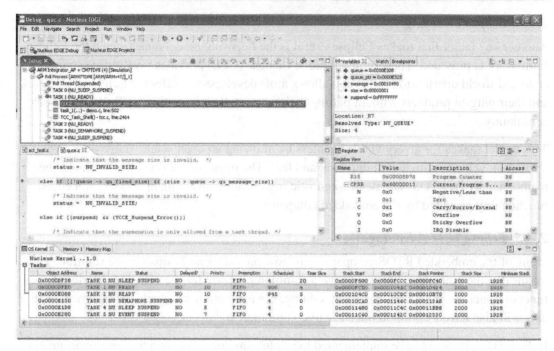

Figure 2.7
Eclipse debug perspective

A nice feature of Eclipse is that plug-ins can cross perspectives. For example, when debugging, the editor and project manager can be made visible and used to make changes to the code if bugs are found and to navigate around the project to fully understand the context of the debugged code. An example of an Eclipse debug perspective is shown in Figure 2.7.

2.4.7 Nonembedded Plug-Ins

Another advantage of Eclipse is the ability to plug in development productivity tools that are not specific to embedded systems. Currently available Eclipse plug-ins provide the following functionality:

- Modeling
- Bug tracking
- Code generation
- Graphics/drawing
- Source analysis/testing
- Project/team management
- Source control (such as CVS, ClearCase, and Visual SourceSafe).

Eclipse provides many advantages for embedded developers; it is now in the hands of the embedded solutions providers to make it a reality.

2.5 Embedded Software and UML

Embedded software is a fashion-conscious business. From time to time, different products, technologies, and methodologies are "flavor of the month." That's not to say that these fashionable ideas don't last—it's just that they generate a surge of interest and then settle into their place. As the twenty-first century began, the UML showed that surge, and everyone is interested in what it can do. It appears to address real problems that many developers are experiencing on real projects. Clearly a practical approach to its use for embedded software is needed—embedded developers are very definitely pragmatic people. In a series of articles in NewBits in 2004 and 2005, Stephen Mellor (with some help from Alasdar Mullarney) described an approach that makes practical sense. This article is based upon that series.

CW

2.5.1 Why Model in UML?

Yes—why? For all the usual reasons: to reduce costs, time to market, and unnecessary redevelopment and to increase productivity, quality, and maintainability. How can UML models do all that?

That depends on what a model is, and how it relates to the systems development process. There are at least three meanings of "model," and each meaning has different uses and implications. Let's take a look at each meaning.

One meaning for the word "model" is a "sketch." For example, we might sketch out a hardware configuration on the back of a beer mat, showing a few boxes for processors and lines for communication or adding a few numbers to indicate bandwidth or expected usage. The sketch is not precise or complete, nor is it intended to be. Often, a sketch of this nature is "talked to" by pointing at various boxes to explain what is happening there and how it relates to other elements. The purpose is to communicate a rough idea, or to try one out just to see if it will work. The sketch is neither maintained nor delivered.

A second meaning for "model" is "blueprint"—a classical example is the set of plans for a house. The blueprint lays out what must be done, describing properties needed to build the real thing, as determined by an architect. Because blueprints are intended to be plans for construction, they often map closely to the artifact that is to be built, so for each important element in construction, there is a "symbol." Because software is a complex beast, the set of symbols—the vocabulary—can become quite large, and without standards, chaos can ensue.

Enter the Unified Modeling Language (UML), which is a language that can be used for building software blueprints. (It has other uses too, as we shall see.) The UML is the result of an effort to reduce needless differences between different systems development methods and establish a common vocabulary for software modeling.

Why would you want to use UML? For all the reasons we outlined previously. Thinking about what you intend to build carefully—to the point of defining it exactly so that someone else can build it—will reduce costs and decrease defects, similar to the efficiency of writing a detailed, reviewed shopping list that avoids all the effort involved in returning a wrong item and getting the right one.

But as anyone who has built a house knows, the blueprint is rarely followed to the letter. Instead, as the builder (in contrast to the architect) constructs the house, the facts on the ground cause some modifications to be made.

The same argument can be applied to models: as we write code, we discover that our design wasn't as clever as we thought. This critique has led to the deprecation of models as "paper mills" that deliver pictures but not working systems. Instead, it has been argued, we should just hack—sorry, write—code because it executes.

Execution is important because it closes the verification gap between a concept on paper and a reality that either works or does not. Code either runs right or it doesn't. You can't be certain of that one way or the other with a blueprint, even with the best review team in the world.

The third meaning for "model" then is an "executable." When we build an executable model, we have described the behavior of our system just as surely as if we had written a program in C. Indeed, when you have a software model that can be compiled and executed, there's no need to distinguish between the model and the "real thing." It is the software.

So does this mean we should "program in UML"? And, if so, why should that reduce costs, time to market, and unnecessary redevelopment, as well as increase productivity, quality, and maintainability?

The answer to the first question is "Yes, but at a higher level of abstraction." For example, when you declare an association between two classes, you do not say whether that will be implemented by a pointer, a reference, or a list (just as when you program in C, you don't think about allocating registers). So while you are "programming," when you build an executable UML model, you don't have to think about a lot of things you normally worry about when programming in a language at a lower level of abstraction.

This approach reduces costs (the first of our reasons for modeling) because the cost of writing a line of code is the same irrespective of language. Studies as far back as 30 years ago showed that, on average, a developer produces 8–12 lines of assembly code, or C, or FORTRAN, or whatever per day. These numbers are "fully loaded," meaning that we're taking into account the time we spend in meetings, unjamming the printer, dealing with performance reviews, fighting the configuration management system, running tests, and all that other stuff. Although some programmers are much more productive, their productivity is also the same irrespective of language.

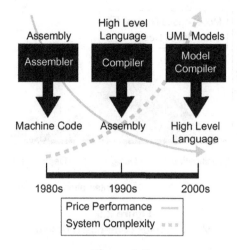

Figure 2.8
The evolution of software development

When we program in an executable UML, we write at a higher level of abstraction, thus reducing costs and increasing productivity. One user of executable UML generates 7–10 lines of C++ for each line of logic written in UML; the amount of code would be greater if this user's projects were written in C. Because the number of lines of code per day is the same, this translates directly into a decrease in time to market and an increase in productivity.

For these reasons, we developed higher-level programming languages as sketched in Figure 2.8.

Using an executable UML also increases quality. Not only is the number of defects reduced, but the errors are found earlier, providing time to react. It is better to know you have a problem when you have 6 months to go on the project than 6 weeks! Figure 2.9 shows the effect of applying this methodology.

Early error identification is achieved by building test cases and running them against the executable model. Because the model is executable, we can provide real values and get real results immediately, using a model executer that interprets the models. You and other experts can see immediately whether the model is doing the right thing. If it is not, you make the change in the models, then and there, and run the tests again.

We must emphasize that model testing occurs early in the life cycle, thus removing downstream defects. In turn, this reduces the effort involved in implementing the wrong thing, just as with that shopping list. The combination of removing defects early and avoiding wasted effort implementing the wrong thing reduces costs and time to market, while increasing productivity and quality.

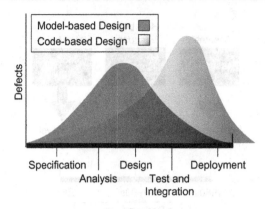

Figure 2.9
Model- versus code-based design

Models are also more maintainable than code because it is easier to manipulate concepts at a high level of abstraction than a lower one. The careful reader will have noted that we have discussed all of our reasons to model except reduction of unnecessary development, or—putting it in the positive—maximizing reuse. While it is certainly easier to reuse models than code (that higher level of abstraction argument again), the main reason you can reuse models is the same reason you are more likely to reuse a C program than one written in ARM assembly language—namely, you can port the C program across multiple hardware platforms.

The same concept applies to executable models. When we built an association, we did not specify whether it was implemented as a pointer, a reference, or a list. This allows us to decide later, once we better understand the speed and performance constraints of our system. In other words, executable models confer independence from the software platform, just as writing in C made us independent of the hardware platform. We can then redeploy the executable model onto different software platforms and implementation environments. This is actually something of an understatement. Models can be translated into just about any form, so long as their application behavior, as defined in the executable model, is preserved.

This brings us to a concern. When we moved from assembly code to C, we lost control of, for example, register allocation, which could lead to a reduction in the performance of the system—a killer concern in an embedded system. The keyword here is "control." If the compiler does a good enough job we don't care, but if the compiler doesn't know enough about our environment to make sound decisions, and we have no control over those decisions, we're in trouble.

For this reason, models need not only to be executable, but also to be translatable onto any software platform, and you, the developer, have to have control of how that translation process takes place, reducing performance concerns to zero. After all, if you can write the code, you can also describe how to go from a concept, as expressed in an executable model, to that code. A translatable UML also offers complete control of how that code is produced.

It is for this reason that we support executable and translatable UML, or xtUML, for short. xtUML models are both executable and translatable to any target software platform in an open manner. We do this by using a trick that differentiates blueprint models from executable models: separation of application from architecture.

2.5.2 Separating Application from Architecture

The separation of the application from the architecture differentiates blueprint-type models from executable ones. To understand that, we first need to understand how "blueprint" model-driven developers do their work.

Blueprint Development
After some initial requirements work, which can be supported by models such as use cases, the blueprint developer builds an analysis model, in UML, that captures the problem domain under study. This model will use various elements of the UML, but there is no universal agreement as to what those elements should be. As a simple example, the UML allows for attributes to be tagged with a visibility (`public`, `private`, etc.). Should an analysis model include this information? That depends upon taste—some do, some don't. Everyone is agreed that an analysis model should not contain design details, but there is little agreement on what that means exactly.

The next stage is to build a design model that does incorporate all that design detail. The design model is a blueprint that captures the software structure of the intended implementation. The work of transforming the analysis model to a design model exercises embedded systems design expertise. For example, we know, as embedded system designers, that a good way to store fixed-size data elements in a memory-limited environment is to pre-allocate memory—or whatever your expertise tells you to do. This expertise is applied to the analysis model to produce the design blueprint.

The next step is to code it up from the blueprint. Putting aside possible errors in the design, this means filling in code bodies. There are two ways to do that. One is to add the code directly to the model and have a tool generate code according to the software structure. Another way is to code up the software structure suggested by the blueprint, adding in coding details. The process is sketched in Figure 2.10.

What's Wrong with That?
Nothing, if you like doing all that work over and over every time the technology—and therefore the software structure—changes. And if you like reinventing and reapplying the same programming constructs when you add new system functionality.

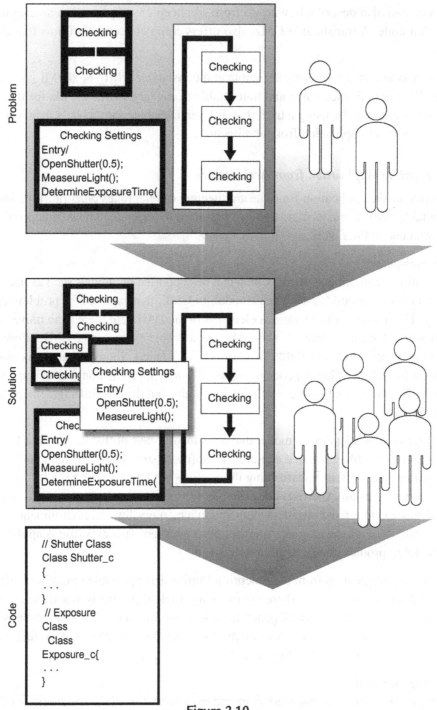

Figure 2.10
The blueprint development process

This approach to software development is rather like using C-like pseudo-code to outline your design, then hand-coding the assembler. Each time you add new application functionality, you have to decide over again how to pass parameters to a function, how to allocate registers to compute an expression, and so on. Each time you port to a new hardware platform, you have to work out what the assembly code meant (you wouldn't trust the pseudo-code, would you?) and rewrite it for a new processor.

At root, you have failed to leverage and capture the embedded systems design expertise represented by going from analysis to design to code. Or, to use the pseudo-code analogy, you have failed to leverage and capture the expertise involved in assembly coding.

Model Compilers

The solution, of course, is to build a compiler from a more formalized pseudo-code (which we may call C) for each of the various processors. Certain parts of the compilers are common, such as building an abstract syntax tree. Others are specialized to the target processor, though they may share common techniques for register allocation, expression ordering, or peephole optimization. The expertise is captured in an artifact (a compiler) which can be reused as required.

The same concept applies to xtUML model compilers. We can build model compilers for each software platform. (Note the adjective: each *software* platform.) A software platform is simply that set of technology that defines the software structure, such as choice of data structure and access to it; concurrency, threads, and tasking; and processor structure and allocation. All these details are filled in by the model compiler, just as a programming language compiler fills in all the details of register allocation, parameter passing, and so on, as determined by the hardware platform.

This approach captures the expertise involved in making embedded software design decisions and allows you to leverage it across a project and across many projects. Model compilers, like programming language compilers, can be bought.

Sets, States, and Functions

Figure 2.11 illustrates the separation between application and architecture. The element to focus on is the dotted line that separates the two.

When we build an xtUML model, it is represented in a simple form for translation, as sets of data that are to be manipulated, states the elements of the problem go through, and some functions that execute to access data, synchronize the behavior of the elements, and carry out computation. The UML is just an accessible graphical front end for those simple elements. When you build a "class" in xtUML, such as CookingStep in a microwave oven, it represents a set of possible cooking steps you might execute, each with a cooking time and power level. Similarly, when you describe the life cycle of a cooking step using a state chart diagram, it follows a sequence of states as synchronized by other state machines (when you open the

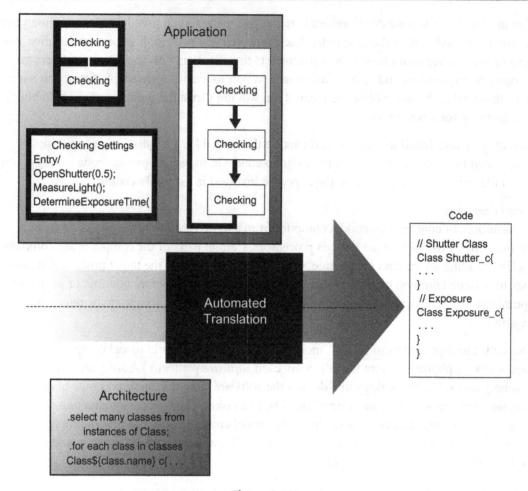

Figure 2.11
Separation of application and architecture

microwave door, it had better stop cooking!), external signals (such as a stop button), and timers. And in each state, we execute some functions.

Naturally, it's a bit more complicated than that, but the point is that any xtUML model can be represented in terms of these primitive concepts. And once that's done, we can manipulate those primitive concepts completely independently of the application details.

Rules
The ability to perform this independent manipulation allows us to write rules. One rule might take a "class" represented as a set `CookingStep (cookingTime, powerLevel)` and

produce a C++ `class` declaration. Crucially, the rule could just as easily produce a `struct` for a C program, or even a `COMMON` block in FORTRAN. Similarly, we may define rules that turn states into arrays, lists, `switch` statements, or follow the state pattern from the Design Patterns community. (This is why I put "class" in quotation marks. A "class" in an executable model represents the set of data that can be transformed into anything that captures that data; in a blueprint-type model, a class is an instruction to construct a class in the software.)

These rules let us separate the application from the architecture. The xtUML model captures the problem domain graphically and represents it in terms of sets, states, and functions. The rules read the application as stored in terms of sets, states, and functions, and turn that into code. This leads to the process shown in Figure 2.11.

The value here is that application models can be reused by applying different sets of rules (a different model compiler) to target a new software platform. Similarly, the model compiler can be reused in any project that requires the same architecture. The applications and the model compilers can each evolve separately, reducing costs, and increasing productivity and reuse.

Open Translation

There is one critical difference between today's programming language compilers and model compilers. With a programming language compiler, you have limited control over the output. Sure, you can apply a few flags and switches, but if you truly dislike the generated code for any reason, you're out of luck unless you persuade the vendor to make the changes to the compiler you require. With a model compiler, the translation rules are completely open. If you can see a better way to generate code because of the particular pattern of access to data, say, you can change the rule to generate exactly what you want. This completely removes any concerns about optimization. It is totally under your control.

I should emphasize that you rarely need to change the model compiler, still less write one of your own. But the knowledge that you can change it should increase your confidence in the technology. Another analogy to programming languages: when they were new, people were concerned about the quality of the output and having some control over it. Over time, of course, those concerns have diminished, even in the embedded space.

2.5.3 xtUML Code Generation

We will now take a look at the code that will be produced from xtUML models. Obviously, what we want is executable code. For an example, we will look at the safety-related logic of a simple microwave oven. The oven components are the door, which must be closed while cooking, and the actual cooking element. There will be some code to manage the cooking times and power levels.

Take a look at some representative code for such a microwave oven:

```
struct Oven_s
{
        ArbitraryID_t OvenID;
        /* Association storage */
        Door_s *Door_R1;
        Cooking_Step_s *Cooking_Step_R2;
        Cooking_Step_s *Cooking_Step_R3;
        Magnetron_s *Magnetron_R4;
        /* State machine current state */
        StateNumber_t current_state;
};
```

The C struct captures information about the oven, which has an arbitrary ID (an identifier to distinguish a particular instance) and some pointers that reference its components. The oven struct also has a current_state that captures the—well—current state of the oven.

The "Cooking step" in Figure 2.11 allows the microwave oven to be programmed to cook in steps, each at a different power level for a certain time. Each step describes cooking parameters for the oven. Typical uses are to program a cooking step to defrost by pushing one button (Time 1, say) with low power and a long time, followed by pushing another button (Time 2, for a second unimaginative name, say) to cook ready to eat at high power for a shorter time. There are twin steps because there are two buttons.

Here is the code for Cooking_Step_s:

```
struct Cooking_Step_s
{
        i_t stepNumber;
        i_t cookingTime;
        i_t powerLevel;
        Timer_s *executionTimer;
        /* Association storage */
        Oven_s *Oven_R2;
        Oven_s *Oven_R3;
        /* State machine current state */
        StateNumber_t current_state;
};
```

This structure includes a step number (used also as an identifier), cooking time, and power level. In the scenario described previously, there could be two instances of this struct, say:

Step Number	Cooking Time	Power Level
1	10 min	20%
2	3 min	100%

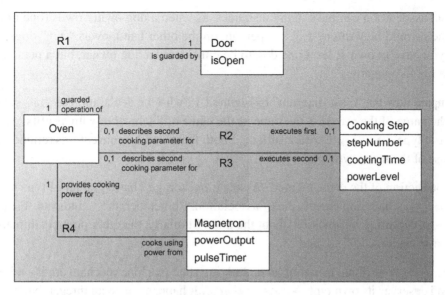

Figure 2.12
Microwave oven class diagram

An execution timer is also used to refer to one of several potential timers. This reference is required so the timer can be interrogated, reset, or deleted. In addition, association storage refers back to the oven. There are two associations because we can program two cooking steps.

Again, the cooking step has a current state attribute to capture whether the cooking step is ready (i.e., has been programmed), executing, or complete.

We need to understand the conceptual entities in the problem and how they are described by data. Figure 2.12 shows a so-called class diagram, which does just that.

This diagram declares four conceptual entities (the oven and the cooking step we have already discussed, plus a door interlock and a magnetron tube), with several associations identified with "R numbers." The numbers are simply a way to identify each association uniquely; the "R" comes from the real-world relationship captured by the associations. The associations also have names that capture the real-world relationship. R2, for example, is read: `Oven` executes first `CookingStep` and `CookingStep` describes first cooking parameters for `Oven`.

The associations also have a multiplicity that indicates how many instances participate in each association. For R2, each oven may have zero or one (0, 1) first cooking steps. And each cooking step may or may not be the first cooking step for this oven (hence the 0, 1 again). The association (R4) between the oven and the magnetron is "1" in both directions because one `Oven` *houses* one `Magnetron`, and one `Magnetron` *is housed in* one `Oven`. In general, an association can have many instances, as when a dog owner owns (one or more) dogs, which would be written "1..*". A person, on the other hand, owns "0..*" dogs, because you have to own at least one dog to be considered a dog owner, but a person is free to choose not to own any.

Let's compare now the "class diagram" of Figure 2.12 with the declaration of Oven_s. We can see that the name of the struct is the same as the name of the box but with a suffix _s, and the OvenID has a type ArbitraryID with a suffix _t; both these coding conventions remind us of the purpose of the name symbols.

The second section of the code, marked "Association storage," has a pointer of type Door_s, named Door_R1, which implements the R1 association. Remember, the _s indicates the pointer is to a struct capturing information about the door. Similarly, the other pointers implement the other associations, R2, R3, and R4.

The current_state attribute is part of an underlying state machine mechanism; it's a little special and deserves its own type—state_number_t. It happens to be an integer, but because we know something about its likely values, we may choose to implement it as an unsigned char, for instance.

The primary observation to make here is the close correlation between the oven diagram and the corresponding declaration. Note that the declaration of Cooking_Step_s and the Cooking Step "class" have the same close correspondence too.

So, why do we put quotation marks around "class"? We do so because the class oven isn't a class at all. It's a C struct! This goes back to the separation between application and architecture we discussed earlier. The application model describes the fundamentals of the solution, while the architecture defines the mapping to the implementation—just as we showed here in this extended example.

Moreover, the C we showed illustrates the point we made right at the beginning. The models we build are executable and translated into the implementation. This is in contrast to "blueprint-type" models that are intended to direct the implementation. There, a "class" means that we should build a class in the code. Here a "class" simply declares important data and houses behavior as defined by a state machine, the trace of which is the current-state attribute.

2.5.4 Conclusions

Modeling and the use of the UML mean different things to different people; the terminology can be confusing and is often misused or abused. There are various possible goals that come out of the use of modeling, but an approach that reduces rework and leverages the diverse expertise of the embedded development team must be a clear winner.

> **A Note on Terminology**
>
> A **software platform** is analogous to a hardware platform. It is the set of technologies that we rely on to make our software work: linked lists, middleware, operating systems, IPC mechanisms, and so on. These technologies don't know about an application, although the choice of software platform will have an impact on system performance.
>
> An **architecture** (more precisely, an application-independent software architecture) is an abstract representation of how to map an application to a specific target platform. (It includes the software platform by reference.)
>
> A **model compiler** is the physical realization of the architecture; it is a program like a programming language compiler.
>
> These terms have analogies in the programming language compiler world: a compiler embodies choices about the architecture based on, and relying on, the facilities provided by the hardware platform.

2.6 User Interface Development

> *In recent years, we have seen evidence of the "iPhone effect"—users are beginning to expect an intuitive, touch-sensitive graphical user interface. As suitable LCD panels are becoming cheaper all the time, this demand can be readily fulfilled, but this has software challenges. With this in mind, Geoff Kendall and I wrote a white paper in 2008 that broadly reviewed user interface design. I used that paper as the basis for this article.*
>
> *CW*

In this article, we are going to take a look at user interfaces in embedded systems. But, before we get started, it is always useful to make sure that all the terminology is straight. What is an embedded system? Wikipedia suggests: "An embedded system is a special-purpose system in which the computer is completely encapsulated by or dedicated to the device or system it controls." This is not bad, but perhaps it would be better to think of one as being any piece of electronics, which includes a microprocessor, but would not normally be regarded as a computer. The definition of a "user interface" (UI) in this context is somewhat easier: the facilities and mechanisms whereby an embedded system interacts with a user.

2.6.1 User Interface Diversity

Since embedded systems are very diverse, it is unsurprising that the range of possibilities for UIs is very wide.

No UI

Some devices do not interact with users in any direct way, but simply interface with other equipment with which they communicate electronically. A good example here would be a disk drive controller.

Buttons and Lights

At a very simple level, an embedded system may just have a few buttons or switches for input and some lights (LEDs) for output. This can result in a surprising level of functionality. An LED, for example, can be on, off, or flashing; the flash rate may be varied to convey different information; the color may also be controllable. A typical example of this kind of UI is a modern domestic washing machine.

Buttons and Simple Alphanumeric Displays

The next step from a few LEDs to display information is the provision of a simple alphanumeric display which can show one or more lines of text. The input buttons may be expanded to become a keypad or even a full keyboard. Examples include driver information systems in cars and intruder alarm systems, which typically employ these facilities.

Buttons, Lights, and a Network

Some specific kinds of device may have a superficially simple lights and buttons display but also utilize a network connection to interact with users in a more sophisticated way. Clearly this is likely to apply to networking equipment like routers and residential gateways.

Buttons and Simple Graphical Displays

As small graphical LCD displays—both monochrome and color—have become very economic, and they are widely used. The obvious typical application is in cell phones.

Buttons, Pointing Devices, and Simple Graphic Displays

It is only one small step to provide a pointing device, which behaves rather like a mouse on a desktop computer. A pointer may just be moved about by the software under the control of buttons or a joystick type of device or the screen may be touch-sensitive. More sophisticated cell phones—"smart phones"—are good examples.

Buttons and Full-Size, High-Resolution Displays

For nonportable equipment, the display size is not limited and this brings a plethora of UI possibilities. The best examples are devices which have a functional need for large displays, as they handle video information, e.g., set-top boxes and personal video recorders.

2.6.2 Implementing a User Interface

Having established the diverse nature of UIs in a wide range of embedded systems, we can give some consideration to how they might be implemented and what the software development requirements and challenges might be.

Lights and Buttons

Processing buttons and switches is quite straightforward. It is simply a matter of polling an interface frequently enough or responding to an interrupt. There are some subtleties: a switch or a button being pressed may be seen as a transition from "1" to "0," not vice versa, as might be expected; switch contacts have a tendency to bounce, which, without care, can be interpreted as a sequence of actuations. Long technical articles have been written on dealing with switch bounce in software.

Lamps are equally simple. They are typically illuminated by setting a bit in a device register. This does have a small challenge: these registers are typically "write only," so the software cannot read back the current state; a shadow copy of the data must be kept in RAM and used with care. Again, articles have been written on the processing of write-only ports.

Flashing lamps need some timing mechanism. This might be handled by a simple clock interrupt service routine or a real-time operating system may be employed for the main application, which can accommodate the timing requirements.

Simple alphanumeric displays are handled in much the same way as lamps.

Network Interfaces
If a device has a network interface (typically wireless or Ethernet), there is the opportunity for the embedded software to communicate with users via other computers on the network, with all the flexibility that the computer's GUI affords. There is the option of writing a specific UI to do this job, or implementing an HTTP server (a "Web server") in the device. The latter approach is very flexible, as it enables the UI to be defined by means of a series of hyperlinked HTML ("Web") pages. There is also the option of making the pages smart by using scripting (like JavaScript) or even the Java language. An HTTP server really requires a real-time operating system environment under which to run.

Menu and Information-Oriented UIs
On devices with graphical screens—either full size, like a TV screen, or small, like a cell phone—the functionality may be defined quite precisely. There are also some common features that may be leveraged. However, there are some common problems that present serious challenges to developers.

Functionality:

Display elements (pages) of such a UI fall very much into two basic types:

• A series of hierarchical menus
• Tabular data (like a set-top box program guide).

Of course, clicking on entries in either type of display may result in the appearance of the other display type. For example, in a personal video recorder, you may have a system menu, which leads you to the program guide; clicking an entry (a show) in the guide produces another menu giving you options to schedule recording, record a series, or show information.

All of the data in menus and tables may be static—"hard wired" into the program—or dynamic—derived or obtained when required.

Common Operation Features:

The way that such displays work may be analyzed and shows a very simple set of behavior patterns. The user selects something and one of three things happens:

- Another menu or table is displayed.
- Some specific action is taken (i.e., some software is started, like making a call on a cell phone).
- Some simple data entry is facilitated (like keying in a number on a cell phone).

Problem Definition:

The implementation of sophisticated UIs brings a number of challenges. This leads to a tension between the cost and difficulty of implementation against the quality of the user's experience. In the consumer marketplace, the user's experience of the UI can be make or break for not only the product, but the vendor's whole product line, present and future.

There are three challenge areas:

- Such complex devices typically have a wide range of functionality, resulting in their implementation as a number of different applications. A smart cell phone, for example, will have Web browser, SMS, and address book applications (and a whole lot else besides) along with the regular phone functionality. It is typical for each application to be developed separately—maybe even by different companies. The result is that each has its own UI. In every application, the UI is doing much the same job as in the others, but in slightly different ways, with a different look and feel. The result is that more work was done in development (creating lots of UIs) and the user experience is degraded (because the UIs are not consistent).
- The usual implementation of UIs takes significant programming skill and effort to implement and, furthermore, to maintain and adapt to future needs.
- There are very limited opportunities to add vendor customization to a device without excessive programming effort. This is a capability demanded by service providers for cellular networks or cable TV providers, for example. These companies regard the phone or box to be simply the physical manifestation of the service they provide and want to reinforce this message at every opportunity.

The good news is that all three challenges can be readily addressed by taking a different approach to UI development.

2.6.3 A Rationalized UI Solution

The way to create a UI for a complex embedded system is to implement it as a separate layer of software, which may be hooked into all the application components of the device. The example considered here is Inflexion Platform UI from Mentor Graphics. This software provides all the display graphics and user interaction control. The developer simply configures the UI using an interactive design tool (which generates an open format output: XML).

The great benefit to the developer is the simplicity of the methods involved in creating a sophisticated, attractive UI. No programming or scripting required—the UI can be completely defined, with a rich set of options being realized through very small parametric adjustments.

Designing an Interface

Sophisticated UI layouts are possible because the visual appearance of any on-screen element (e.g., an individual icon or caption) can be controlled to a high degree. Any element may have a location, scale, and orientation in three dimensions. Device screens are of course two-dimensional, but Inflexion Platform UI allows compelling three-dimensional effects by applying appropriate perspective and rotational effects automatically at run time, according to the layout parameters specified.

Such effects can be applied to any element and at multiple levels of abstraction (e.g., individual items or a whole menu), to enable extremely sophisticated and diverse layouts.

Support for 3D Rendering

As noted, Inflexion Platform UI enables the layout of an interface's elements to be specified in three dimensions. Many visual effects (such as smooth fading, scaling, and zooming) can be achieved even if the target hardware only has basic 2D graphics APIs. However, the 3D capability of Inflexion Platform UI really comes to life—with full rotation, solid models, texture mapping, and lighting—if OpenGL/ES is available. OpenGL/ES is the open standard API for rendering 3D graphics on embedded systems. OpenGL/ES support can be embedded in a device through either a software library or—for maximum performance—a hardware-accelerated graphics processor.

Usage Models

Given that a sophisticated device may run a wide range of applications, selected as required by the user, there are really three ways that a rationalized UI solution, like Inflexion Platform UI, may be deployed:

- The UI can be applied to all the applications in the device immediately. This is the ideal approach for a brand new system, as it simply requires all the developers to take a unified approach. It is a rather more challenging prospect if all the applications already exist. However, this approach would yield the best user experience, as it results in all the applications having a common look and feel and any customizations (color, backgrounds, "skins," etc.) apply system-wide.
- Developers may take advantage of the UI technology as and when possible. This approach results in new applications, and those being updated, acquiring the rationalized UI. Over time, the complete system will be encompassed.

- The UI can be used very easily—and with minimal effort—to deploy service-provider-specific facilities in a device. This could be an additional application (e.g., a portal to promote downloadable content accessible across a network), or an encapsulation (and "badging") of facilities already present in the device.

2.6.4 Conclusions

The user interfaces of embedded systems may be as varied and diverse as the systems themselves and require various approaches in their design. As soon as a device has a sophisticated graphical display, the challenge in UI design is increased drastically. The attractiveness of a device to potential purchasers is strongly influenced by the UI, which can thus be the key determining factor in the success of the product.

2.7 Software and Power Consumption

In early 2011, it was very clear that device power consumption was becoming a very important issue for many developers. Although, in the past, this would have been regarded as a purely hardware issue, the role of software was becoming more significant. As Mentor Graphics is the only EDA company with a significant involvement in embedded software, we thought that we should bring this broad knowledge base to bear on this topic. So, I collaborated with two colleagues, Russell Klein and Shabtay Matalon to write a paper looking at the balance of hardware and software contribution to minimizing power. This article is based on that paper.

CW

2.7.1 Introduction

The power consumption of devices and the issues around designing for low power are hot topics at this time. This chapter looks at the issues from a system-wide perspective and gives guidance on design strategies that encompass both hardware and software development.

State of the Art—Hardware Versus Software Effort

In the early days of embedded systems, most of the design effort was expended on hardware; software was little more than an afterthought. Over time, design teams began to include software specialists working alongside the hardware engineers. In recent years, the size of the average embedded software team has outstripped that of their hardware counterparts by a factor of 2, 5, or even 10.

The embedded software team is not homogeneous. The expansion of teams has not been simply a matter of adding manpower; there has been a focus on having a wide range of expertise within the team. Some software engineers are likely to be very well-versed in working close to the hardware, i.e., they are classic embedded software specialists. Others, however, may be much more focused on the application and have very little awareness of the embedded environment.

Why Power Matters

There are many reasons why a designer of an embedded system is increasingly driven to stay within tight system power budgets. Battery life in portable electronics has driven awareness and optimization of system power. It is obvious that Apple has a huge sales advantage over their competition in the tablet market. To beat them, someone needs to build a better iPad. One area where Apple sets the bar pretty high for their competitors is battery life. CNET measured almost 12 h of video playback, and this is for a device that weighs in at just slightly over one pound.

For portable devices, an increase in power consumption obviously results in a decrease in battery life. Although great progress has been made in battery technology over the last three decades, it has not kept pace with the demands of modern electronics. This has obvious effects upon the consumer satisfaction with such devices, where, for many potential customers, convenience is a strong driver in the selection process. This has a nonobvious knock-on effect. In the early days of the iPhone, Apple's stock price was adversely affected by user disappointment with the device's battery lifetime; rectification of this issue resulted in a restoration of the share price.

While the connection between the benefits of low power consumption and battery-powered devices is clear, power matters in many other areas as well. When Google was deciding on a location for their new server farm, power was an important consideration. Google is pretty tight lipped about the details, but a local paper (Beck, Byron. "Welcome to Googleville." Willamette Week. Portland, Oregon, June 4, 2008) reported that their new building will have a 6 mW connection to the power grid. Assuming that each of their servers will burn about 500 W, and that they might have 10,000 servers packed into one building, i.e., 5 mW of power needed just to drive the servers. Power costs 6 cents per kilowatt hour in The Dalles, Oregon. That same paper reported that Google got a break from the power company by pre-negotiating their rates down to 3 cents per kilowatt hour. Commercial power rates in Mountain View, California, are about 16 cents per kilowatt hour. Let's do the math—assuming that all those servers are going to be crunching full time on that secret Google search algorithm: 10,000 servers at 500 W each, 365 days per year, 24 h per day is 43.8 million kilowatt hours. At 16 cents per kilowatt hour, the tab in Mountain View is just a bit over $7 million. The tab in The Dalles, Oregon—after some haggling with the power company—is about $1.3 million. There's more power needed to run the air conditioners needed to dissipate all that heat. Google reaps more than $5 million a year in operational savings.

It's not all about cost. The vast majority of power generation is produced with fossil fuels, adding environmental motivation to reduce power in nearly every system. Many items of equipment feature a standby mode, which facilitates fast start-up when required. There is a strong incentive to minimize the power consumption under these circumstances.

Some things are worth even more than money. Implanted medical devices need to run for years off little more than a watch battery. Each time a battery needs to be replaced, it means opening up the patient. Despite the marvels of medical science, a small percentage of folks never recover from that type of major surgery—so any improvement in battery life can literally be a life saver. Pacemaker batteries have a capacity of several hundred milliamp hours (mAh), a small fraction of a laptop battery. They need to function for at least 10 years; 300 mAh divided by 10 years, 365 days per year, and 24 h per day means that the pacemaker itself needs to run on 19 microamps. Even lower power consumption would be desirable, as that means less frequent invasive surgeries.

The skill in designing an embedded system today is to use strategies that find the best compromises between performance and power consumption and utilize every possible efficiency measure to deliver the desired design goals. There are two key strategies that must be adopted:

- Recognize that designing for optimum power consumption requires a system-wide approach—both hardware and software designers need to be involved.
- Having the right tools to enable power consumption estimations to be made, as early in the design cycle as possible, is essential to avoid time-consuming redesigns late in the day.

2.7.2 Software Issues

Software Architecture
Any nontrivial embedded application is likely to be built on an operating system (OS) of some kind. This code may be developed in-house but is commonly provided by a software IP provider, such as Mentor Graphics. It could be a simple kernel or it may be a full featured OS. One of its primary functions is to provide the software developer with a multithreading environment (using a thread or process model).

In broad terms, embedded applications are either real time or non real time. A real-time application is one where the timing of actions and responses to events is critical and a Real-Time Operating System (RTOS) would normally be used. An RTOS is usually fast, but its key characteristic is predictability—such an OS is termed "deterministic."

For an application where timing is not a critical issue, an OS adapted from the desktop may be desirable, and a common choice is some variant of the Linux® development platform. The Linux OS offers a familiar programming environment for many developers and a large range of available middleware offers many opportunities for software reuse. Android is another option which provides a clean programming environment, particularly suited to applications where post-deployment software components ("Apps") are a requirement. Linux and Android present a significantly larger memory footprint than a conventional RTOS.

The choice of OS is driven by a number of factors—primarily the specific embedded application. Mentor Embedded solutions encompass the complete spectrum from the Nucleus® RTOS, to Mentor Embedded Linux and Android.

Power Issues
Historically, the power consumption of a device would have been regarded as the sole province of the hardware designer. In recent years, the influence of the software design on power consumption has been recognized. Some key areas, where software has an influence on power consumption, may be identified:

- **Display:** Sophisticated displays are becoming increasingly common on a variety of embedded systems. Their size and resolution is steadily increasing and touch sensitivity is widely implemented. This hardware is a major power drain. It is essential that the driving software monitors utilization very carefully and dims or shuts down displays when they are not in use. Passive displays that do not emit light, such as those used on a number of popular e-book readers, minimize power consumption. Backlit displays, like those used on tablets and laptops, are far more power hungry.
- **Wireless Peripherals:** Embedded devices have become increasingly connected over the last decade and that connectivity is often wireless. A variety of technologies are available: Wi-Fi, Bluetooth, ZigBee, and cellular technologies such as GSM are all contenders, along with RFID and Near Field Communication (NFC). In almost all cases, the software may have the opportunity to turn off the wireless interface when it is not in use or optimize the transmission power for current circumstances.
- **USB:** For an embedded developer, USB may appear in two guises. If the device is a USB peripheral, its software should respond correctly to requests from the host to shut down or suspend. If the device draws power from the host, this must be strictly limited (to 500 mA with USB 2.0 and 900 mA with USB 3.0). An embedded system may also be a USB host, in which case the host software must monitor USB peripheral utilization and shut down or put to sleep unused devices. If USB 3.0 is implemented, full use of its more advanced power management capabilities should be implemented at either end of the bus.
- **CPU Utilization:** The execution performance of the code may be quite significant— minimizing the number of instructions that need to be executed to perform a given task reduces the number of Watt-Hours required. Both the application code and the additional IP have an influence. The application code execution speed may be affected by the design and the development tools employed. An RTOS typically supports a multithreading environment much more efficiently, in terms of CPU utilization, when compared to Linux and other general-purpose OSes. Studies have shown that the power consumed by

a device, performing identical tasks using different operating systems, may vary widely. In any case, an OS may be tuned to minimize overheads. Low power consumption is closely related to algorithm performance. Anything that can be done to reduce the number of clock cycles needed to execute an algorithm can be applied to reducing power consumption.

• **Memory:** The size of the code and live data of an application obviously affects the amount of memory required for the design. Although it's unlikely to be a significant factor, memory does consume power, and minimizing the memory footprint of the software may be desirable. A design with just enough memory would be short-sighted, as possible future software enhancements may be precluded. The size of an embedded application is broadly affected by two factors: the application code itself and any additional software IP. The size of the application code may be minimized by careful design and the use of high-quality development tools. Additional software components may be chosen with memory footprint being included in the selection criteria. Typically, an RTOS is much more compact than an implementation of Linux. RTOSes are usually scalable—only the required components are included in the memory image and redundant code is minimized or eliminated altogether. Care is needed though, because with some RTOSes, the kernel itself may be scalable, but additional middleware (networking, file system, etc.) is not. This isn't the case with the Nucleus RTOS, where all components are easily scalable.

The key to addressing the issue of power consumption from the software perspective is *power visibility* and *analysis*. This requires the right tools or, rather, a set of tools. These tools would offer hardware simulation and probing that offers precision monitoring of power consumption and relates it directly back to the code.

2.7.3 Power Control in Software

Broadly, developers have two facilities to control system power consumption via software: (1) sleep/suspend and (2) Dynamic Voltage and Frequency Scaling (DVFS).

Sleep/Suspend
Historically, developers only had a limited range of core processor sleep states. New System-on-Chip (SoC) devices continue to expand the array of mechanisms available for a software developer to leverage in-power management. Numerous low power states (sleep, doze, hibernate, etc.) for the CPU and on-chip peripherals have become increasingly sophisticated. Coordinating the broad array of states across CPU and peripherals is a major task.

Low power consumption is closely related to algorithm performance. Anything that can be done to reduce the number of clock cycles or the clock frequency needed to execute an algorithm can be applied to reducing power consumption. Where a platform supports sleep or various power down modes, if a system can complete the work in fewer clock cycles, it can then remain in a lower power mode longer, thus consuming less power.

Dynamic Voltage and Frequency Scaling

Some systems provide DVFS facilities. Rather than entering a sleep or suspended mode, where DVFS is available, the program can reduce the clock frequency and voltage of the system to conserve power.

There are trade-offs involved in deciding which approach to take, a suspended mode or a reduced clock frequency where both are available. When considering the power used by a processor as it executes, running at the highest clock frequency and voltage will result in the highest power consumption. Recall that switching power consumption in a processor is proportional to the clock frequency and the square of the voltage. As the clock frequency is lowered, the power consumption is proportionally lowered. But as the voltage is lowered, the power drops much faster. Therefore, if it is possible to meet the performance requirements of the system at a reduced clock frequency and a reduced voltage, this will usually be preferred, as it will result in the lowest power consumption.

For example, consider a task which requires 1 ms to run on a 1 GHz processor at 1.4 V. Running the clock at one-half the frequency (0.5 GHz) will increase the time required to complete the task to 2 ms. Reducing the clock frequency to 0.5 GHz will reduce the power consumption by one-half, and the reduction of the voltage by about one-half, which will result in a 4 times further reduction in power. Therefore, if the performance requirements of the system can accept a 2 ms run time for this algorithm, by reducing the clock frequency and voltage it can consume one-eighth the power.

The alternative of going into a suspended mode saves less power. By running the system at full power for 1 ms and then suspending for 1 ms, we'll burn 100% of the power for half the time, then in sleep mode we will not drop to no power consumed, but in many cases we will burn just a few percent of the running mode of the processor. Thus, using a power down strategy to save power will result in a reduction of a bit under one-half. Where performance goals can be met, DVFS will result in a more power optimal solution, but higher response times and throughput can be achieved by using power down modes.

One of the challenges with either DVFS or power down modes comes as software complexity increases. If there is only one software task or process occurring at one time, that task will have complete knowledge of the current requirements for power and performance. It will know when to power off a peripheral or enter a sleep mode. However, as the software becomes more complex, there may be more than one (in fact, there may be many) processes running concurrently. They will all have differing requirements for the performance at the same time. It is no longer appropriate for any one process to suspend the processor, reduce the clock, or turn off a peripheral. A framework is needed to allow all processes to report their minimum performance needs, and a facility which would monitor those needs and keep those minimum performance levels, while turning down or turning off power where it is no longer needed.

Of course, the execution performance of code is radically affected by the processing power of the CPU. A powerful processor executes many millions of instructions per second, but runs with a very high clock frequency, which directly affects its power consumption. Multiple, lower-power CPUs, running at lower frequencies may be able to offer a similar execution throughput, while consuming less power. However, this introduces additional complexity into the software, which must be configured for multicore.

2.7.4 Multicore

An increasing number of embedded systems are being implemented with multiple CPU cores. This may be a number of identical cores (homogeneous multicore) or cores of multiple architectures (heterogeneous multicore). In either case, new challenges are presented to the software developer.

In a homogeneous multicore system, there may be the possibility to use an OS that supports Symmetrical Multi-Processing (SMP), where a single instance of the OS runs across all the cores and manages distribution of work between them. A well-designed multithreaded application ports readily to an SMP environment. Operating systems that support SMP include Linux and Nucleus.

Another approach, Asymmetrical Multi-Processing (AMP) is when a separate OS instance runs on each processor. Each OS may be selected to be optimal for the function of a particular core. A standardized inter-core communication mechanism needs to be adopted and Multicore Communications API (MCAPI) provides a viable option for many applications.

Multicore systems offer the potential of significantly reduced power consumption for a processor-based system. It may not be obvious, but smaller and simpler processors are far more power efficient than larger more complex ones. This can be true even if the complex processors have sophisticated power management facilities which are used effectively.

As an example, consider the ARM Cortex A9 when compared with the ARM Cortex R4. When implemented on a 65 nm process, the Cortex A9 delivers 2075 DMIPS and has a power efficiency of 5.2 DMIPS/mW and runs at 830 MHz. While the ARM Cortex R4 implemented on the same process delivers 1030 DMIPS and has a power efficiency of 13.8 DMIPS/mW and runs at 620 MHz. (For more detailed information, visit www.arm.com/products/processors/cortex-a/cortex-a9.php performance tab, single core 65 nm, optimized for performance.)

Recall that Watts are Joules/Second and MIPS is Millions of Instructions/Second. Dimensional analysis allows us to multiply the numerator and denominator by Seconds and get "Millions of Instructions/Joule." Inverting this we get Joules/Instruction—or the amount of energy needed to accomplish a given workload on the core. For the A9 we get 1/5.2 or

0.192 MilliJoules/Million Instructions, and for the R4 we get 1/13.8 or 0.072 MilliJoules/ Million instructions—or a difference of a factor of 2.65. Thus, you will burn almost 3 times the energy implementing an algorithm on an A9 as compared with the R4. This is energy and does not include any time aspects. If we look at power (Watts/Second), you need to include the speed of the processor. Looking at the power differential between these processors, the A9 burns 400 mW, while the R4 burns a mere 75 mW. That's 5.3 times more power, but just twice the compute throughput. Two R4s will deliver the same compute capability while consuming only 37% of the power. And the story gets more compelling as the processors get smaller and simpler.

The challenge, of course, is to enable algorithms to take advantage of the two cores. But the benefit of going to multiple smaller cores, in terms of power consumption, is significant.

2.7.5 Hardware Issues

Hardware Architecture

Early validation of software on an early virtual prototype of the hardware allows design teams to readily change the hardware design topology and influence the RTL design specification before RTL specs are finalized and implemented. At this design stage, it's still easy to add or remove compute resources, add a hardware accelerator block for a performance critical function, and optimize the design for low power.

Hardware component selection includes choosing the processors, peripherals, and memories that will compose the design. A key decision is choosing the right processor. Some processors have been optimized for low-power operation in portable battery-operated devices and some are suitable for desktops and servers. There is also a wide range of memories with different performance/power characteristics. In some designs, memories may account for up to half of the design gate count and may consume large amounts of power. And last, the most advance devices need to support numerous interfaces connecting them to the outside world communicating video, audio, text messages, and control signals in a variety of formats. Peripherals support data handling, computation, and communication through these interfaces and their large number may impact power.

Hardware topology defines how the design components are connected with each other usually using standard busses using standard protocols. As some busses are required to support fast communication speeds and various arbitration schemes, and others may need slower speed or dedicated communication resources, the bus architecture can become quite complex. It's common to see various busses and even bus layering schemes providing the internal and external communication resources for the components. The hardware topology impacts important attributes such as latency and throughput of the data flowing through the design and thus impacts directly the level of power consumed.

Power domains are sections in the design that can be controlled independently from others to conserve power. Such sections may not be required to operate all the time and may be shut down during idle times of certain functions in the design. Others may be controlled to run slower under given conditions conserving on power. Power domains are usually associated with power states that define when to apply a specific power mode to the design and a power controller unit that can control the power domains during operation.

Factors That Affect Power Consumption

The power consumed by a device is composed of static and dynamic power. Both static and dynamic power can be influenced at the various design stages starting at the system level and continuing at the implementation level. System level design entails architecture level design and optimization of power under the control of the application software. This stage presents the opportunity to reduce power by up to 80%.

As static and dynamic power are additives, each needs to be considered independently. Static power, which is also defined as "leakage," is consumed in the absence of any design activity. It is highly associated with the current flowing through the transistor in idle state determined by the transistor attributes. Dynamic power is associated with the activity in the design influenced by the volume of data that the design has to process in a given time unit by switching the transistors between on and off states.

The design operating frequency is usually determined by the frequency of the clocks operating in the design. The faster the clocks run, they increase the switching frequency of the transistors resulting in larger dynamic power consumption. Reducing the clock frequencies to the minimal level required to meet performance will result in power savings.

Implementation entails all the stages for transforming the design from architectural representation into RTL, and from the RTL to gates using RTL synthesis, and even further down through place and route up to design tape-out. In some cases, implementation may start earlier when high-level description in C/C++/SystemC is synthesized to RTL. Various implementation techniques can be used to reduce power during implementation. For example, clock gating may be used to block ("gate") the clock from switching the gates in a design block when input data does not change and unblock the clock when new data needs to be processed by the design block.

Voltage determines both static and dynamic power. The larger the voltage, the higher the power consumption. In particular, voltage may significantly increase dynamic switching power. Reducing the voltage may cause performance reduction, but at a certain point may prevent the design from operating properly.

The process technology has traditionally been the primary factor determining power as moving to smaller process geometry allowed reducing Vdd voltage and reducing transistor leakage. However, moving to the next process technology also resulted in increased operating frequency and increased gate count. This resulted in requiring a reduction in the number of transistors that can be active in any given time. In fact, the CTO of ARM has said, "In a decade, 11 nm process technology could deliver devices with 16× more transistors running 2.4× faster than today's parts. But those devices will only use a one-third as much energy as today's parts, leaving engineers with a power budget so pinched they may be able to activate only 9% of those transistors."

As simply riding the process technology curve could no longer drive the design power down and meet the design requirements, new designs combined with advanced power analysis and optimization techniques had to emerge to tackle the power problem.

2.7.6 Virtual Programming

The best way to meet all of these challenges lies in the emergence of hardware aware virtual prototyping. Virtual prototyping techniques provide an alternative to hardware prototyping by using abstracted functional models of the hardware.

Virtual prototyping opens the way to concurrent development of hardware and software and continuous analysis and optimization of the design for performance and power.

2.7.7 Conclusions

Minimizing power consumption of an embedded device is a major undertaking—there are no simple solutions. A rich mixture of software and hardware design factors affect power consumption, which can only be effectively addressed using an integrated hardware/software approach.

Programming

Chapter Outline

In this chapter, the articles focus on specific programming problems encountered in embedded systems.

3.1 Programming for Exotic Memories

I wrote the original version of this article for NewBits in 1991, when I was working for Microtec Research in the United Kingdom, and it required little revision for inclusion in this book. The issues discussed in this article continue to be of concern. It is interesting that, even back when I first wrote the article, solving a problem by hiding it in a C++ class seemed like a good idea. I was even referring to multichip embedded systems, which were far from common. This article may be usefully read in conjunction with another one I wrote, "Self-Testing in Embedded Systems," later in this chapter.

CW

3.1.1 Exotic Memories

A palm-fringed beach with waves washing tendrils of white foam from the deep blue ocean onto the even whiter sand. The sun beating down on your head and an ice cold bottle of Mexican beer in your hand. The hint of suntan oil in the air and the appetizing aroma of lunch cooking on the barbecue in the nearby bar.

A memory of last summer's vacation? No, mine wasn't like that either.

Anyway, the kind of exotic memory I want to talk about is the much more prosaic kind: embedded systems memory storage, other than regular RAM and ROM. Nonvolatile RAM and shared memory both present specific challenges to the embedded systems programmer.

Embedded Software: The Works. DOI: 10.1016/B978-0-12-415822-1.00003-9

3.1.2 Nonvolatile RAM

In the good old days, real computers (the ones with rows of lamps and switches on the front) had magnetic core memory. We still use the terms "core memory" or "core dump" today, even though the technology has passed into the history books. Core memory had two significant characteristics that differ from those of modern semiconductor memory. The first was size. Core memory was generally packaged in slabs, several inches square, which typically held 4 K words. (I heard it suggested that a 4 by 4 matrix of slabs, i.e., 64 K—could be termed a "patio," but I'm not sure this terminology ever escaped the lab in which I worked at the time.) The second feature was nonvolatility; the data was still there even after the power had been cut. This tended to be exploited during debugging. At the end of the day, the engineer would shut down the computer and go home. The following day, he would switch on the computer and carry on his work from where he left off. (This assumes that the core had not experienced any significant mechanical shock—such as being dropped—because it tended to corrupt the memory, which was a source of concern for military and aerospace applications.)

Nowadays, most RAM is volatile and engineered to be nonvolatile only for specific applications. Without assistance, the data is lost on power down, and the RAM is filled with garbage on power-up. Typically you could get nonvolatility by providing a battery backup power supply to low-power RAM chips. In an embedded system, nonvolatile RAM is typically used to hold setup or configuration data, which must be retained across power shutdowns.

Why should these requirements present programming problems? Can't we treat nonvolatile RAM just like regular RAM? The answer to the second question is generally yes; while you can treat it like normal RAM, a little care is needed in two specific respects.

First, it is common practice to carry out initialization and self-test of RAM on start-up. It may be obvious that initialization of NVRAM (which should contain useful data retained from earlier) may well be illogical, but self-test sounds reasonable. After all, failure of RAM chips (like most electronic devices) almost always occurs on power-up, so a test before beginning to use them seems only prudent. Of course, the battery-backed NVRAM hasn't been powered up, but it has been pulled out of a low-power "sleep" mode, and that amounts to the same thing.

The problem is that a self-test of RAM necessarily corrupts (normally only on a temporary basis) the RAM locations under test. If you carried out a test of NVRAM on start-up, and it was interrupted (e.g., by a bouncing power supply switch), the data is corrupted. To overcome this problem, you can exploit the fact that memory chip faults almost never result in the failure of a single byte of storage; most likely a particular bit in a whole sequence of bytes gets stuck. Consequently, you only need to test a small number (maybe just one) of the locations, which are dedicated to this use. Testing one byte per chip is reasonable. It would not be too hard to figure out suitable addresses. If you have information about the architecture of the chips, you can consider a test of a representative cell on each row and/or column, but that might be just a little "over the top."

The second problem is the classic "chicken and egg" dilemma. Since the NVRAM is supposed to survive power failure, the data should not be initialized on start-up. The first time that the equipment is powered up, the NVRAM will contain random data, just like regular RAM. How can you recognize that the data has not been set up? An additional consideration is the possibility of battery backup failure or a power glitch, which corrupts the data. There are two possible corrective measures:

- Include a nonrandom byte "signature" to indicate the NVRAM is set up.
- Use a checksum to ensure the data has not been corrupted.

The signature can be used under any circumstances; the checksum is most appropriate when the data is not updated too often (e.g., setup parameters) since it must be recalculated each time the data is updated. Care should be taken with the checksum calculation to avoid any locations that are dedicated to self-testing, as described previously.

Here is an example testing code:

```
#define SIG 0
#define CHK 4
#define NVR_SIZE 256
extern unsigned char nvram[NVR_SIZE];
int nvr_check()
{
    int i;
    unsigned char sum = 0;
    if( nvram[SIG] ! =0 ||  /*              check signature */
        nvram[SIG+1] !=0xff ||
        nvram[SIG+2] !=0x55 ||
        nvram[SIG+3] !=0xaa )
            return (-1);                   /*failed */
    for (i=CHK; i<NVR_SIZE; i++)           /*check checksum */
        sum +=nvram[i];
    if (sum !=0)
    {
        nvr_init(); /* failed */
            return (-1);
    }
    else
            return (0);                    /* passed */
}
```

The C function, `nvr_check()`, is designed to check the NVRAM signature (00, $FF, $55, $AA) and verify the checksum's integrity. The external array `nvram[]` maps to the NVRAM.

No attempt is made to accommodate self-test cells. If the NVRAM passes the tests, 0 is returned. If it fails, the function `nvr_init()` is called and −1 is returned.

Updating the checksum on an area of NVRAM can be a chore, but C++ can make it very straightforward. All that is required is the definition of a new data type (i.e., class), which looks like a regular array to the user. By overloading the appropriate operators, the updating of the checksum could be totally automatic. In addition, the constructor function can take care of the verification of the signature and checksum.

Persistent memory is commonly provided by use of flash. This is very cost effective and reliable but brings a slew of other programming issues, which are beyond the scope of this article. The upcoming technology is MRAM (magnetic RAM), which promises to replace both flash and SRAM, having the nonvolatility of one, with the speed of the other.

3.1.3 Shared Memory

In an embedded system that includes a number of microprocessors, a common form of interface between them is an area of shared memory, often called "dual port RAM." This area provides a high-speed communications medium with a high degree of flexibility. Although simple in concept, at least three problems must be addressed:

- Since both microprocessors are running simultaneously, some data may be read by one before it has been written in its entirety by the other.
- The data representation (e.g., byte ordering and floating point format) may not be the same when two different processors are involved.
- The address of the memory may not be the same from the viewpoint of both processors. For example, a 4 K shared memory area may extend from $1000 to $1FFF on one side and from $23,000 to $23FFF on the other.

All three problems can be readily overcome. The first one can be addressed by defining a message-passing scheme. Perhaps a single byte is set nonzero when a valid message has been completely loaded elsewhere. The receiving chip clears the byte when all the data has been read, thus completing the handshake. Some devices used to implement dual port RAM provide the alternative of using an interrupt-driven handshake.

To solve the second problem, it is necessary to agree which representation will be used in the shared memory; one method is for the "losing" chip's software to convert all transferred data. I encountered such a problem many years ago when I was using a PDP11 minicomputer linked by shared memory to a TMS9900 microprocessor. Those two CPUs have different byte ordering (i.e., the address of the most significant byte of a word may be lower or higher than that of the least significant) and differing floating point formats. I settled on the 9900 format for data in the shared RAM even though it could be argued that it was the stranger one. (In the manual for that chip, the bits were numbered 0 for the most significant and 15 for the least; I guess the author hadn't heard of powers of two!) Byte-ordering differences are still common today; just try connecting an x86 and a Coldfire. Most RISC chips even have the ability to run in two different modes, to facilitate compatibility with other devices. The modern term for memory architecture is "edianity" (or "endianness")—a device is either "big-endian" or "little-endian."

The different addresses mentioned in the third problem are simply a linking issue and are really both unsurprising and a source of only a little confusion.

3.1.4 Conclusions

Handling specialized, exotic types of memory need not be a problem when the relevant issues are correctly addressed in the software. As always, the choice of development tools may be critical. At the simplest level, being able to locate appropriate data structures at the correct addresses is necessary. At a more complex level, the use of C++ classes enables special memory to be treated quite transparently.

3.2 Self-Testing in Embedded Systems

This article is based upon one that I wrote for NewBits in 1993. It is fair to say that failure in an embedded system is at least as likely now as it was then—maybe more so. So the issues discussed in the original article are still important.

CW

Ever since the early microprocessors were first used in the design of embedded systems, code has been incorporated to ensure hardware integrity. Logically, if a complex board has some intelligence, it should be able to perform self-testing. With careful design, diagnostic software can detect most of the likely failures in an embedded system. Only the unlikely failure of the CPU or ROMs storing the program precludes any testing being performed at all.

3.2.1 Memory Testing

A clear candidate for testing is the system memory; its integrity is critical to system operation. Since memory is readily accessible to the CPU, implementing tests is quite straightforward. Some thought is required to design appropriate and safe tests, which may be applied in each of a variety of situations.

RAM—Start-up
RAM chips are like light bulbs (and most electrical and electronic components): they most commonly fail on power-up. The fact that the RAM does not contain any useful data on start-up indicates an ideal opportunity to perform a thorough test. The most common form of test used at this time is a "moving ones" (or "moving zeros") test. Specifically:

1. Clear all bits to zero.
2. Set a single bit to one and check all the other bits to ensure they are still set to zero.

3. Clear the bit and check all bits to ensure that they are still set to zero.
4. Repeat the test for each bit.

The following code implements this algorithm:

```
#define RAM_SIZE 0=1000
extern char RAM [RAM_SIZE];
register int bit, ramloc, count;
register char mask;
for (count=0; count<RAM_SIZE; count++)
     RAM [count]  = 0;
for (ramloc=0; ramloc<RAM_SIZE; ramloc++)
{
     mask = 1;
     for (bit=0; bit<8; bit++)
     {
        RAM[ramloc] = mask;
        if (RAM[ramloc] !=mask)
            fail();
        for (count=0; count<ramloc; count++)
            if (RAM[count] != 0)
                fail();
        for (count=ramloc+1; count<RAM_SIZE; count++)
            if (RAM[count] != 0)
                fail();
        RAM[ramloc] = 0;
        for (count=0; count<RAM_SIZE; count++)
            if (RAM[count] !=0)
                fail();
        mask <<=1;
     }
}
```

It is assumed that the array RAM has somehow been located over the RAM space and that the constant RAM_SIZE correctly reflects the size of the memory. The function fail() is called if the test fails.

However, these tests have two problems:

- The testing code must itself not use any RAM. All variables must be held in machine registers, and all functions called by the memory test must be inlined to avoid using the stack. Both these requirements may be met by use of a modern optimizing compiler, which will make good use of machine registers and perform inline optimization of small functions.
- Testing takes a long time. Even a high-performance microprocessor can take minutes or hours to complete such a test. Time can be saved by testing each RAM chip separately. Failures that cause a bit to affect other bits in the same chip are not likely to affect bits in other chips. Only memory bus problems cause these failures, which prevent the program from running anyway.

RAM—Testing on the Fly

Once the application code is running, it is clearly impossible to perform any RAM tests that would corrupt the data held there. The only feasible kind of testing is to check each byte, word, or long word in turn, having saved its data in a machine register first so that it may be restored. This test detects two types of failure:

* Sticky bits, which stay set to zero or one, whatever is written to them.
* Adjacent bit crosstalk, which can occur with certain memory architectures.

The following code performs such a test:

```
#define RAM_SIZE 0×1000
extern char RAM [RAM_SIZE];
const char tests[] = { 0, 0xff, 0x55, 0xaa };
int ramtest()
{
    register int testnum, ramloc;
    register char save;
    for (ramloc=0; ramloc<RAM_SIZE; ramloc++)
    {
        save = RAM[ramloc];
        for (testnum=0; testnum<sizeof(tests); testnum++)
        {
            RAM[ramloc] = tests[testnum];
            if (RAM[ramloc] != tests[testnum])
                return (1); /* test failure */
        }
        RAM[ramloc] = save;
    }
    return (0); /* test success */
}
```

Test values of $00 and $FF are used to check for sticky bits; $55 and $AA are used to check for crosstalk. Once again, all variables must be placed in machine registers, and no function calls are made during the test. This test can be run from time to time or may be initiated by an operator action. In a multithreading environment, it can be run continuously as a background task; see the reentrancy considerations discussed in the section on multithreading below.

PROM—Checksum

In-service failure of program memory (ROMs, PROMs, or flash memory devices) is unusual, but they may be subject to human error. If there is a set of memory chips containing application code and constant data—and there is usually more than one—the possibility exists that one or more devices will be of the wrong revision, be incorrectly programmed, or be inserted the wrong way around or in the wrong sites. If the device containing the start-up code is correct, a power-up self-test can perform a checksum verification on the memory and detect many problems.

The checksum algorithm must be chosen carefully. Most would not detect devices in the wrong sites, because their contents would still add up. A cyclic redundancy check (CRC) or carefully designed diagonal parity algorithm are possible solutions.

The insertion of the checksum into the memory image has to be done after the linker has generated the absolute file. A utility can be written that reads the S-records (or Intel hex or whatever), calculates the checksum and generates an additional record to locate the value at a known memory address.

Special Memory Types

Special kinds of memory require special care in the design of test algorithms. With nonvolatile RAM, conventional RAM tests cannot be performed because the system may be powered down while the memory is corrupt. The slow access time of flash RAM can present other problems.

In addition, shared RAM cannot be tested conventionally because its corruption would most likely interfere with the interprocessor protocol. The solution is to incorporate a self-test sequence into the protocol, letting each CPU write and read from the shared memory. In addition, such a sequence enables each CPU to verify the integrity of the other—two tests for the price of one.

3.2.2 Input/Output Devices

Since all I/O devices are unique, it is hard to make any useful general comments. The testing of an I/O device generally requires a loop back facility to be included in its design. Loop back permits the value sent to an output port to be read back in and verified or for test input to be sent to input ports.

3.2.3 Multithreading Issues

In a multithreading environment, where a real-time kernel, RTOS, or interrupt service routines are used, additional care must be exercised in implementing self-tests. With an RTOS, additional tests may be possible.

RAM Testing Reentrancy

The ongoing testing of RAM in multithreading environments is a problem when an interrupt occurs while a location under test is corrupt. The only solution is to disable interrupts for the short period of time that the location does not hold valid data.

Stack Limits

When using an RTOS, each task typically has its own stack. The size of each stack must be specified but is hard to determine. If you can calculate how much stack space the deepest nested function call requires (assuming there is no recursion) and add to that the maximum stack requirement of any interrupt service routines (which might "steal" some stack space), the allocation is likely to be too great. The usual practice is to make educated guesses and increase the allocations after a crash.

A simple self-testing technique that helps prevent crashes is the use of guard words at the top of the stack for each task. These words are set to a particular value and monitored at regular intervals. When these words become corrupted, an error is declared: that stack is on the verge of overflow. The most suitable value for guard words is zero, as mostly nonzero addresses are pushed onto the stack.

3.2.4 Watchdogs

A common way to monitor CPU and software integrity is to incorporate a watchdog timer circuit into the hardware design. These can come in two forms:

- A reactive watchdog asserts a specific interrupt and expects a specific response, which arms it for another period. This type of watchdog tests whether the CPU is functioning but does nothing to monitor the status of the software. The mainline code could be in an unintended loop, interrupted only to service the watchdog.
- A proactive watchdog requires the program to access a particular address at a minimum frequency. Watchdog assertions are placed at suitable strategic positions in the main software.

3.2.5 Self-Test Failures

What action should self-test routines take when errors are located? What should a watchdog do if it times out?

The answers to these questions will vary from one embedded system to another. A common practice is to perform a reset—return the code to the power-up state. While this practice may be effective in correcting a random RAM glitch caused by a power spike, for example, it hides the details of a number of other faults. An indication of the nature of the failure should be reported or displayed. If this is not possible, at least an error code should be stored somewhere, so that an engineer using an infield debugger can inspect the value and identify the fault.

3.2.6 Final Points

To write self-testing code, it is critical that the C code is translated into very efficient machine code and that special requirements (such as register usage, function inlines, and memory locating) are properly accommodated by your development tools.

3.3 A Command-Line Interpreter

For some years before joining Microtec, I had been interested in the Forth programming language. One of the first embedded systems that I worked on used this language, and

I remained fascinated by its possibilities. In this article, which is based upon one I wrote for NewBits in 1996, I applied some of the key ideas of a threaded interpretive language to a user interface.

CW

Two terms are commonly confused in the world of embedded systems: "debugging" and "diagnostics." The reason for the confusion is quite subtle, and we will come to that shortly. First of all, let's define the terms. *Debugging* is the process of verifying that code performs the function for which it was designed, and if not, fixing it accordingly. *Diagnostics* are components of an application program that enable the user to confirm or ensure the well-being of the software and/or hardware after deployment.

The two terms should not be confused because debugging tools should be present only during the development phase. After deployment, they should leave no trace; just the application code (including any diagnostics) should be taking up space on the target. The reason for confusion is that some development tools do not conform to this ideal and consume valuable target resources long after they have ceased to be useful. It is in the interests of the vendors of such tools to propagate this confusion.

In this article, we are going to look at the implementation of diagnostics on an embedded system. It must be emphasized that this process has very little to do with the debugging phase; it is a means of implementing a post-deployment testing facility.

3.3.1 Embedded Systems Diagnostics

With a simple embedded system, a malfunction is easy enough to identify—the system no longer works. The cause of the fault may or may not always be apparent, but the fact that a fault exists is quite clear.

More complex systems present a greater challenge. With a high level of functionality, a system can develop a potentially catastrophic fault, without showing any immediate symptoms. It is essential that such systems incorporate protection against such a mishap.

The subject of self-testing was covered in the previous article in this chapter, so we will not dwell on it here. Suffice it to say that self-testing takes three forms:

1. Tests executed on reset
2. Tests executed continuously or periodically during the operation of the system
3. Tests initiated by the operator—i.e., diagnostics.

Diagnostics concern us here.

Communication with the operator is an obvious requirement. If the system has a screen/keyboard, then these components may be used. Otherwise, a connection to an external terminal (or computer) is required. Maybe a spare serial line is available, or one could be included in the design specifically for this purpose. Another possibility is that a serial line, which is normally employed by the application, can be used when the system is in "diagnostic mode."

3.3.2 An Embedded System Comes Alive

A difficulty (psychological, at least) with developing an embedded system is that an incomplete application, even thought it may well be executing correctly, does not appear to be doing anything. A much greater feeling of satisfaction is achieved when the software "says its first word," which is a major argument for implementing diagnostics early in the development cycle. Apart from this, working diagnostics can also help with debugging and integration.

3.3.3 A Command-Line Interpreter—Requirements

There are broadly two ways that a keyboard/screen user-interface to diagnostics may be implemented:

- **Single key action:** Commands are implemented as single keys, which are actioned immediately. Other data that is required is obtained by prompting the operator. This approach may be good for very simple requirements, because a "state machine" approach to the implementation may be taken. The downside is that the structure is not intrinsically extensible.
- **Mnemonic commands:** Complete command lines are entered by the operator and only actioned when the Return key is pressed. This approach has greater flexibility because it may be extended easily. It has the additional advantage that the programmer is not burdened with line editing, because the library functions can take care of these matters. A possible downside is that interpretation of the commands may require complex programming.

Although the first approach is valid, we will address the issues involved in the second one in this article.

As with many common software requirements, the inclusion of a command-line interpreter (CLI) presents the designer with two options: buy it or build it. It is almost always better to buy a software component, if such a product exists. For larger embedded systems, a good approach is to include a UNIX-style shell as the CLI—these are available for popular RTOSs. For a smaller system, the overhead of such a sophisticated solution may not be bearable. The only option is to implement a simple, custom CLI.

3.3.4 Designing a Command-Line Interpreter

The operation of a CLI occurs in three phases:

1. Recognize the command
2. Collect and process any required parameters
3. Execute the command.

Phase 1 is quite straightforward. All that is required is a table of valid commands, against which a sequence of string comparisons is made. A little care with case sensitivity is the only real precaution that is necessary. Phase 3 is also quite easy. A table of function addresses (i.e., pointers to functions) is constructed, with entries corresponding to each command. This table is indexed as a result of the string comparisons performed in phase 1. The only challenge to the programmer is to untangle the intractable C syntax!

Most problems occur in phase 2. Here are some of the issues:

- How many parameters (if any) are required?
- What kind of data is required for each parameter?
- Are any parameters optional?
- How much flexibility in entry format can be provided to the operator?
- How are parameters passed to the function that implements the command?

The remainder of this article will propose a design for a CLI that addresses these issues, resulting in the complete code for a real implementation.

3.3.5 A CLI Implementation

The CLI described here exhibits the following behavior:

- When ready to accept a command line, it prints a $ prompt.
- It accepts text "blindly" until the Enter key is pressed.
- An empty command line (no characters or just spaces) is ignored and a new prompt issued. This means that repeatedly pressing Enter just causes more prompts. This is useful when checking the communications line.
- A command line is considered to be a sequence of tokens, separated by spaces. (Incorporating support for other or additional separators is straightforward.)
- When a line has been received, the tokens are processed, in turn, from left to right. (Although, if preferred, you could readily adapt this approach to parse right to left.)
- Each token is checked to find out if it is a valid command. The first three characters and the length of the command are considered significant. So PRINT and PRIME are considered the same, whereas PRINTER is considered different. The comparison is not case sensitive (all processing is performed using lowercase), and any nonspace ASCII characters may be used. If the command is valid, its corresponding function is executed.
- If a command is invalid, an attempt is made to process it as a number. If this attempt is successful, the value is pushed onto the "parameter stack," where it may be retrieved by subsequent commands. If the token does not represent a valid number, an error message is displayed, and the remaining tokens on the line are not processed.

This behavior yields a very flexible operator interface. Here are some possible command lines, with the processing sequence:

6 LED-FLASH: The 6 results in this value being pushed onto the parameter stack. The function corresponding to the command LED-FLASH is executed and presumably expects a value on the stack.

5000 5015 DUMP: The 5000 and 5015 each place a value on the stack. The function corresponding to DUMP is executed and expects two parameters on the stack: at the top it expects the last address and below it the first address or a range to be dumped in hexadecimal format.

MEM DUP 15 + DUMP: The MEM causes a function to be executed that places an address on the stack. DUP causes a function to be executed that makes a copy of the top stack item. 15 pushes the value on the stack. + adds together the top two stack items. DUMP functions as before.

The user interface may be used in a simplistic manner by an operator with limited training. A more sophisticated user can take advantage of the flexibility.

Readers with knowledge of the Forth programming language may find this approach familiar. Although this CLI is far from being a full-blown Forth interpreter, it could form the basis for the design of such a tool.

3.3.6 CLI Prototype Code

All the code for a prototype CLI is included in this section. The program is written in ANSI C and should build with any compliant compiler. In tests, using an efficient 68000 compiler, the entire program resulted in less than 1 K of code. This represents a very small overhead to gain a powerful facility.

Of course, there are library functions too, but they may well be present in the application anyway. The most significant one is printf(), but its use may be circumvented if the memory hit is too great.

Here is the command table, which may easily be extended:

```
struct dtable                        /* command dictionary entry */
{
    union
    {
        struct
        {
            unsigned char len;
            char word[3];
        } bytes;
        unsigned long bits;
```

```
        } id;
        void (*func)();
    };
    #define ENTRIES 10          /* command dictionary definition */
    void dot(void);             /* command functions */
    void store(void);
    void at(void);
    void query(void);
    void plus(void);
    void dump(void);
    void quit(void);
    void v1(void);
    void v2(void);
    void v3(void);
    struct dtable dictionary[ENTRIES]  =
    {
        {1, '.', 0, 0, dot},                    /* . command */
        {1, '!', 0, 0, store},                  /* ! command */
        {1, '@', 0, 0, at},                     /* @ command */
        {1, '?', 0, 0, query},                  /* ? command */
        {1, '+', 0, 0, plus},                   /* + command */
        {4, 'd', 'u', 'm', dump},               /* dump command */
        {4, 'q', 'u', 'i', quit},               /* quit command */
        {2, 'v', '1', 0, v1},                   /* v1 command */
        {2, 'v', '2', 0, v2},                   /* v2 command */
        {2, 'v', '3', 0, v3}                    /* v3 command */
    };
```

Each entry includes the first three characters and the length of the command, coded into an
`unsigned long` (32 bits) by means of a `union`. This technique is portable because it makes no
assumptions about the layout of structures.

Here is the `main()` function:

```
    void main()
    {
        char str[100];
        char *tok;
        unsigned long ident;
        unsigned match, error;
        unsigned i, j;
        struct dtable scratch;
        while (1)
        {
            printf("\n$ ");         /* prompt */
            gets(str);              /* get line of input */
            tok =strtok(str, " ");
            if (tok)
                do                  /* compute 32 bit identifier: */
                {
```

```
                  scratch.id.bytes.len  =
                      (unsigned char)strlen(tok);
                  for (i=0; i<3; i++)
                      if (i<strlen(tok))
                          scratch.id.bytes.word[i]  =
                              (char) tolower(tok[i]);
                      else
                          scratch.id.bytes.word[i] = 0;
                  ident = scratch.id.bits;
                  match = 0;                        /* scan dictionary: */
                  for (i=0; i<ENTRIES && !imatch; i++)
                  {
                      if (ident == dictionary[i].id.bits)
                      {
                          /* match - execute function */
                          match = 1;
                          (*dictionary[i].func)();
                      }
                  }
                  error = 0;/* no match -try a number */
                  if (!imatch)
                  {
                      if (!trypush(tok))
                          error = 1;
                  }
              } while ((tok = strtok(NULL," ")) && !error);
      }
  }
```

An outer loop displays a prompt and accepts command lines. The inner loop breaks down the line into tokens and decodes them into commands and parameters. The parameter stack definition, which is simply an `int` array, is shown here, along with the associated utility functions.

```
#define STACKSIZE 100              /* local parameter stack */
int stack[STACKSIZE];
int sp  = 0;                       /* stack pointer */
/*** Utility Functions ***/
/* trypush() - try to convert token to number a push on stack returns true on
success */
int trypush(char *str)
{
    unsigned i;
    int val = 0;
    for (i=0; i<strlen(str); i++)
    {
        if (isdigit(str[i]))
            val = val * 10 + str[i] - '0';
        else
```

```
                {
                        printf("%s???\n", str);
                        return (0);
                }
        }
        push(val);
        return (1);
}
/* push() - push a value onto the stack */
void push(int n)
{
        if (sp == STACKSIZE)
        {
                printf("Stack full!!!\n");
                exit(-1);
        }
        stack[sp++] = n;
}
/* pop() - pops a value off of the stack [else returns 0] */
int pop(void)
{
        if (sp == 0)
        {
                printf("Stack empty!!!");
                return (0);
        }
        return (stack[--sp]);
}
```

Here is the code for commands implemented in this example:

```
/*** Command functions ***/
/* . command - pop value off the stack and display */
void dot(void)
{
        printf("%d ", pop());
}
/* @ command - pop address off the stack and push the data at which it points */
void at(void)
{
        push(* (int *)pop());
}
/* ! command - pop an address and some data off the stack and store the data at
the address */
void store(void)
{
        int addr;
        addr = pop();
        * (int *)addr = pop();
}
```

```
/*?command - effect the @ and . commands in sequence */
void query()
{
     at();
     dot();
}
/* +command - sum top two stack items and push the result */
void plus(void)
{
     push(pop() +pop());
}
/* dump command - pop last and first address and display that range of memory
addresses in hex */
void dump(void)
{
     int last, i, j;
     last =pop();
     for (i=pop(); i<last; i+=8)
     {
          printf("%04X ", i);
          for (j=i; j<i+8; j++)
               printf("%02X", *(char *)j);
          printf("\n");
     }
}
/* quit command - pop code and exit if correct */
void quit(void)
{
     if (pop() ==999)
          exit(0);
}
/* v1 command - push the address of the first user variable */
void v1(void)
{
     push((int)&vars[0]);
}
/* v2 command - push the address of the second user variable */
void v2(void)
{
     push((int)&vars[1]);
}
/* v3 command - push the address of the third user variable */
void v3(void)
{
     push((int)&vars[2]);
}
```

It should be noted that these commands, which allow very free access to the application's memory space, would be very unlikely candidates for inclusion (or at least documentation) in a real system. The actual commands would be much more application-specific. These commands are simply presented as examples:

. (dot): Removes the top value from the parameter stack and displays its value (in decimal).

@ (at): Removes the top value from the parameter stack and, treating it as an address, pushes the data found at the location to which it points.

! (bang): Removes the top value from the parameter stack and, treating it as an address, removes the next value off the stack and stores it at that address.

?: Combines the effect of the @ and . commands: it removes the top item from the stack and, treating it as an address, obtains the data to which it points and displays its value (in decimal).

+: Removes the top two items from the stack and pushes their sum.

dup: Makes a copy of the top item on the stack and pushes it.

dump: Removes the top value from the parameter stack and treats it as the last address of a range. The next item is removed and treated as the first address of a range. The range of memory addresses is then displayed in hex.

quit: Removes the top value from the stack and, if it has the value 999, terminates the CLI (otherwise, it does nothing).

v1, v2, and v3: Push the address of the first, second, and third user variables, respectively.

3.3.7 Conclusions

The implementation of a flexible command-line interpreter facilitates easy-to-use post-deployment diagnostics. The code may readily be written in ANSI C, which by use of a modern optimizing compiler results in a very small memory overhead.

3.4 Traffic Lights: An Embedded Software Application

When I wrote the original article on this topic for NewBits in 1996, I was responding to a common request: "Can you show me an embedded application that does not require reams of code and does something that I understand." I figured that we all cross the road sometimes.

CW

The objectives of this article are ambitious: to address a number of key techniques in the development of a simple embedded system. Issues to be covered include:

- Basic application design
- Building and debugging the program
- Programming and debugging interrupts.

3.4.1 The Application

The example program is a simplified traffic-light-controlled pedestrian crossing system. These systems work in various ways in different countries. This example is based upon the system used in the United Kingdom:

1. The walker presses a button.
2. The traffic lights cycle through yellow to red, and the crossing light changes from "Stop" to "Go."
3. After a delay (while the walker crosses the road), the yellow light and "Go" are set flashing.
4. After a further delay, the traffic light is restored to green and "Stop" is illuminated.
5. Then the button (the one the walker pushed in step 1) is inoperative for a period of time (to avoid too frequent interruption of traffic).

In this example, some details are omitted for simplicity: traffic sensing, pedestrian detection, sound generation, and so on.

3.4.2 Hardware Configuration

The controller is based upon a microcontroller, which has two memory-mapped I/O ports. At $1000 is the output port that controls the (five) lights. The allocation of bits is illustrated in Figure 3.1; setting a bit illuminates the bulb. The most significant three bits should not be set.

At $2000 is the input port that corresponds to the press button; bit 0 is set (and latched) when the button is pressed, and it may be cleared by writing 0 to the port. A 250-mS timer interrupt is set up on a vector at address $100.

3.4.3 Program Implementation

The logic of the program consists of two parts: a main loop, which awaits the button pressing and sequences the lights, and an interrupt service routine, which deals with time delays and lamp flashing.

Figure 3.1
Output port at $1000

3.4.4 Main Loop

The structure of the main loop closely maps to the previous behavioral description; the pseudo-code is shown is as follows:

```
initialize lights to Green/Stop;
initialize button;
repeat
    wait for button to be pressed;
    clear button register;
    set lights to Yellow/Stop;
    delay 2 seconds;
    set lights to Red/Stop;
    delay 2 seconds;
    set lights to Red/Go;
    delay 5 seconds;
    set lights to flashing Yellow/Go;
    delay 3 seconds;
    set lights to Green/Stop;
    delay 1 minute;
end
```

Here is the actual C code:

```
/* Bit patterns for lights */
#define GREEN 1                          /* traffic */
#define YELLOW 2
#define RED 4
#define FLASH_YELLOW 0 × 200
#define GO 8                             /* pedestrian */
#define STOP 0 × 10
#define FLASH_GO 0 × 800
/* Control word for lights */
volatile int lights;
/* Button read/clear definitions - bit 0 of $2000 */
#define READ_BUTTON ((*(int *) 0 × 2000) & 1)
#define WRITE_BUTTON (*(int *) 0 × 2000)
/* Timing facilities */
volatile int countdown;
#define WAIT(s) countdown = s*4; while(countdown);
main()
{
    /* Initialisation */
    lights  = GREEN | STOP;       /* set lights */
    WRITE_BUTTON  = 0;            /* init button */
    asm("MOVE #$2000,SR\n");      /* enable interrupts */
    /*** Main loop ***/
    while (1) /* go round for ever */
    {
```

```
        /*** Wait for button ***/
        while (!READ_BUTTON)
            ;                           /* just hang */
        WRITE_BUTTON = 0;
        /*** Change lights to allow crossing ***/
        lights = YELLOW | STOP;
        WAIT(2);
        lights = RED STOP;
        WAIT(2);
        lights = RED | GO;
        WAIT(5);                        /* crossing now */
        /*** Change lights back ***/
        lights = FLASH_YELLOW | FLASH_GO;
        WAIT(3);
        lights = GREEN | STOP;
        /*** Let traffic flow for a while ***/
        WAIT(60);
    }
}
```

This code is quite simple. There are a couple of points to note:

- Access to the button is by means of two "memory variables": READ_BUTTON and WRITE_BUTTON. These macros handle the syntactical untidiness of accessing a memory location (I/O port) directly in C.
- A significant amount of work is done by the timer interrupt (which is discussed in more detail in the section that follows). The main loop communicates with the interrupt service routine (ISR) by means of two global variables: lights and countdown. The variable countdown is in turn accessed by the macro WAIT().

3.4.5 Interrupts

The only interrupt in this system is a 250-mS timer. The ISR supports two facilities:

- Countdown timer for performing time delays in the main loop
- Output to the lamp-driving hardware, taking care of flashing.

Here is the code for the ISR:

```
/* Lights port - $1000 */
#define LIGHTS (*(int *)0 × 1000)
interrupt void timer()
{
    int flash_bits;
    if (countdown != 0)             /* effect timer */
        countdown--;
```

```
        flash_bits = lights >> 8;           /* update lights */
        lights ^ = flash_bits;
        LIGHTS = lights & 0xff;
}
/* Interrupt vector */
#pragma asm
        ORG $100
        DC.L _timer
#pragma endasm
```

Note that the function is declared as an ISR by using the keyword `interrupt`, which is a common extension to ANSI C. This ensures that context-saving code is generated and the function ends with an `RTE` instead of `RTS`. The function is necessarily declared to be of type `void`, with no parameters.

Also shown is the definition of the interrupt vector, using an assembly language insert. There are several other methods by which this may be achieved. For example, using an array of pointers to functions, which is located at link time. The method chosen has the advantage of brevity and simplicity. Anything to do with interrupt vectors and interrupt control is necessarily very processor-specific.

3.4.6 Time Delays

The time-delay facility is implemented very simplistically. A global variable, `countdown`, is monitored by the ISR. If it is nonzero, it is decremented by one on each tick. To effect a delay, the code in the main loop simply loads this variable with the number of 250-mS intervals that it wished to delay for and waits for it to become zero. The `WAIT()` macro takes care of the details. Note that countdown is declared `volatile`; this ensures that the compiler generates code to actually access the variable's memory location each time it is referenced (instead of optimizing it into a register).

This timing mechanism yields a mean timing precision of $+/-125\,mS$, which is quite acceptable for this application.

3.4.7 Lights

The ISR has exclusive access to the lamp-driving hardware. The main loop uses the global variable `lights` (again declared `volatile`) to indicate its requirements. The least significant byte of this variable maps directly onto the output register (as shown in Figure 3.1). The most significant byte is in the same format but indicates lamps that are to be flashed on each timer tick. The ISR code updates the lamp-driving hardware every tick, having toggled the bits of the flashing lamps accordingly.

3.4.8 Using Global Variables

The technique of using global variables as a means of communication between tasks or between mainline code and ISRs is widely considered to be unreliable. To use it safely, certain conditions must be met:

- The application must be simple—this one is.
- The variables must be declared volatile—they are.
- The protocol for their use (i.e., which code writes and reads them and when it can do it) must be clearly defined—it is.

C Language

Chapter Outline

C is still the most commonly used programming language for embedded software applications. Although a small language, it continues to challenge software engineers. Many of these challenges arise from the origins of the language, which were quite a different context from that in which it is now applied. The articles in this chapter address the topic from various perspectives.

4.1 C Common

This article is based upon one written for NewBits way back in 1990 by Ken Greenberg. It addresses an issue that was—and still is—a common area of confusion among C developers. Over the years, embedded tools have become more like those used for host native development in both their capability and the standards to which they adhere, but some issues, such as the handling of weak externals, persist.

CW

In this article, we will discuss the concept of C common, sometimes referred to as *weak externals*. A C common data object is a global variable that has not been explicitly declared to be either static or external and has not been initialized in any module. Thus, a declaration of the form:

```
int currentTask;
```

may be in C common. You can't tell by looking at only one module whether a given data object will or will not end up in C common since it may have been initialized elsewhere.

It is now common—maybe even standard practice—for software engineers to prototype their applications (or at least as much of the applications as practical) on their development platform (host). Ideally, the program should work the same way in the host-prototyping environment as in the target. The use of C common makes an embedded compiler work more like a host compiler and enables the programmer to transfer code from native to cross-compilers without making changes.

The concept of C common is closely tied to the idea of a defining point. In C, every data object is to have only one defining point. In practice (as defined by C compilers on UNIX), it is also possible to have no defining point at all, with the linker taking care of those objects not defined elsewhere. A defining point is most commonly determined by an explicit initialization. If you coded:

```
int currentTask=0;
```

then you create the defining point for `currentTask`, and no other module is allowed to initialize it. If two modules initialize the same variable, a link-time error (multiple definitions) results. Initializing a variable tells the linker that this object is an "external definition"; it also has the side effect of placing it in a separate program section from the uninitialized variables. Suppose you had a declaration like:

```
extern int currentTask;
```

then you would, in effect, tell the linker that this is an "external reference" to a data object defined elsewhere. If all of your modules had identical declarations, this would be a link-time error of a different sort. If all modules tell the linker that this variable is somewhere else, the variable ends up being undefined.

Variables declared as `static` do not export their names to the linker, so they are treated differently. Since they can be accessed only from within the module in which they are declared, they don't really need a defining point.

C common data objects have no defining point at all. They are not initialized in any modules. They may include the storage class `extern` in some modules (but this is essentially superfluous; the `extern` keyword is, in this context, arguably obsolete). The behavior expected by most (host) C programmers is that declarations will refer to the same variable. If all modules simply declare `currentTask` to be a variable of type `int`, then it is up to the linker to establish that they refer to the same `int`. Since the variable `currentTask` is not initialized in any modules, then the linker is responsible for allocating space for it in the uninitialized variables section. Further, if the variable is declared as:

```
long currentTask;
```

in one module, then it still refers to the same object. The linker must allocate space based upon the longest declaration found. It should be emphasized that this is *not* good programming practice!

For embedded systems developers, it has been traditional to take a somewhat different approach. In the past, most C compilers have not supported C common. Any variable declared as external is a reference; any variable not declared as external is a definition. The burden is on you to make sure that all declarations are external except one.

However, the techniques used in embedded development have been changing over the last few years. Embedded tools increasingly conform to the standards of host tools, and support for C common is an example of that conformance.

In summary, Figure 4.1 illustrates the valid code constructs for the traditional and the C common approaches; Figure 4.2 illustrates the possible error conditions.

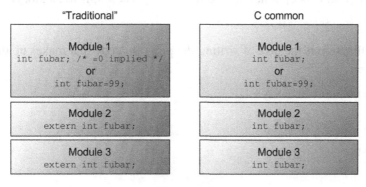

Figure 4.1
Valid syntax

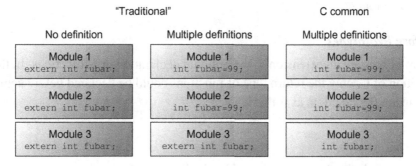

Figure 4.2
Error conditions

4.2 Using C Function Prototypes

This article is based upon one that appeared in NewBits in 1990 under the title "Using ANSI C Function Prototypes"—there is no record of the identity of the original author. At that time, the ANSI standard for the C language was new, and compiler support was just beginning to appear.

CW

When the ANSI C language standard was first published (after a long wait), it incorporated much more than a consolidation of the language features in use at that time. A number of entirely new capabilities were added, among which were function prototypes. This feature allows programmers to specify more information about functions for improved error detection during compilation.

In discussing function prototypes, it is important to understand a three-way distinction among function declaration, definition, and calling.

- **Declaring** a function establishes a form or *prototype* for subsequent use of the function. Function declaration is optional in C (unlike C++ where it is mandatory). For example:

  ```
  int my_function(int, char);
  ```

- **Defining** a function is the task of writing the program code that makes up the function. For example:

  ```
  int my_function(int x, char c)
  {
      ...
      if (c=='y')
          return (x*47);
          ...
  }
  ```

- **Calling** a function involves its actual use as a program element, either from other functions or, in the case of recursion, from the function itself. For example:

  ```
  z = my_function(22, 'y');
  ```

4.2.1 Before Prototypes

Function prototypes are an extension to the optional declaration of C functions. In the original K&R definition of C, only the return type of functions could be declared before actually defining or using the function, as follows:

```
return_type my_function();
```

where `return_type` is the data type returned by the function. In the absence of `return_type`, function type defaults to `int`. This seems odd, because `void` is a more logical choice.

However, `void` was not included in the language standard until ANSI, so there were backward compatibility issues.

Note that such a declaration causes the compiler to check the return assignment of a function but makes no effort to check the types of parameters passed in a call to that function. In the absence of such parameter "prototypes," the compiler generates code assuming that all parameters are correct, leading to errors that are difficult to diagnose at runtime.

4.2.2 Applying Prototypes

Function prototypes extend function declaration to include the formal parameters.
For example, if the `my_function()` function, discussed in the previous section, requires two integer parameters, the declaration could be expressed as follows:

```
return_type my_function(int x, y);
```

where `int x, y` indicates that the function requires two parameters, both of which are integers. Note that the variable names `x` and `y` are optional, formal parameters. The actual parameter names in the function definition and subsequent calls need not be the same; only the types must match. Using meaningful formal parameter names is helpful in documenting the functionality.

Because the declaration includes parameters, the compiler can check whether calls to this function are made properly. If, for example, the following call is made:

```
a = my_function("hello world");
```

the compiler generates an error informing you that `my_function` was called with an incorrect number of parameters (one instead of two) and an incorrect first parameter (i.e., an integer was expected but a string was received instead).

Note that C automatically performs some forms of type conversions to ensure that actual parameters match the declared type of a function's formal parameters. Character variables (`char`), for example, are automatically promoted to integer (`int`) for parameter passing and arithmetic.

4.2.3 Prototypes in Use

Using function prototypes in include (header) files can be a convenient method of preparing a specification of your code module and its external interface for others to read. Users of your code will be less reliant on browsing your source code to determine the parameters that your functions require.

The following are several examples of function prototype declarations. Note that the inclusion of parameter names is optional:

```
int uart_init(int baud, char bits, char parity);
void make_buffer(int size);
void kill_list(struct list *head);
char *look_ma(int, float, char); /* no parameter names */
```

In summary, C provides software developers with a standard language syntax to express functions' declarations in more detail. Function prototypes increase the readability of include files and allow compilers to support improved compile-time error checking.

4.3 Interrupt Functions and ANSI Keywords

This article is closely based upon a piece by Ken Greenberg in NewBits in 1990. ANSI C was new then, but the points and principles discussed in this article still hold today.

CW

4.3.1 Interrupt Functions

I would like to start this article with a discussion of interrupt functions. As the term implies, interrupt functions are used primarily as interrupt handlers. Traditionally, an embedded system designer would write a short routine for each interrupt in assembly language. The addresses of these routines are then inserted into the interrupt vector table. These routines save registers as necessary, call a C function to do any processing required, restore registers on return from the C function call, and return with the special "return from exception" instruction required by the specific processor. The code for each routine, and all functions that it calls, is collectively known as the Interrupt Service Routine (ISR).

Why are these short assembly language routines necessary? There are two reasons. First, every compiler has a set of registers that it is free to use as scratch registers and other registers that it is obligated to save through function calls. I like to call the first set the *scratch* set and the second the *preserve* (or *save*) set. For example, the preserve set for a particular 80 × 86 compiler includes the registers CS, DS, SS, SP, BP, SI, and DI; the scratch set includes ES, BX, CX, DX, and the return value register AX. If you wrote your ISR entirely in C, it is safe to assume that the preserve set registers would be as they were at the time an interrupt occurred when you were ready to resume execution of the interrupted code. However, it is just as safe to assume that the scratch set would be completely destroyed by the ISR. For this reason, these short routines will push the contents of the scratch set registers before calling a C function and will pop them before returning.

The second vital function performed by these short assembly routines is the generation of an appropriate return from the ISR. When an interrupt occurs, many processors push state information on the stack along with the return address. For example, the 80×86 family processors push a "flags" word, the contents of the CS register, and the contents of the IP register on the stack. Clearly, a normal return instruction would not clean up the stack properly, so the `IRET` instruction must be used instead.

Ideally, C programmers like to write everything in C, including their ISRs. With modern compilers, it becomes simple to get rid of the short routines for interrupt handlers that you used to write in assembly language. All you need to do is use the `interrupt` keyword when declaring a function. The compiler will then generate a special function prologue and epilogue. The function prologue will save any registers from the scratch set that will be used, along with the preserve set registers; the epilogue will restore the contents of all registers (both preserve set and scratch set) and generate an appropriate "return from exception" instruction. The actual behavior of the prologue is somewhat compiler-dependent. On the Motorola 68000, which has a "save multiple registers under mask" instruction, only those registers that are actually used are likely to be saved; on the Intel 80×86 family, it is probable that all the scratch registers will be saved.

What does such a C interrupt function look like? Here's an example of a keyboard interrupt handler:

```
void interrupt far kbdintsvc()
{
    register int c;
    register DEVICE d;
    c = inport(0);
    ainbuf[ainptr++] = c;
    if (ainptr>=IBUFSIZ)
        ainptr = 0;
    d = deviceTable[D_CONIN];
    if (d->interruptQueue)
        SignalTask(&d->interruptQueue);
    else
        ++(d->status.dcbsema);
}
```

Don't worry too much about the work done by the handler; the form of the C interrupt service routine is what you need to understand at this point. Note that all interrupt functions must be of type `void` and have no arguments. Since the address of this function will be placed in the processor's interrupt vector table, no mechanism exists for passing any arguments to the function. In other words, this function is never actually called. Since it has no caller, there is nobody to return a value to, so void is an appropriate type. Also, this should be a `far` function for the 80×86 family. Leave the `far` keyword out for processors that do not support (need) it (i.e., processors with flat address space, like Coldfire, PowerPC, and ARM).

It should also be noted that in the example, the `kbdintsvc()` function makes two function calls: `inport()` and `SignalTask()`. These types of function calls indicate an important concept: interrupt functions may call "normal" (non-interrupt) functions with no interesting side effects. Of course, any assembly language functions must adhere to the C calling conventions for preserving registers. Since a copy of the scratch registers has been preserved on the stack, it isn't necessary to keep saving them. In fact, you may not call another `interrupt` function, since `interrupt` functions are not designed to be called. Thus, it is illegal to explicitly call any function declared with the interrupt keyword.

4.3.2 ANSI C const Keyword

As long as we're discussing declaration modifiers like `interrupt`, I should say a few things about the `const` keyword for ANSI C compilers. The basic concept is easy to understand: do not allow modification of any data object declared as `const`. The syntax can be a little tricky, however. Just remember that C is a left-associative language in most cases. That is, "`int errno`" says that `errno` is of type `int`. Each token refers to the token to its left. (Unfortunately, the `far` and `near` keywords don't follow this rule, since they were modeled after Microsoft's keyword usage.)

The `const` keyword also modifies the token to its left. Thus, "`char const * pcc`" indicates that `pcc` is a pointer to a constant `char`. The pointer itself is not constant, but the character at which it points is. Alternatively, we could have a declaration like "`char * const cpc`," which states that `cpc` is a constant pointer to a `char`. The pointer itself is not allowed to vary, but the character that it points to can be modified. A few examples of how `const` is used may be helpful.

```
static char * const days[] = {"Sun", "Mon" "Tue" "Wed" "Thu" "Fri" "Sat"};
```

In this example, we declare `days` to be an array of constant pointers to `char` with `static` scope. There really is no reason for these pointers to be anything else but constant, since the order of days in a week is unlikely to be modified at runtime. The benefit from declaring the pointers to be `const` is that they will be placed in a separate section by the compiler, and that section may then be located in memory as desired. Since the pointers will never change, they can be placed in ROM, for example. The characters themselves are allocated to the strings program section by the compiler and may also be placed in ROM. Thus, we have the advantage of being able to use pointer types (which are quite efficient) without wasting any RAM space for them. We also won't need to copy them from ROM to RAM at program start-up, which is an advantage if your system is ROMable.

I'd like to give one more example of how `const` can be used in an embedded system. Suppose you have an input routine that obtains values in decimal or hexadecimal from the system operator. You will need to convert these values from their ASCII representation to a form your program can use. The following code converts characters representing numbers in any base up to 16 into binary form:

```
while (c = *s++)
{
    for (index=0; index< base; index++)
        if (c == hextable[index])
            break;
    value = value * base + index;
}
```

The table that is used to look up the individual characters may well be a const, since the table can be placed in ROM with other data that doesn't change at runtime. We can achieve this by declaring it as:

```
static char const hextable[] = "0123456789abcdef";
```

Note that this is an array of constant characters, where the previous example used constant pointers.

4.3.3 ANSI C Volatile Keyword

ANSI C programmers tend to think of const and volatile together, since there are similarities in how they modify storage. Syntactically, they are identical; semantically, they are nearly opposites. Whereas a const data object can never change, a volatile one is assumed to have changed since its last usage. This means that declaring something volatile will have an effect upon optimization. The keyword tells the compiler to refrain from optimizing accesses to a given variable, since external events can alter the variable's contents.

In the embedded systems world, this typically applies to memory-mapped I/O devices. It may also apply to data objects in regions of memory that are subject to direct memory access (DMA) transfers, or are used for data shared by multiple tasks in a multitasking environment or between mainline code and ISRs. Volatile variables never end up as register variables.

The volatile keyword is typically applied to pointer variables. For example, a memory-mapped I/O device may be declared as:

```
char volatile * const receivedData = 0xfeed;
```

This means that receivedData is a constant pointer to a volatile character. We will now use the pointer in a loop:

```
while (*receivedData !=3) /* wait for CTRL-C */
    ;
```

The code generated for this loop will perform a fetch through the pointer on each pass through the loop. If we do not declare `receivedData` as a pointer to a `volatile char`, then the optimizer is free to fetch through the pointer once and repeatedly compare the result to the constant 3. This is meaningless but perfectly legal optimization. The addition of the `volatile` keyword to the C language provides a means to communicate your intentions to the compiler. If a variable is not safe to optimize because it is really a memory-mapped I/O port, you can convey this information in a standard, portable way.

Using `const` for Data Protection

A variable with the type modifier `const` cannot be modified for the scope of its declaration, and, if the variable is static or global, will likely be placed in ROM. The ANSI C keyword `const` can also serve another function: data protection. Suppose that your program includes a data structure initialized and modified by one module—the "implementer" module—that must be accessible as read-only to some other "consumer" modules, and for the purposes of this discussion, a procedural interface is considered inefficient. You can declare the data structure as `const` in the consumer modules, while retaining the ability to modify it (non-`const`) in the implementer module. As long as the definition of the data structure is in the implementer module (non-`const` declaration), the data will not end up in the `const` section, yet the consumer modules will not be permitted to modify the data.

- Module a—the implementer:

```
int sacred_data = 0;
void implementer(int x)
{
    sacred_data = x + 2;
}
```

- Module b—the consumer:

```
extern const int sacred_data;
char consumer()
{
    display(sacred_data);
}
```

4.4 Bit by Bit

This piece is based upon one that I wrote for NewBits in 1992, when I was with Microtec UK. Even today, C programmers find bit fields challenging.

CW

In this article I plan to get down to basics and take a look at the issues and challenges associated with memory in embedded systems. At the lowest level, memory is just lots of bits: numerous logic states represented by 1s and 0s. The grouping of these bits into bytes and words is really quite arbitrary; sometimes it becomes necessary to deal with memory bit by bit or in small groups of bits. I will look at how this is accomplished from the C programmer's perspective.

4.4.1 Bitwise Operators

C is generally considered a good language for embedded systems programming for a number of reasons, one of which is its ability to perform bitwise arithmetic. All the primary bit operations are available: inversion (using ~) to change 1s to 0s and vice versa; AND (using &) to test bits; OR (using |) to set bits; exclusive OR (using ^) to all toggle-specific bits; shift right and left (using >> and <<) by one or more bits.

Using these operators, the programmer can perform any required bit manipulation. However, the notation does not yield very easily readable code. This situation is exacerbated by the need for the expression of bit patterns in hexadecimal (or octal). C does not have a means of including binary constants. For example, it is not intuitive that the following code:

```
bytevar |= 0xc0;
```

results in the setting of the top 2 bits of the 8-bit variable `bytevar`.

4.4.2 Binary Constants

There is a way to provide what appear to be (8-bit) binary constants in standard C. All that is necessary is a series of macro definitions in a header file (`bits.h` perhaps), like this:

```
#define b00000000 ((unsigned char) 0x00)
#define b00000001 ((unsigned char) 0x01)
#define b00000010 ((unsigned char) 0x02)
...
#define b11111110 ((unsigned char) 0xfe)
#define b11111111 ((unsigned char) 0xff)
```

4.4.3 Bit Fields in Structures

Since working with single bits or small groups of bits is useful, the C language was designed to include integers of any bit size in structure definitions. In ANSI C, these integers may be signed or unsigned and may be as large as the widest available base type (typically 32 bits).

The mechanism for allocating memory to bit fields is governed by the rules and definitions of the C language and the design of the specific compiler. An understanding of the latter is important, since, without care, it is easy to write compiler-specific code. As Kernighan and Ritchie wrote in the second edition of *The C Programming Language*, "Almost everything about (bit) fields is implementation-dependent."

Bit fields may be allocated from the top of the first word of memory or, less commonly, from the bottom. Unnamed fields may be used for padding. Bit fields are allocated through the first word until one of a number of criteria cause the next word to be utilized:

- Insufficient bits remain in the current word, and word boundary straddling is not supported by the chip/compiler.
- An unnamed field with width 0 is encountered.
- A new base type is encountered.

The last criterion has interesting effects. Compare the following two structures, `alpha` and `beta`:

```
struct
{
    char a : 2;
    char b : 3;
} alpha;
struct
{
    char a : 2;
    int b : 3;
} beta;
```

These two apparently similar structures use different amounts of memory. The structure `alpha` requires a single byte of memory, whereas `beta` requires 8 bytes.

Remember that the C bit fields are just a convenient notation for the programmer. The compiler must still generate the instructions to perform the necessary ANDs and ORs. However, beyond their notational convenience, C bit fields also give the well-designed compiler an opportunity to generate particularly efficient code.

4.4.4 Microprocessor Bit-Field Instructions

Recognizing the importance of the C language and the usefulness of the bit-field notation, a number of microprocessors include instructions to operate on bit fields of arbitrary width. Typical chips with this capability are the Freescale Coldfire and the Intel Pentium family in native 32-bit mode. Here is some compiler output that shows bit fields in use:

```
;struct
;{
; unsigned a : 3;
; unsigned b : 4;
;} blob;
```

```
;main()
;{
     XDEF _main
_main:
; blob.b = 1;
     moveq #1,d0
     bfins d0,_blob{3:4}
;}
     rts
```

In the same way that the use of bit fields in C gives the compiler a chance to produce optimal instructions, the use of a microprocessor's bit-field instructions enables the chip to perform well. Although the same memory read/write cycles are required, a significant speed increase is possible.

4.4.5 I/O Devices and Bit Fields

A commonly cited example of the application of bit fields is for accessing input/output device control registers. It is very common for I/O devices to have control registers that are segmented into a number of fields of various bit lengths. An obvious concept is to map a C structure onto such an I/O device control register. Each of the individual fields can then be read and/or written as required. There are potential problems with this technique, if the ports are write-only; I have addressed this issue from the C++ standpoint in another article— "Write-Only Ports in C++" in Chapter 5.

It is generally difficult to find examples of code used typically in an embedded system that you, the reader, can experiment with. This is simply because such examples make assumptions about the available target hardware, assumptions that may not be fulfilled. However, in one suitable example, the target is a standard PC and the I/O device the program addresses is the screen. The screen of a PC, in text mode, appears to the program to be an area of memory, where each word is divided into 2 (8-bit) bytes, the first of which contains the ASCII code of the character to be displayed; the second is divided into a number of bit fields. Specifically, the latter is divided into a 1-bit, a 3-bit, and a 4-bit field, which specify flashing, background color, and foreground color, respectively. Here is the definition of a structure that reflects this format:

```
struct cell
{
     unsigned sym : 8;
     unsigned fg : 4;
     unsigned bg : 3;
     unsigned flash : 1;
};
```

In the following code, the variable p is a pointer to type struct cell and is initialized to point to the screen memory. The code of main() simply writes a pattern on the screen using the bit-field references.

```
struct cell *p = (struct cell *) 0xb8000000;
void main()
{
    int row, col=0, inx;
    for (row=0; row<25; row++, col+=3)
    {
        inx = row*80 + col;
        p[inx].sym = 1;
        p[inx].fg = 1;
        p[inx].bg = 0;
        p[inx].flash = 0;
    }
}
```

This code does, of course, make certain assumptions about the compiler—i.e., the allocation of fields across a word is compiler-dependent.

4.4.6 Conclusions

For embedded systems, the manipulation of bits of memory is often necessary, and the C language bit-field notation is useful for this application. If you are using a high-performance target chip that supports bit-field manipulation directly, you need a compiler that can take full advantage of this facility. Since the use of bit fields, in general, introduces compiler dependencies, the documentation should carefully identify these dependencies.

4.5 Programming Floating-Point Applications

This article uses a test case from a piece in NewBits in 1992 by Sarah Joseph-Bigazzi, who was a software engineer with Microtec. The original article was very product-centric, but the example code has much wider application.

CW

Floating-point values are used widely in mathematical and number-crunching applications. Capable of representing extraordinarily large and small numbers, floating-point numbers offer a numerical range that is essential to many embedded systems designs. Yet, despite the importance of these numbers, few developers are fully aware of the intricacies and pitfalls in their use. In this brief article, we will look at an example of some floating-point code that looks fine on the surface, but contains a subtle, but important, error.

4.5.1 A Test Case

Here we have a program using floating-point operations:

```
void compare(double x, double y, int i)
{
    if (x==y)
        printf("Compare #%d worked\n", i);
    else
        printf("Compare #%d failed\n", i);
}
main()
{
    const double PI=ASM(double, "fldpi");
    double d_array[5], d_value=PI;
    float f_array[5], f_value=PI;
    register int i;
    for (i=0; i<5; i++)
    {
        d_array[i]=d_value / (i +1);
        f_array[i]=f_value / (i +1);
        compare(d_array[i], f_array[i], i+1);
    }
}
```

This program simply fills d_array (a double array) and f_array (a float array) with the floating-point values π, $\pi/2$, $\pi/3$, $\pi/4$, and $\pi/5$ (the ASM instruction fldpi extracts the value π from the 8087 coprocessor—we are assuming an x86 processor, but the same principle can be applied for other devices). Each element of d_array and each element of f_array will be passed to compare() as double-precision parameters. The parameters x and y are then compared. If the values are equal, the program will display a message in the form:

```
Compare #n worked
```

for values of n from 1 to 5.

If the values are not equal, the program will display:

```
Compare #n failed
```

4.5.2 Running the Test Case

Since both arrays are loaded with equal values, you should expect all the comparisons to succeed and the following output to appear:

```
Compare #1 worked
Compare #2 worked
Compare #3 worked
Compare #4 worked
Compare #5 worked
```

But that is not what happens. (Try it!) What you actually see is:

```
Compare #1 failed
Compare #2 failed
Compare #3 failed
Compare #4 failed
Compare #5 failed
```

So what is going on here?

4.5.3 Troubleshooting

If you have a debugger available, step through the code and have a look at the two parameters, x and y, of compare(). You will observe that they are, indeed, slightly different. To find out why they are not equal, we need to go back to the code in main() where the values were derived.

Upon examination of the declaration of arrays d_array and f_array, we find the source of the problem: the first is declared double and the second is float. Thus, the values π, $\pi/2$, $\pi/3$, $\pi/4$, and $\pi/5$ were calculated in d_array and f_array using different precisions. Since the values in f_array were calculated using a single-precision floating-point format (float), f_array remained less precise than d_array even after the values in f_array were promoted to double-precision prior to comparison. Because of this difference in precision, in every call to compare(), x and y were inherently unequal. Using the == operator reflects a classic programming error, namely, an attempt to test for equality between two values of different precision.

The comparison (x == y) should have been expressed alternatively as (fabs (x-y) <delta), where fabs() is a C library function that returns the absolute value of its real argument, and delta denotes the maximum difference acceptable for the two values to be considered equal. In this case, a value of delta of 0.0005 is appropriate.

4.5.4 Lessons Learned

Although this was a purely fabricated example, two lessons may be learned and applied to real embedded floating-point applications.

First, it is very easy to "play fast and loose" with data types in C—in other words, allowing automatic conversions to occur and not worrying about the consequences. This practice can cause enough difficulties with integral data types (sign extension, etc.), but with floating point, the bugs can be very subtle and hard to locate.

The second lesson is to dismiss all thoughts of "equality" between any two floating-point values. There is always the possibility of very slight rounding errors, even if the data type issues are taken care of. It is normally not even safe to look for a precise zero value—even though there is an exact representation for zero. You should never code ($x == 0.0$); it should always be something like: `(fabs(x) < TINY)`, where `TINY` is defined to be a value small enough to represent a rounding error in the floating-point format for your processor.

4.6 Looking at C—A Different Perspective

This article was written as a result of my experience training engineers in C programming for embedded applications. It appeared in a more expanded form in NewBits in 1993, including some topics that are covered in articles elsewhere in this book. I have also added some new examples.

CW

Even back in the mid-1980s, when many other languages were significant for microprocessor software development (assembler, Pascal, even FORTRAN), C was emerging as the clear winner and the choice for the 1990s.

Nowadays, it's no contest. Everyone uses C (or its derivative C++) for embedded systems work. It's easy to forget that C was not designed for this purpose. Of course, when Kernighan and Ritchie first defined the language, microprocessors had not been invented, and electronic control systems were either hard-wired or controlled by minicomputers.

The experience of training is always a two-way flow of information; the teacher generally has something to learn from the pupil. In C language training, that something is a different perspective on aspects of the language that we all take for granted.

4.6.1 Static Things

The keyword `static` is often a source of confusion for those learning C for the first time. The reasons for the confusion are that the keyword really performs two separate functions (three in C++), and some variables may be stored statically but are not declared as `static`.

The word *static* means "unmoving or stationary." Static variables are ones that are allocated space at compile time rather than being dynamically allocated machine registers or stack space at runtime. All variables outside of C functions are stored statically. Inside a function, local variables may be declared as static (using the `static` keyword as a qualifier on the declaration of the variable), if it is required to preserve a value from one call to the next. This capability is useful, but results in the function not being reentrant.

Variables outside of a function need not be declared static, but if they are, they are rendered local to the module (file) in which they appear, instead of being global, as is otherwise the case. This is a useful facility because it provides a means of communication between a group of related functions. However, it is really a different use of the `static` keyword. The same idea can be applied to functions. Functions are normally global. By declaring one as static, it is made available only in the module in which it is defined.

So the static keyword has two distinct meanings: it can specify memory allocation, and it affects the scope of a variable or function. I have seen the suggestion that you could #define a new "keyword" local, which would just be static in disguise.

4.6.2 All Those Semicolons

When someone first learns C programming, he or she usually has problems with using the semicolon. After all, many of the semicolons do seem unnecessary. As a result, lots of compiler diagnostics about missing semicolons are generated. Ironically, the effect of missing a semicolon can cause errors several lines down in the program, which is confusing for the beginner.

The usual result is a big swing the other way. The C novice starts putting in more semicolons left, right, and center. This solves the problem of the missing ones but introduces some more subtle difficulties at runtime. The following loop, for example, gets executed only once (for a value of i of 5):

```
for (i=0; i<5; i++); printf("%d", i);
```

instead of the five iterations that the programmer intended.

This string-character-counting loop never finishes (or never actually gets started):

```
while (*str); str++, len++;
```

Under some circumstances, you want a loop to be empty. This code, for example, is waiting for a device to return a nonzero value:

```
while (*devptr == 0)
    ; /* wait */
```

It is a good practice to put the lone semicolon (the empty statement) on a line by itself and include a comment, as shown in the example. And let's hope that devptr is declared as a pointer to a volatile int, or we will be in real trouble.

4.6.3 Pointers and Pointer Arithmetic

For many, pointers are thought of as strange and mysterious entities and operations on them akin to a black art. This topic is far-ranging enough to deserve its own article, "Pointers and Arrays in C and C++," later in this chapter.

4.6.4 When Being Clever Is Not Being Smart

A certain kind of programmer likes to demonstrate how clever they are at every opportunity. I guess that we have all met similar people in many walks of life. Such programmers specialize in arcane code that *might just* be a tiny bit more efficient. A little thought about what "efficient" might mean can often give some useful insight.

Here is an example of some "clever" code:

```
void bin(int x)
{
        if ((x/2) != 0)
                bin(x/2);
        printf("%d", x%2);
}
```

What does it do?

To save you having to spend ages trying to analyze the code, I will tell you: it prints a number in binary format.

What is wrong with the code? Two things. First, it is hard to understand, and since a large proportion of programming time is spent on maintenance, such lack of clarity is short-sighted. Second, the code is recursive, and while it is valid in C for an embedded application, recursion should be used with great caution and only when absolutely necessary. It is easy for the code to consume more resources (i.e., stack space) than may have been anticipated.

Consider this alternative code (which does almost exactly the same thing):

```
void bin(int x)
{
    unsigned mask=0×80000000;
    int i;
    for (i=0; i<32; i++, mask >>=1)
            if ((x & mask) != 0)
                    printf("1");
            else
                    printf("0");
}
```

This implementation may seem simplistic, but it is certainly much easier to understand.

So, did the "clever" code have any virtues or advantages? It may have been faster, but, for a user I/O function like this, speed is not an issue. What about size? We compiled both pieces of code for a 68 K target using "optimize for size." I expected the results to be similar and to argue that a few extra bytes are a small price to pay for extra clarity. I did not expect the result that I got: both produce exactly the same amount of code—60 bytes.

4.6.5 Conclusions

Although it is the most used language for embedded systems work and is well suited for this purpose, care is still required when writing in C. Another precaution to take is the selection of software development tools that are specifically designed for embedded systems development.

4.7 Reducing Function Call Overhead

Embedded engineers are always concerned about resources (usually time and memory) and want to be sure that they are working efficiently. This quest for efficiency can often be at the expense of good programming practice. In this article, based upon one that I wrote for NewBits in 1994, I look at how modern compilers address some of the efficiency issues without conflict.

CW

It sometimes seems that the life of a programmer is cursed. There never seem to be any good breaks. If code is small, it probably runs slowly. Fast code is generally too big. Worst of all, code that is easy to read (i.e., structured code) is probably large and slow. Why is this the worst case? Because, although it is commonly believed that programming is all about making a microprocessor do something, it is really a task performed to communicate your ideas to another programmer (or yourself at a later date). Since at least 90% of programming time is spent on maintenance, it is financial common sense to write clear, easy-to-read code.

4.7.1 Compilers and Structured Code

Recognizing these needs, modern compilers are designed to process structured code well. Broadly, this goal involves translating the intentions of the programmer into machine code rather than simply making the executable map directly onto the original source code. The following loop provides a simple example:

```
char v[4];
for (i=0; i<4; i++)
    v[i] = 0;
```

which an efficient compiler may compile into a single 32-bit move of 0 into memory.

This presents a minor debugging challenge but not really a difficulty for a debugger designed to cope with fully optimized code. The improvement in both code size and speed is well worth it. The biggest challenge in efficiently processing structured code is function calls. An easy-to-read program consists of a modular collection of smaller functions that call one another as required. This approach to program design achieves the goal of more maintainable and flexible code. However, the cost in runtime performance may be severe, because every call/return sequence carries overhead.

4.7.2 Inline Functions

Before considering what a function call overhead really means, it is worth considering how function calls may be avoided. One solution might be the C++ `inline` keyword. This language extension is supported by a number of modern C compilers. The `inline` keyword offers advice to the compiler that communicates that the indicated (small) function could be usefully inlined. For each call to the function, the compiler inserts the actual code of the function instead of the call. Instead of the `inline` keyword, some compilers have command-line switches that the programmer can use to indicate which functions to inline.

Both of these techniques have a drawback: the programmer must determine which functions to inline. The best approach is for the compiler to be "smart" enough to make the decision itself, as long as the programmer has indicated that execution time is a priority.

4.7.3 Function Calls

The process of executing a function call can be broken into four discrete stages:

1. Evaluating the parameters (if any) and preparing them for transfer into the called function.
2. Constructing the stack frame; space is likely to be required for local variables (which have not been allocated machine registers) and for internal workings.
3. Executing the function code.
4. Returning a value from the function.

Each of these stages (except stage 3) represents a real or potential overhead beyond the processing necessary to perform the job required of the function.

4.7.4 Parameter Passing

In C, the way parameters are handled is clearly defined: they are evaluated from right to left and pushed onto the stack in turn. In principle, the parameters could be passed by other means (e.g., via machine registers), but studies tend to indicate that such other means do not yield the expected performance benefits.

Why are the parameters dealt with from right to left? (Were Kernighan and Ritchie left-handed?) This question is commonly asked in C language training courses. It seems a reasonable query because a number of other languages deal with parameters the more "logical" way—left to right. The reason is simple: C permits a given function to accept a variable number of parameters. The only way such a function can determine just how many parameters have actually been passed is from information in a mandatory (typically the first) parameter. For this to work, that parameter must be at a predictable location, such as the top of the stack. An example of this technique is the standard `printf()` function.

The initial (mandatory) parameter is a string (char*) containing format descriptors that indicate how many parameters follow. Of course, programmer error may result in the failure of this mechanism, but a disadvantage of such great flexibility is some lack of robustness.

Can the knowledge of parameter passing be exploited by the programmer? If the ability to handle a variable number of parameters is used, it is significant that the first parameter is on top of the stack. By taking its address, indexing off of it can provide access to all the others. But, of course, this practice will probably result in nonportable code.

4.7.5 Local Storage

Once a function has been called, the parameter values are pushed onto the stack with the return address on top. The next concern is the provision of local work space for the function. This includes memory allocated to local (automatic) variables and storage space for intermediate values during expression evaluation.

The need for intermediate storage is a misconception that often results in less-readable code. For example, the complex statement:

```
v[i]=w[p + q-> t];
```

may be used out of fear that using a variable to store an intermediate value would result in unwanted overhead. In fact, recoding the statement as a series of simple steps:

```
j = p + q-> t;
v[i] = w[j];
```

is unlikely to have any effect at all upon the resulting machine code.

In a small function, both local variables and intermediate storage may be accommodated by machine registers, but this depends upon the target microprocessor architecture and the degree of compiler optimization.

4.7.6 Stack Frame Generation

In most functions, local memory storage is a requirement, and the stack is the obvious place to accommodate it, which is logical since indexing off of the stack pointer is efficient for most 16/32 bit processors. Also, functions are inherently reentrant, since a new area can be allocated on each call, and the code may be shared between tasks (or between mainline and interrupt code). This chunk of memory on the stack is known as the *stack frame*.

The generation of a stack frame is quite a simple process. The compiler determines how much local storage will be required, and the stack pointer is incremented, or more often decremented, by the appropriate amount. At the end of the function, the stack pointer's previous value is restored so that the return address is on top again.

Because high-end microprocessors are invariably programmed in a high-level language, support for stack frame generation is incorporated into the instruction sets of such devices. The instructions that perform the stack frame construction and destruction are generally called LINK and UNLK (i.e., unlink), respectively. The idea is quite simple. A suitable (address) register is nominated to be the *frame pointer* (FP). For the Coldfire family, the frame pointer is generally A6 because A7 is used as the stack pointer (SP). The FP points to a linked list of stack frames to facilitate their easy de-allocation by the UNLK instruction. The LINK instruction builds the frame in three simple steps:

1. The current FP is pushed onto the stack.
2. The FP is set to point to the pushed (old) value.
3. The stack pointer is advanced by an appropriate amount to allocate the required space.

The structure of the stack at the beginning of the "real" function code is shown in Figure 4.3.

The UNLK instruction undoes the frame structure in two stages:

1. The SP is set to the value of the FP.
2. The old FP is popped off of the stack.

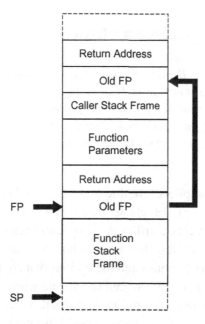

Figure 4.3
Stack structure after a function call

Once the stack frame is established, the function parameters may be accessed at (known) positive displacements (for most processors) from the FP. Local variables and work space are at negative displacements.

In principle, the LINK/UNLK sequence has the ability to dynamically adjust the size of the local work space during the execution of the function while still being able to locate parameters easily. However, the C language does not really exploit this capability.

4.7.7 Return Values

In C, any non-void function is assumed to return a value. For most functions, this is an integral type (int, char, short, etc.), a floating-point value, or a pointer (address). Almost without exception, a machine register (or pair of registers) of the appropriate type is employed. There is little scope for improved efficiency here.

4.7.8 Conclusions

Programmers who are particularly concerned with the runtime performance of their programs can apply a knowledge of function call overheads to their program design and use modern compilers that combine efficient local optimization with automatic inlining capabilities.

4.8 Structure Layout—Become an Expert

This article has a long and complex history. It is based upon a piece in the Spring 2004 issue of NewBits by Antonio Bigazzi, the leader of the compiler team at Accelerated Technology, with valued assistance from Meador Inge, which, in turn, was an updated version of an article he had done at Microtec, also for NewBits, back in 1993. It illustrates a fine example of something that seems quite simple and can be taken for granted but, in the embedded world, must be understood in more detail.

CW

Modern embedded C/C++ compilers give fine-grained control and a wealth of options for determining how C structures (or C++ classes) are laid out. The result is that any arbitrary layout can be attained. This article describes the simple and consistent rules that many compilers use in laying out structures, showing you how to tailor these rules to your needs. You will need to check whether the rules and the options that control them are implemented in a specific compiler, and we have endeavored to indicate where options may be unusual. These rules are described with reference to Freescale 68 K, ColdFire, and PowerPC, but the same ideas apply to all target architectures. We will discuss these rules for C, with the understanding that the same rules apply to C++.

The article will help you to:

- Determine the memory layout of any structure, whether packed or unpacked; determine the offset and alignment of members.
- Force any desired layout; for instance, force cpu32 (e.g., 68332—a bus16) and cpu32p (e.g., 68340—a bus32) structures to have identical layouts.
- Achieve maximum space or time efficiency for accessing the structures in your program.
- Learn a number of useful techniques and test yourself; a test is included at the end of this article.

4.8.1 Key Concepts

To understand structure layout fully, you first must be familiar with the concepts of data bus width and natural boundaries. The compiler uses these concepts in the basic algorithms that compute offsets and total size when allocating structures.

Data Bus Width

The 68-K family includes many variants, which are divided into two groups:

- Those that utilize BUS16: 68000/08/10, 68302, 68330/1/2/3, 68340, 68ec000, 68hc000/1, cpu32.
- Those that utilize BUS32: 68020/30/40/60, 68ec020/30/40/60, cpu32p.

All ColdFire devices are BUS32 (the 5204, now obsolete, was a BUS16).

PowerPCs are either BUS32 or BUS64. For example, PPC403GA is BUS32, while PPC740 is BUS64.

BUS16s have a 16-bit data bus width, meaning that a memory cycle can access a maximum of 16 bits. Multibyte quantities can be properly accessed only if they reside at an address that is a multiple of two. For instance, fetching a short integer (which takes up 2 bytes), located at an odd address, causes an address exception. Since it is not, in general, possible or practical to recover from such an event, the compiler must avoid this kind of situation by properly accessing misaligned data in a bytewise, piecemeal fashion.

BUS32s have a 32-bit data bus width, meaning that a memory cycle can access a maximum of 32 bits. Multibyte quantities can be properly accessed at any address, but if they are not properly aligned, performance will be degraded because the hardware must supply extra memory cycles. In other words, the piecemeal access is provided by the hardware itself—at a cost. BUS64s have a 64-bit data bus width, meaning that a memory cycle can access up to 64 bits.

Efficiency considerations are similar to BUS32. Eight-byte quantities (C doubles) should be aligned to a multiple of 8 for maximum efficiency.

Natural Boundaries

Each C type has a natural boundary—i.e., a number N such that a variable of that type is accessed best if aligned to an address multiple of N. As a general rule for scalar types, the natural boundary is determined by:

N = min (bus width, data size)

The following table specifies the sizes and natural boundaries for all C types:

C Type	Size	BUS16		BUS32/64
char	1		1	
short	2		2	
int	4	2		4
long	4	2		4
pointer	4	2		4
enum	4	same as int		
packed enum	1	same as char		
packed enum	2	same as short		
packed enum	4	same as int		
integral type : bit(m)	-	same as integral type		
float	4	2		4
double	8	2		4/8
struct	-		same as most demanding member	
union	-		same as most demanding member	
array	-		same as component	
packed struct	-		1	

Unpacked Structures

Memory layout for unpacked structures is governed by Golden Rules 1 and 2:

- **Golden Rule 1:** The natural boundary for an unpacked structure is the natural boundary of its most demanding member. Padding bytes may be needed between members to guarantee proper aligning.
- **Golden Rule 2:** The size of an unpacked structure must be a multiple of its natural boundary. Trailing bytes may be needed at the end of the structure to satisfy this rule.

These two simple rules guarantee that, should a structure become a substructure or a component of an array, the natural alignment of its members is preserved.

The table below shows the notation used in the code examples that follow:

Notation	Description
A...Z	Uppercase letters denote a nonbit structure member. The letter is repeated to show how many bytes the member occupies (one letter per byte). For example, AAAA represents a 4-byte structure member.
{ }	Curly braces surround an unpacked structure. For example, {AAAABBBB} represents an unpacked structure containing 4-byte members A and B.
( )	Parentheses surround a packed structure. For example, (BCCD) represents a packed structure containing a char-shot-char.
Ø	This character represents a padding byte or trailing byte. For example, {AØBBCØ} shows a padding byte between A and B and a trailing byte after C.
a..z	Lowercase letters denote a bit field. The letter is repeated to show the number of bits the member occupies (one letter per bit). For example, aaa represents a 3-bit field.
I	A bar emphasizes byte boundaries when bit fields are represented. For example, aaaaaaab|cccccccc indicates 2 bytes.
Δ	This character represents a trailing bit. For example, {abΔΔΔΔΔΔ} means 6 trailing bits follow bit-field member b.

Consider the following declarations:

```
struct simple
{
    char A;
    int B;
    char C;
};
struct simple vect[2];
struct outer
{
    char X;
    struct simple Y;
};
struct point
{
    char C;
    double X, Y;
};
```

Memory layout on BUS16:

```
struct simple {AØBBBBCØ}
array vect [{AØBBBBCØ}{AØBBBBCØ}]
struct outer {XØ{AØBBBBCØ}}
struct point {CØXXXXXXXXYYYYYYYY}
```

Memory layout on BUS32:

```
struct simple {AØøøBBBBCØøø}
array vect [{AØøøBBBBCØøø}{AØøøBBBBCØøø}]
```

```
struct outer {Xøøø{AøøøBBBBCøøø}}
struct point {CøøøXXXXXXXXXYYYYYYY}
```

Memory layout on BUS64:

```
struct simple {AøøøBBBBCøøø}
array vect [{AøøøBBBBCøøø}{AøøøBBBBCøøø}]
struct outer {Xøøø{AøøøBBBBCøøø}}
struct point {CøøøøøøøXXXXXXXXXYYYYYYY}
```

On BUS32, the natural boundary for simple is 4, which means its size must be 12. The most demanding member in `simple` is the integer B, whose natural boundary is 4 on BUS32. By following the Golden Rules, the compiler guarantees that B is properly aligned in every possible case, as shown by the memory layouts for structure `simple`, array `vect`, and structure `outer`. In all instances of structure `simple`, B retains an offset that is a multiple of 4. For BUS16, the same code fragment would yield a natural boundary of 2 and a size of 8 for `simple`. In all cases, B retains an offset that is a multiple of 2.

Removing any of the padding or trailing bytes would make B improperly aligned in some cases (try to lay out `vect` if `simple` has no trailing bytes). It is worth noting that the `size of` operator must always yield the same value when applied to a type. Adding padding or trailing bytes only when `simple` is used as a component of a structure or of an array would make its size ambiguous. Let's state this as:

Golden Rule 3: The size of a structure cannot depend on its context.

Packed Structures

With packed structures, the compiler ignores the natural boundaries of C types. There are never padding or trailing bytes between members of packed structures. See the example that follows.

Memory layout on any bus:

```
packed struct simple {ABBBBC}
array of packed struct simple [{ABBBBC}{ABBBBC}]
packed struct outer {X{ABBBBC}}
packed struct point {CXXXXXXXXYYYYYYY}
```

Unpacked Inside Packed

Unpacked structures are never packed, even when enveloped by packed structures. If unpacked structures were packed when nested inside packed structures, we would reach the following paradox:

```
struct alpha
{
    char A;
    short B;
```

```
        char C;
    } solo;

{AøBBCø} sizeof(alpha) == 6

packed struct omega
{
    char X;
    struct alpha Y;
} pair;

(X{ABBC}) sizeof(alpha) == 4 ?!
```

4.8.2 Bit Fields

There is no `bit` keyword in C. Bit fields are declared through `int` or `unsigned int`. In many compilers, however, you may use any integral type to declare bit fields: `char`, `short`, `long`, or even a packed enumerated type. The integral type determines the byte offset of the first bit, whether the bits are signed or unsigned and whether they come in groups of 8, 16, or 32. Notice that the keyword `packed` affects the alignment of the bitfield containers only. Packing of bits always happens.

Examples of memory layout for bit fields (assuming big endianity):

```
struct ua
{
    unsigned char a : 4, b : 4;
};

{aaaabbbb}

struct ub
{
    char A;
    short b : 3, c : 3, d : 3;
    short:0;
};

{A|ø|bbbcccdd|dΔΔΔΔΔΔΔ}

struct uc { enum color a : 12, b : 12; };
{aaaaaaaa|aaaabbbb|bbbbbbbb|ΔΔΔΔΔΔΔΔ}

packed enum UBYTE {x1 = 0u,y1 = 0xffu};
packed enum UWORD {x1 = 0u,y1 = 0xffffu};
packed enum ULONG {x1 = 0u,y1 = 0xffffffffu};
packed struct pa
{
```

```
        char A;
        UBYTE b : 3;
        UBYTE c : 4;
    };
```

{A|bbbccccΔ}

```
packed struct pb
{
    char A;
    UWORD b : 4, c : 4, d : 4;
};
```

{A|bbbbcccc|ddddΔΔΔ}

```
packed struct pc
{
    char A;
    ULONG b: 4, c : 4, d : 4;
    ULONG e : 20
};
```

{A|bbbbcccc|ddddeeee|eeeeeeee|eeeeeeee}

4.8.3 Tips and Techniques

Quick Detection of Size

A simple trick to detect the size of any type is to "make a mistake" and pass the type as a parameter to any function. Your compiler may reveal the size. For example:

```
1 #pragma options -p68000
2 struct tag
  {
3       char A;
4       int B; };
5 int foo(int x) { foo(struct tag); }
                         ^
>> (W) type used as argument; replaced with its sizeof: 6
```

Detection of Offsets and Alignments

A more general approach to discover sizes, offsets, and alignments is to use an inspector program as follows:

```
#include < stddef.h > /* get offsetof macro */
#include < stdio.h >
#define alignof(type) offsetof(struct{char any;type x;},x)
#define print_offset(tag,member) \
    printf("\noffset of %s.%s\t%d", \
    #tag,#member,offsetof(tag,member))
```

```
#define print_alignment(type) \
    printf("\nalignment of %s\t%d",#type,alignof(type))
struct x
{
    char a;
    long b;
};
struct y
{
    char a;
    struct x b;
    long c;
    char d;
};
void main(void)
{
    printf("size of x: %d\n", sizeof(struct x));
    print_offset(struct x, a);
    print_offset(struct x, b);
    print_offset(struct y, a);
    print_offset(struct y, b);
    print_offset(struct y, c);
    print_offset(struct y, d);
    print_alignment(char);
    print_alignment(int);
    print_alignment(struct x);
    print_alignment(struct y);
}
```

Tight Packing

Structures come in three flavors: explicitly unpacked (rare), explicitly packed, and plain. For example:

```
unpacked struct unpk {...
packed struct pk {...
struct plain {...
```

Plain structures are, by default, unpacked. They can all be made packed via the -KP compiler option. If you have fairly large arrays of structures, this option may save some space, but it trades off memory with runtime efficiency. In these examples, the additional keywords packed and unpacked are illustrated; many compilers achieve the same result with #pragma directives.

Inflating Structures

Two minor options may be available and can be used to affect the size of structures—namely −Zn<num> for unpacked and −Zm<num> for packed. Their defaults are −Zn1 and −Zm1; i.e., make the size of unpacked and packed structures a multiple of 1 (trivially true). In some cases, you may want to adjust sizes to be a multiple of a given value. For instance, to confirm a memory override during debugging, you may want to specify larger values of −Zm and −Zn. For instance, −Zm8 will make the size of all your packed structures a multiple of 8.

Managing Alignment

Perhaps you have an array of structures, and efficiency reasons dictate they should be 8 bytes long. For instance:

```
typedef packed struct {char head; long load;} MSG; MSG arry[1000];
```

To make sure that array also starts at a multiple of 8, you can inject an alignment directive just before the array declaration, as follows:

```
#pragma asm
.align 8 # PPC assembler
#pragma endasm
MSG arry[1000];
```

An alternative is to give the array a "roommate" in an unpacked union, as follows (assuming BUS64):

```
unpacked union
{
    double alignMeToMultipleOf8;
    MSG arry[1000];
} u;
```

As the alignment requirement of the union is that of its most demanding element (the double), the whole union will be aligned to a multiple of 8. Recent compilers are likely to provide an option (-Zd<num>) to the same effect:

```
#pragma options −Zd8
MSG arry[1000]; /* aligned to 8*n */
```

Forcing Identical Layout Across BUS16 and BUS32

If you are writing an application that is meant to be portable across all variants of the 68 K family, you may want the layout for all structures in your program to be identical across processors.

For unpacked structures, your compiler may support the options −Za2 and −Za4 to alter the natural boundaries. The option −Za2 forces all types with a natural boundary of 4 to "lower" their demands to 2. The option −Za4 "raises" the demands of the types with a natural boundary of 2 (and a size greater than 2) to 4. Substantially, −Za2 and −Za4 allow you to define the bus width.

An example:

```
struct
{
    char A;
    int B;
    float C;
};
```

```
68000 {AøOBBBBCCCC}
68020 {AøøøBBBBCCCC}
```

Specify −Za4 to make the 68000 version identical to the one for the 68020. You could also use −Za2 to make the 68020 structure comply with the 68000 one, with a saving of 2 bytes at the expense of runtime efficiency for the 68020.

```
−Za2 {AøBBBBCCCC}
−Za4 {AøøøBBBBCCCC}
```

Forcing Identical Layout Across BUS32 and BUS64

If there are no doubles, the layout is identical. If there are doubles, it should be pretty obvious that "doubling" the −Zas will obtain the desired result:

```
struct
{
    char A;
    double B;
};
```

```
ppc403ga {AøøøBBBBBBBB}
ppc740   {AøøøøøøøBBBBBBBB}
−Za4     {AøøøBBBBBBBB}
−Za8     {AøøøøøøøBBBBBBBB}
```

Be aware that many PowerPC compilers use an alignment of 8 for doubles regardless of the bus width. Ideally, doubles are aligned to 4 for BUS32 variants and to 8 for BUS64 variants.

Test Your Knowledge (Difficult)

Test your knowledge of structure alignment. Assign yourself a point for each correct answer. Check your answers with the solutions at the end of this article and compute your score and expertise as follows:

Score	You Are
0–7	Well …
8–15	Structure Apprentice
16–22	Structure Cadet

23–29   Structure Master
30   The Compiler

1. Is the layout of the following structure the same for BUS16 and BUS32 and for BUS32 and BUS64?

```
struct
{
    char A;
    float B;
};
```

2. What is the offset of X in the following structure for BUS16, BUS32, and BUS64?

```
struct
{
    char A;
    int X;
};
```

3. Same question for

```
struct
{
    char A;
    double X;
};
```

4. How many trailing bytes can be found in the following BUS32 structure?

```
struct
{
    char A, B;
    int C;
    char D;
};
```

5. How can you avoid trailing bytes in the previous structure? What size would result?
6. How many trailing bits can be found in the following structure? Would it make a difference to remove the keyword `packed`?

```
packed struct { char a : 7; };
```

7. How many trailing bits can be found in the following structure? Would it make a difference to add the keyword `packed`?

```
struct { short a : 7; };
```

8. How many padding bytes can be found between X and Y? Are there any trailing bytes?

```
struct { char X; short Y; };
```

9. How many padding bytes can be found between X and Y? Are there any trailing bytes?

    ```
    #pragma options -pcpu32 -Za4 -Zm4
    packed struct
    {
        char X;
        short Y;
    };
    ```

    In questions 10–16, consider the following structure:

    ```
    packed struct mix
    {
        char A;
        struct
        {
            char B;
            double C;
        } T;
    };
    ```

10. Are there padding bytes between A and T in the structure mix?
11. Are there padding bytes between B and C in the structure mix.T?
12. What is the size of the structure mix for BUS16 and -Zm4?
13. What is the size of the structure mix for BUS32?
14. What is the size of the structure mix for BUS64 and -Zm10?
15. What is the minimum set of options to force the structure mix to have an identical layout for BUS16 and BUS32?
16. What is the minimum set of options that can make the structure mix have an identical layout across BUS16 and BUS32 and also make its size 16 bytes?
17. What is the size in bytes of the following structure?

    ```
    struct
    {
        short a : 7;
        short b : 8;
    };
    ```

18. Would declaring the preceding structure to be packed change its size in bytes? Or the bit offset of b?
19. What is the size in bytes of the following array? Are the bit fields signed?

    ```
    typedef packed enum {F, T} BOOLEAN;
    struct
    {
        BOOLEAN a:1, b:1, c:1, d:1,e:1, f:1, g:1;
        BOOLEAN:0;
    } arry[100];
    ```

20. What is the size in bytes of the preceding array with the following definition of BOOLEAN? Are the bit fields signed?

    ```
    typedef unpacked enum {F=0u, T} BOOLEAN;
    ```

21. Write a C declaration for a tightly packed BUSany structure containing only 8 (1-bit) bit fields followed by a short.

22. Can unpacked substructures exist inside packed structures?

23. What option would make sure the offset of B would be 4 in the following BUS16 structure?

    ```
    struct
    {
        char A;
        int B;
    };
    ```

24. Write a C declaration for a BUSany structure that contains a double and has a size of exactly 9 bytes.

25. Write a C declaration for an array of 100 single unsigned bits and size = 100.

26. Write a union of size 1 containing 8 bits that can be accessed together as a byte or individually as a bit.

27. What is the size of the following structure if BUS64 and −Zm2?

    ```
    struct
    {
        char A;
        struct inner
        {
            char B;
            short C;
        } S;
    };
    ```

28. What offset has sub in the following BUS32 declaration?

    ```
    struct
    {
        char A;
        struct inner
        {
            char B,C,D;
            long E;
        } sub;
    };
    ```

29. Determine the size of the following structure if BUS16/32/64 and if packed or unpacked.

    ```
    struct
    {
        char A;
        double B;
    }
    ```

30. What option will give the following C structure a size of 128 bytes?

```
packed struct {char A;};
```

Now check your answers. Add up your points to find your score. If you are dissatisfied with your results—which is not unlikely because the test is difficult—experiment a bit with the compiler and try the test again. After all, what really matters is how much you know in the end.

Answers

1. No.
   BUS16: {Aø0BBBB}
   BUS32: {AøøøBBBB}

   −Zn2 or −Zn4 would force the same.
   Yes, BUS32 and BUS64 are identical but for doubles.
2. <2, 4, 4>.
3. <2, 4, 8>.
4. Three, to make the size a multiple of C's alignment:
   {ABøøCCCCDøøø}
   Golden Rule number 2.
5. Declare the structure packed. Size 7.
6. One: {aaaaaaa△}.
   No.
7. Nine: {aaaaaaa△|△△△△△△△△}.
   No.
8. One: {XøYY} with any bus. No trailing bytes.
9. None: (XYYø). One trailing byte. -Za4 has no effect on packed structures or on a short type, since its size is 2.
10. Never. A and T are fields of a packed structure.
11. Always. B and C are fields of an unpacked structure, and the byte after B has an odd offset.
12. 12 bytes: (A{BøCCCCCCCC}ø)
13. 13 bytes: (A{BøøøCCCCCCCC})
14. 20 bytes: (A{BøøøøøøøCCCCCCCC}øøø)
15. Three solutions:
   −Za2 (A{BøCCCCCCCC})
   −Za4 (A{BøøøCCCCCCCC})
   −KP (A(BCCCCCCCC))
16. −Za2 −Zm8: (A{BøCCCCCCCC}øøøøø)
   −Za4 −Zm16: (A{BøøøCCCCCCCC}øøø)
17. 2 bytes: {aaaaaaab|bbbbbbb△}
18. No. No.
19. 100: [{abcdefg△}{abcdefg△}...]. Yes, since BOOLEAN is a signed char.
20. 400:
   [{abcdefg△|△△△△△△△△|△△△△△△△△|△△△△△△△△}...]
   Unpacked enums are integers—unsigned integers in this case.
   No, they are unsigned, as is their container.

```
21. packed struct
    {
        char a:1, b:1, c: 1, d: 1, e: 1, f: 1, g: 1; h: 1; short sh;
    };
```
22. Yes. And vice versa.
23. −Za4: {AøøøBBBB}
```
24. packed struct
    {
        char A;
        double B;
    };
```
25. struct {unsigned char x : 1;} a[100];
26. typedef struct {char a:1, b: 1, ..., h: 1;} BITS;
    union { BITS bits; char allBits; };
    Use a typedef to increase readability. Don't give yourself a point if your answer included an unnecessary packed.
27. Six: {Aø{BøCC}}. For any bus, -Zm2 is immaterial since the structures are not packed.
28. Four bytes. According to Golden Rule 1, sub has 4 as its natural boundary—i.e., the natural boundary of its most demanding field, the long E. {Aøøø{BCDøEEEE}}.
29. Unpacked BUS16:10 {AøBBBBBBBB}
    Unpacked BUS32:   12 {AøøøBBBBBBBB}
    Unpacked BUS64:   16 {AøøøøøøøBBBBBBBB}
    Packed BUSany: 9 {ABBBBBBBB}
30. -Zm128.

4.9 Memory and Programming in C

In the mid-1990s I wrote a regular column in New Electronics magazine for a while. The brief was to "educate" hardware engineers in some of the concepts of embedded software. Of course, it didn't last. One of the other embedded software vendors was aggrieved that I was writing this column and complained, even though I had been careful not to write a "sales pitch." This article is closely based upon my first "Byte Site" column.

CW

In this article, we focus on various facets of embedded microprocessor software development and address some of the challenges faced by engineers new to this programming environment. Nowadays, the most commonly used programming language for embedded systems applications is C (or C++); assembler being reserved for particularly tricky situations. To an engineer who is familiar with the hardware details of the embedded system, C may appear to be rather abstract and far removed from the "real world." Keeping that in mind, we will take a look at memory usage by programs written in C.

4.9.1 Memory

Typically, an embedded system uses a number of different types of memory, including read-only memory (ROM), read/write memory (RAM), and CPU registers. In turn, the different types of memory may be used in different ways: ROM contains both code and constant data; RAM contains variables and stack space; registers may be used for different variables at different times. Add to this the complication that input/output devices may be mapped to memory addresses. Memory usage is clearly visible to the assembly language programmer, but the C programmer needs to understand how the development tools accommodate memory usage.

4.9.2 Sections

The most powerful facility that enables the programmer to control memory usage is support for multiple *program sections*. A *section* is a logical area of memory that is given an arbitrary name and may be located at specific addresses at link time. The simplest possible case might contain two sections, one called "CODE" and another "DATA," each assigned to the addresses of ROM and RAM, respectively.

A cross-compiler (one designed specifically for embedded systems development) will automatically segment the code and data into a number of sections. A typical compiler might, for example, generate up to seven sections for different categories of information. These sections will most likely have default names that may be overridden by the programmer, as required.

As each module (file) is compiled, the resulting relocatable object module contains parts of each section that will be contained in the final executable. The task of the linker is to assemble all the parts of each section and correctly locate the sections in memory.

Fundamentally, this approach works because the programmer does not (ever) need to know the precise address of a variable or a piece of code. It is sufficient to be sure that the code and data have been placed in the right type of memory and that the debugging tools can locate them symbolically, when required.

Figure 4.4 illustrates this concept, showing the memory map for an application. In this case, ROM is located at address $00000000 and RAM at $80000000. Three modules *alpha*, *beta*, and *gamma* are included, each contributing fragments of the CODE and DATA sections.

4.9.3 Conclusions

Clearly, for a software developer to easily map code and data to the right type of memory and then to debug the result, tools must be designed specifically for embedded systems software development and include facilities to enable flexible handling of program sections. Only in this context will the programmer find all the control required to complete the task.

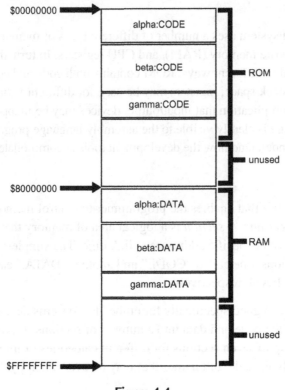

Figure 4.4
Typical memory map

4.10 Pointers and Arrays in C and C++

This article is based on another of my "Byte Site" columns in New Electronics in 1996.

CW

4.10.1 Pointers and Pointer Arithmetic

For many, pointers are thought of as strange and mysterious entities, and operations on them are considered akin to a black art. It is surprising that even programmers with years of assembler experience, where manipulating addresses is second nature, can come unstuck with pointer arithmetic.

This last point is really the key to the secret. A common question from C language students is, "What is the difference between a C pointer and an address?" The answer is that, although the value of a pointer may indeed be an address, it is itself a distinct data type, the characteristics of which are influenced, in turn, by the type of data to which it points.

When the C language was first defined, pointers had "target-specific" types for two reasons. The first was to permit type-specific arithmetic, which we will consider again shortly. The second was the notion that the addresses of different kinds of data might indeed be different in nature.

From one perspective, this may seem to have been an unnecessary precaution, because modern computers tend to have simple, single address spaces (if you ignore `near/far` in 80 × 86, I/O space addressing, and the multiple address spaces of 680 × 0 and SPARC). However, other chips used for embedded systems designs do have distinctly different ways of accessing different kinds of data. The most graphic example is the 8051 series, with its five address spaces.

In C, as with most high-level programming languages, all the arithmetic operations are defined in type-specific ways. For example, the process of adding two integers is quite different than that applied to floating-point numbers, but the programmer need not normally be concerned with this underlying complexity. With an embedded system, manipulation of addresses may be necessary, so an appreciation of how the C language does arithmetic on pointers is likely be useful.

Here is an example of some pointer arithmetic:

```
int *ptr;
...
ptr = ptr + 1;
```

Apart from the passing thought that this is rather Pascal-like in its style (a C programmer would normally use ++ or +=), the code is quite simple. However, a program line just like this resulted in a support call a few years ago, "When I add one to the pointer, why is it that the code generated by my compiler adds four?" On the surface this query seems reasonable. The reason for the addition of four is that integer variables occupy 4 bytes of memory (with a compiler for a 32-bit processor anyway). The idea is that when a pointer points at a value in memory, incrementing the pointer should result in it pointing to the next variable of that type. In other words, C automatically scales the arithmetic according to the size of the data type that is pointed to. This even works with user-defined data types (`struct` and `typedef`).

If you need a way to deal with addresses directly in C, the best approach is to use pointers to single-byte data types. A good choice would be a pointer to `unsigned char`. Although this approach will make the arithmetic "easy," accessing nonbyte data will require a cast.

The idea of a generalized pointer, capable of containing the address of anything, was introduced in ANSI C: the `void` pointer. No operations may be performed on a `void` pointer until it has been cast into a "normal" pointer type.

4.10.2 Arrays and Pointers

Another common query in C courses is, "Just what is the relationship between arrays and pointers—they seem kind of similar, but different?" The fact is that the relationship is very close. An array is just an area of memory with a name (and data type). The name is simply a pointer to the memory area and can be treated just like any other pointer. The only difference is that it is a constant, so it cannot be incremented, assigned to, or modified in any way. The usual way that array elements are accessed is by means of an index in brackets after the array name.

What is not obvious is that these brackets are simply an operator (look at the table of operator precedence in Kernighan and Ritchie).

To place the value 99 in the third element of an array z, you would probably write:

```
z[2] = 99;
```

However, by treating z as a pointer, the same code may be written:

```
*(z12) = 99;
```

The brackets notation is just an alternative way to show array indexing, which is much clearer and easier to understand. No rules designate one notation as better than the other. A good guide to writing clear code is to treat "real" pointers as pointers and arrays as arrays. For example, to get the data at the location pointed to by ptr into the variable x, you would normally write:

```
x = *ptr;
```

There is nothing to be gained from the (equally syntactically valid) form:

```
x = ptr[0];
```

The one exceptional circumstance is when you want to take offsets from a pointer that has been passed to a function. Within that function, treating the pointer like an array is reasonable.

4.10.3 Conclusions

Although the use of pointers causes 70% of the bugs in C/C++ programs, pointers are a necessity for most applications. With care, their use may not be hazardous, and the intelligent use of arrays may simplify matters further.

4.11 Using Dynamic Memory in C and C++

In 2009, I did a Web seminar on C++ for embedded. One of the big benefits of this medium is that attendees have plenty of opportunity for questions and feedback, which often gives me useful input for further sessions, articles etc. On this occasion, there were many questions about dynamic memory, so I developed a specific Web seminar on that topic, which also spawned a white paper. I used that as the basis for this article.

CW

4.11.1 C/C++ Memory Spaces

It may be useful to think in terms of data memory in C and C++ as being divided into three separate spaces:

1. **Static memory:** This is where variables, which are defined outside of functions, are located. The keyword static does not generally affect where such variables are located; it specifies their scope to be local to the current module. Variables that are defined inside of a function, which are explicitly declared static, are also stored in static memory.

Commonly, static memory is located at the beginning of the RAM area. The actual allocation of addresses to variables is performed by the embedded software development toolkit: a collaboration between the compiler and the linker. Normally, program sections are used to control placement, but more advanced techniques, like Fine Grain Allocation, give more control. Commonly, all the remaining memory, which is not used for static storage, is used to constitute the dynamic storage area, which accommodates the other two memory spaces.

2. **Automatic variables:** Variables defined inside a function, which are not declared static, are automatic. There is a keyword to explicitly declare such a variable—auto—but it is almost never used. Automatic variables (and function parameters) are usually stored on the stack. The stack is normally located using the linker. The end of the dynamic storage area is typically used for the stack. Compiler optimizations may result in variables being stored in registers for part or all of their lifetimes; this may also be suggested by using the keyword `register`.

3. **The heap:** The remainder of the dynamic storage area is commonly allocated to the heap, from which application programs may dynamically allocate memory, as required.

Figure 4.5 illustrates a typical memory layout for the three C/C++ data memory spaces.

If a real-time operating system is used, each task normally has its own private stack.

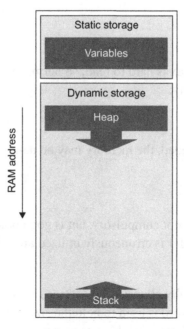

Figure 4.5
A typical memory layout for the three C/C++ data memory spaces

4.11.2 Dynamic Memory in C

In C, dynamic memory is allocated from the heap using some standard library functions. The two key dynamic memory functions are `malloc()` and `free()`.

The `malloc()` function takes a single parameter, which is the size of the requested memory area in bytes. It returns a pointer to the allocated memory. If the allocation fails, it returns NULL. The prototype for the standard library function is like this:

```
void *malloc(size_t size);
```

The `free()` function takes the pointer returned by `malloc()` and de-allocates the memory. No indication of success or failure is returned. The function prototype is like this:

```
void free(void *pointer);
```

To illustrate the use of these functions, here is some code to statically define an array and set the fourth element's value:

```
int my_array[10];
my_array[3] = 99;
```

The following code does the same job using dynamic memory allocation:

```
int *pointer;
pointer = malloc(10 * sizeof(int));
*(pointer13) = 99;
```

The pointer de-referencing syntax is hard to read, so normal array referencing syntax may be used, as [and] are just operators:

```
pointer[3] = 99;
```

When the array is no longer needed, the memory may be de-allocated thus:

```
free(pointer);
pointer = NULL;
```

Assigning NULL to the pointer is not compulsory, but is good practice, as it will cause an error to be generated if the pointer is erroneously utilized after the memory has been de-allocated.

The amount of heap space actually allocated by `malloc()` is normally one word larger than that requested. The additional word is used to hold the size of the allocation and is for later use by `free()`. This "size word" precedes the data area to which `malloc()` returns a pointer. There are two other variants of the `malloc()` function: `calloc()` and `realloc()`.

The `calloc()` function does basically the same job as `malloc()`, except that it takes two parameters—the number of array elements and the size of each element—instead of a single parameter (which is the product of these two values). The allocated memory is also initialized to zeros. Here is the prototype:

```
void *calloc(size_t nelements, size_t elementSize);
```

The `realloc()` function resizes a memory allocation previously made by `malloc()`. It takes as parameters a pointer to the memory area and the new size that is required. If the size is reduced, data may be lost. If the size is increased and the function is unable to extend the existing allocation, it will automatically allocate a new memory area and copy data across. In any case, it returns a pointer to the allocated memory. Here is the prototype:

```
void *realloc(void *pointer, size_t size);
```

4.11.3 Dynamic Memory in C++

Management of dynamic memory in C++ is quite similar to C in most respects. Although the library functions are likely to be available, C++ has two additional operators—`new` and `delete`—which enable code to be written more clearly, succinctly and flexibly, with less likelihood of errors. The new operator can be used in three ways:

```
p_var = new typename;
p_var = new type(initializer);
p_array = new type [size];
```

In the first two cases, space for a single object is allocated; the second one includes initialization. The third case is the mechanism for allocating space for an array of objects.

The `delete` operator can be invoked in two ways:

```
delete p_var;
delete[] p_array;
```

The first is for a single object; the second de-allocates the space used by an array. It is very important to use the correct de-allocator in each case.

There is no operator that provides the functionality of the C `realloc()` function.

Here is the code to dynamically allocate an array and initialize the fourth element:

```
int* pointer;
pointer = new int[10];
pointer[3] = 99;
```

Using the array access notation is natural.

De-allocation is performed thus:

```
delete[] pointer;
pointer = NULL;
```

Again, assigning NULL to the pointer after de-allocation is just good programming practice.

Another option for managing dynamic memory in C++ is to use the Standard Template Library. This may be inadvisable for real-time embedded systems.

4.11.4 Issues and Problems

As a general rule, dynamic behavior is troublesome in real-time embedded systems. The two key areas of concern are determination of the action to be taken on resource exhaustion and nondeterministic execution performance. When considering dynamic memory utilization, there are two likely problem areas: stacks and the use of malloc().

Stacks
A stack is located in memory, set up to be a particular size and the stack pointer initialized to point to the word following the allocated memory. Thereafter, problems occur if stack operations result in data being written outside of the allocated memory space. This may be stack overflow, where more data is pushed on to the stack than it has capacity to accommodate. Or it can be stack underflow, where data is popped off of an empty stack (as a result of a programming error).

In both cases the problems can be very hard to locate, as they typically manifest themselves later in completely unrelated code. It may even result is problems in another, unrelated task.

Use of malloc()
There are a number of problems with dynamic memory allocation in a real-time system.

Firstly, the standard library functions (malloc() and free()) are not normally reentrant, which would be problematic in a multithreaded application. If the source code is available, this should be straightforward to rectify by locking resources using RTOS facilities (like a semaphore).

A more intractable problem is associated with the performance of malloc(). Its behavior is unpredictable, as the time it takes to allocate memory is extremely variable. Such nondeterministic behavior is intolerable in real-time systems.

Without great care, it is easy to introduce memory leaks into application code implemented using malloc() and free(). This is caused by memory being allocated and never being de-allocated. Such errors tend to cause a gradual performance degradation and eventual failure. This type of bug can be very hard to locate.

Memory allocation failure is a concern. Unlike a desktop application, most embedded systems do not have the opportunity to pop up a dialog and discuss options with the user. Often, resetting is the only option, which is unattractive. If allocation failures are encountered during testing, care must be taken with diagnosing their cause. It may be that there is simply insufficient memory available—this suggests various courses of action. However, it may be that there is sufficient memory, but not available in one contiguous chunk that can satisfy the allocation request. This situation is called *memory fragmentation*.

4.11.5 Memory Fragmentation

The best way to understand memory fragmentation is to look at an example. For this example, it is assumed that there is a 10K heap. First, an area of 3K is requested, thus:

```
#define K (1024)
char *p1;
p1 = malloc(3*K);
```

Then, a further 4K is requested:

```
p2 = malloc(4*K);
```

The resulting situation is shown in Figure 4.6—3K of memory is now free.

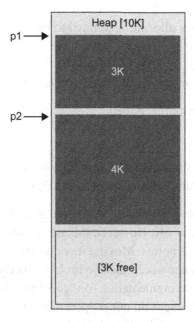

Figure 4.6
Memory fragmentation

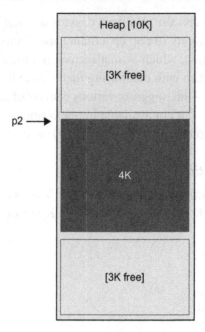

Figure 4.7
Non-contiguous free memory

Some time later, the first memory allocation, pointed to by p1, is de-allocated:

```
free(p1);
```

This leaves 6K of memory free in two 3K chunks, as illustrated in Figure 4.7.

A further request for a 4K allocation is issued:

```
p1 = malloc(4*K);
```

This results in a failure—NULL is returned into p1—because, even though 6K of memory is available, there is not a 4K contiguous block available. This is memory fragmentation.

It would seem that an obvious solution would be to defragment the memory, merging the two 3K blocks to make a single one of 6K. However, this is not possible because it would entail moving the 4K block to which p2 points. Moving it would change its address, so any code that has taken a copy of the pointer would then be broken. In other languages (such as Visual Basic, Java, and C#), there are defragmentation (or "garbage collection") facilities. This is only possible because these languages do not support direct pointers, so moving the data has no adverse effect upon application code. This defragmentation may occur when a memory allocation fails or there may be a periodic garbage collection process that is run. In either case, this would severely compromise real-time performance and determinism.

4.11.6 Memory with an RTOS

Memory management facilities that are compatible with real-time requirements—i.e., they are deterministic—are usually provided with commercial real-time operating systems. A widely used scheme is to allocate blocks—or "partitions"—of memory under the control of the OS.

RTOS Block/Partition Memory Allocation

Block memory allocation is performed using a "partition pool," which is defined statically or dynamically and configured to contain a specified number of blocks of a specified fixed size. For example, using the Nucleus RTOS, the API call to define a partition pool has the following prototype:

```
STATUS NU_Create_Partition_Pool(NU_PARTITION_POOL *pool,
CHAR *name, VOID *start_address, UNSIGNED pool_size,
UNSIGNED partition_size, OPTION suspend_type);
```

This is most clearly understood by means of an example:

```
status = NU_Create_Partition_Pool(&MyPool, "any name",
(VOID *) 0xB000, 2000, 40, NU_FIFO);
```

This creates a partition pool with the descriptor MyPool, containing 2,000 bytes of memory, filled with partitions of size 40 bytes (i.e., there are 50 partitions). The pool is located at address 0xB000. The pool is configured such that, if a task attempts to allocate a block, when there are none available, and it requests to be suspended on the allocation API call, suspended tasks will be woken up in a first-in, first-out order. The other option would have been task priority order.

Another API call is available to request allocation of a partition. Here is an example using Nucleus RTOS:

```
status = NU_Allocate_Partition(&MyPool, &ptr, NU_SUSPEND);
```

This requests the allocation of a partition from MyPool. When successful, a pointer to the allocated block is returned in ptr. If no memory is available, the task is suspended, because NU_SUSPEND was specified; other options, which may have been selected, would have been to suspend with a timeout or to simply return with an error.

When the partition is no longer required, it may be de-allocated thus:

```
status = NU_Deallocate_Partition(ptr);
```

If a task of higher priority was suspended pending availability of a partition, it would now be run.

There is no possibility for fragmentation, as only fixed-size blocks are available. The only failure mode is true resource exhaustion, which may be controlled and contained using task suspend, as shown.

Additional API calls are available which can provide the application code with the means to handle and obtain information about the status of partitions:

NU_Delete_Partition_Pool()—removes a partition pool.

NU_Partition_Pool_Information()—returns a variety of information about a partition pool.

NU_Established_Partition_Pools()—returns the number of partition pools [currently] in the system.

NU_Partition_Pool_Pointers()—returns pointers to all the partition pools in the system.

Care is required in allocating and de-allocating partitions, as the possibility for the introduction of memory leaks remains.

Memory Leak Detection

The potential for programmer error resulting in a memory leak when using partition pools is recognized by vendors of real-time operating systems. Typically, a profiler tool is available which assists with the location and rectification of such bugs.

4.11.7 Real-Time Memory Solutions

Having identified a number of problems with dynamic memory behavior in real-time systems, some possible solutions and better approaches can be proposed.

Stacks

Although it could be a result of a programming error, the most likely reason for stack overflow is that the stack space allocation was insufficient. Specifying stack size is very challenging. It is almost impossible to estimate statically for most systems. The simplest approach is to make measurements on a running system. Allocate a generous amount of space for the stack, fill it with a known value—nonzero, odd numbers make most sense, as they are less likely to be a valid address. Run the code for a reasonable period of time and inspect the stack to determine how much was used.

To monitor at runtime for stack overflow/underflow, a good technique is to place a "guard word" at either end of the stack space, as illustrated in Figure 4.8. These may be initialized to a known value (again a nonzero, odd number is best) and a background (low-priority) task used to monitor the words for changes during execution. Alternatively, a memory management unit may be used to trap an error if the guard words are accessed.

Dynamic Memory

It is possible to use partition memory allocation to implement malloc() in a robust and deterministic fashion. The idea is to define a series of partition pools with block sizes in a geometric progression, e.g., 32, 64, 128, 256 bytes. A malloc() function may be written to deterministically select the correct pool to provide enough space for a given allocation

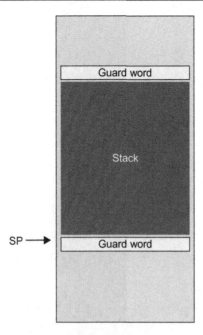

Figure 4.8
Stack guard words

request. For example, if 56 bytes are requested, a 64-byte partition would be used; for 99 bytes, a 128-byte partition. Unfortunately, a 65 byte requested would be satisfied with a 128-byte partition. This approach takes advantage of the deterministic behavior of the partition allocation API call, the robust error handling (e.g., task suspend) and the immunity from fragmentation offered by block memory. The choice of partition sizes, being powers of 2, is to facilitate efficient code using a bit mask.

4.11.8 Conclusions

C and C++ use memory in various ways, both static and dynamic. Dynamic memory includes stack and heap.

Dynamic behavior in embedded real-time systems is generally a source of concern, as it tends to be nondeterministic and failure is hard to contain.

Using the facilities provided by most real-time operating systems, a dynamic memory facility may be implemented which is deterministic, immune from fragmentation and with good error handling.

C++

Chapter Outline

C is still the default language for embedded software development, but most modern engineers have some knowledge of C++, even if they have yet to apply it in this context. In this chapter, articles cover the broad topic of moving from C to C++, some specific language features are looked at in detail, and particular applications of C++ are outlined.

5.1 C++ in Embedded Systems—A Management Perspective

Engineers are often attracted by new technology. They are naturally enthusiastic about being on the leading edge—using the latest programming techniques and languages, for example. Even though embedded developers are, on the whole, somewhat conservative, they ultimately have the same mind set. Managers, on the other hand, have an "if it ain't broke, don't fix it" attitude. In any embedded development environment, a healthy tension exists between these two cultures. I wrote this white paper a few years ago in answer to engineers' request: "How can I persuade my manager to pay for C++? I know it's what we need."

CW

The technical pros and cons of using C++ for embedded systems applications are widely discussed. This paper considers the matter from the viewpoint of the manager. A software team leader is not concerned with the finer points of a programming language; he is interested only in the benefits that its adoption might yield.

5.1.1 Embedded Systems Development Teams

In the early days of embedded systems development, the "team" would be just one engineer, who would design the hardware, build the prototype, and implement the software. He would

Embedded Software: The Works. DOI: 10.1016/B978-0-12-415822-1.00005-2

understand all aspects of the system. The software, which was probably written in assembly language, represented quite a small part of the overall design effort.

Over time, as systems became more complex, separate hardware and software development teams evolved. The program design and implementation represented a much greater proportion of the development investment. Software was increasingly written in high-level languages, typically C.

Nowadays, with very large and complex embedded systems being implemented in ever-shorter design cycles, large teams of software engineers may be involved. This results in some challenging management problems.

5.1.2 Object-Oriented Programming

In recent years, there has been an increase in the popularity of object-oriented programming techniques. The idea is not new. The first object-oriented programming language was implemented in the late 1960s. However, this approach lends itself particularly well to the programming problems of today.

Broadly, object-oriented programming is a technique whereby software is wholly or partly implemented as a set of *objects*. An object is a self-contained package of code and data, with a clearly defined interface to the outside world. *An object may be utilized without the programmer having to understand its inner workings.* This is exactly the same approach that hardware designers use when making use of integrated circuits.

5.1.3 Team Management and Object-Oriented Techniques

At the philosophical level, object-oriented programming facilitates the "encapsulation of expertise": the specific knowledge and experience of one engineer may be placed inside an object and made available for the use of others, without their needing to acquire that expertise.

The alternative, which must be faced when conventional, procedural programming techniques (such as those imposed by the C language) are used, is that all members of the team must have wide knowledge of the complete application, which is inefficient. Object-oriented programming techniques enable the collective expertise of the team to be *divided, distributed, and managed in an efficient manner.*

5.1.4 C++ as an Object-Oriented Language

C++ is an extended variant of the C language. The extensions provide object-oriented programming facilities. Strictly speaking, C++ is not an object-oriented language; it is a procedural language with some object-oriented capabilities. This does not devalue C++ as a useful language for embedded systems development. Although there are other languages that may be described as more purely object oriented, none of them have the backward compatibility (to C) that C++ offers. This is a benefit because it reduces the cost of training engineers and porting existing code. The use of C may continue, with C++ being introduced incrementally, as required. *No sharp learning curve limits continued productivity.*

5.1.5 Overheads

At any mention of C++, many experienced embedded systems developers will make dark comments about excessive memory use and real-time overhead. They are wise to be concerned about these issues. There are always some resource limitations in an embedded system. The same concerns arose when C was first used. However, over time, three things happened:

- C compilers improved in quality and functionality (the overhead was reduced).
- Microprocessor power and available memory increased (the overhead was less important).
- Demand for increased programmer productivity increased (the overhead was a lower priority).

Exactly the same factors apply to C++.

5.1.6 The Way Forward

Given that C++ appears to be an attractive option for the future, a number of steps must be taken:

1. Determine the appropriateness of C++ for your type of application.
2. Select and acquire C++ tools.
3. Develop a migration plan—e.g., staff training, existing code conversion, and update development procedures.

C++ offers clear benefits to the development manager:

- Increased productivity
- Reduced time to market
- Full use of software team members' individual skills.

5.2 Why Convert from C to C++?

Many embedded developers still use C. For many years, there have been ongoing arguments over the benefits of moving to C++. In the early 1990s, Michael Fay (who now teaches at Santa Barbara City College) addressed this topic in a piece for NewBits, which formed the basis of this article.

CW

The well-publicized advantage of C++ over C is that it supports object-oriented programming. Reflecting more deeply on the topic, we can identify several specific advantages of C++ over C.

5.2.1 Hide Implementation Details

In C, one clean way to provide a data structure is to encapsulate it in a separately compiled module in which only functions can access the underlying data. Here is a simplified example of a stack implemented that way:

```
static int stack_array[MAX];
static int stack_top = -1;
void push(int i)
      { ... }
int pop(void)
      { ... }
```

In C++, you can hide implementation details on a finer level, within each object. For example, here is a class that defines simplified stack objects:

```
class stack
{
    int stack_array[MAX];
    int stack_top;
public:
    stack()
    {
        stack_top = -1;
    }
    void push(int i)
        { ... }
    int pop(void)
        { ... }
};
```

In this example, stack_array and stack_top are private class elements, visible only to push() and pop(), whereas in C, stack_array and stack_top are visible to all subsequent functions in the module. (The element stack() is a constructor—i.e., a function invoked when a stack is declared.)

5.2.2 Reuse Class Code

If a program has more than one stack of ints, the single class definition in the preceding C++ example will suffice as the only implementation of int stack operations. You can declare many such stacks, each having its own private data:

```
stack s1;
stack s2;
```

In C, you have to expand the stack implementation to cover multiple `int` stacks.

5.2.3 Reuse Generic Classes

If your only requirement for stacks is for `int` data, the example previously given will suffice. To accommodate a different data type—say `float`—you need to define another, almost identical, class. An alternative is to use a C++ class template, which enables you to outline what a class would look like and leave the compiler to generate code for specific data types as needed.

5.2.4 Extend Operators

You can extend operators to apply to new, user-defined types. For example, if you define a class `complex` to handle complex numbers, you may wish to define (i.e., overload) the + operator to apply this operation to complex numbers in a natural way. Trying to use an undefined operator on a new data type would, of course, result in compile-time errors.

This capability may be used (or abused) creatively. There is no specific reason why the functionality of a redefined operator needs to mirror that of its standard counterpart—although this can lead to endless confusion. An interesting example is the extension of the << operator in the `stream.h` library. By default, << has the same meaning as in C: left shift. However, `stream.h` defines a class called `ostream`, and if the left operand of << is an `ostream`, << means to output the right argument to the `ostream` and return an `ostream` as the result. Since <<, like most operators, is left-associative, you can output a series of variables and constants by connecting them with <<:

```
cout << "Total = " << i << '\n';
```

5.2.5 Derive Classes from Base Classes

Suppose you need to define an object (e.g., a `filename`) that is a special case of a simpler object (e.g., a `string`). A `filename` could have a slightly different concatenation operation (+) and a new `open()` function. In C++, you can define `filename` in terms of `string` by declaring it to be a "derived class." A derived class inherits the properties of its base class. In the following example, adapted from Ref. [1], class `filename` inherits the properties of class `string`:

```
class filename : public string
{
public:
    friend filename & operator+(filename &, filename &);
    FILE *open();
};
```

With this definition, a `filename` allows all the operations that can be performed on any `string` but has its own + operator and `open()` function. (`string` may have these operations; if it does, they will be superseded in `filename` objects.)

The preceding example illustrates other features of C++ beyond inheritance. To avoid copying large strings, the & operators denote call and return by *reference* rather than by *value*. The friend keyword indicates that a function has access to the nonpublic members of objects in the class.

5.2.6 Avoid Errors Through Function Prototyping

C++ requires you to give the types of all parameters and to declare all functions before they are used:

```
extern void func(long 1);
...
func(1, 2); // error!
```

This requirement is necessary so that the C++ compiler can apply various types of conversions. It also prevents you from passing too many parameters or parameters with incompatible types.

Although prototyping is optionally available in C, it is mandatory in C++.

5.2.7 Add Parameters Without Changing Function Calls

In C++, you can specify the default values of parameters in case the trailing parameters are omitted at the calling point:

```
extern void f2(int j, int i=0);
...
f2(3); // j=3; i=0
```

Thus, by giving default values in the function declaration, you can expand the number of parameters of an existing function without changing all the calls. This makes it easier to extend the application of a function.

5.2.8 Using Safer, Simpler I/O

By using the stream.h standard library, you don't have to match printf() format strings with parameters, which is a common source of C errors.

C:

```
#include <stdio.h>
printf ("%d\n", floatvar); /* error: incorrect format */
```

C++:

```
#include <stream.h>
cout << floatvar <<"\n"; // format implicity matches type
```

If you still need `printf()` functionality, a series of functions are available to do simple formatting (`oct()`, `dec()`, `hex()`, etc.). Another useful function is `form()`, which operates rather like `sprintf()`.

5.2.9 *Improve Performance with Fast Inline Functions*

Within the module in which a function is defined, you can declare it to be `inline`, giving a hint to the compiler that the function code should be substituted rather than called. Inline functions are faster than ordinary functions because there is no call or return overhead, and the inlined code is subject to optimization. Inline functions are safer than macros because they can impose type promotion and do not evaluate their arguments more than once. The following example is from the standard include file `defs.h`:

```
inline int max(int a, int b)
{
    return a > b ? a : b;
}
```

When you define class member functions, you have the option of defining the function within the class body or simply declaring it for definition elsewhere. The former method has an implied `inline` declaration and is typically used for simple, one- or two-line function definitions.

5.2.10 *Overload Function Names*

C++ lets you redeclare (i.e., overload) the same function name as long as the parameters' types are not all the same. Overloading is convenient because you do not have to remember different names for functions that perform essentially the same task. For example, the aforementioned `defs.h` provides two inline `max()` functions for other types as well as `int`:

```
inline float max(float a, float b)
{
    return a > b ? a : b;
}
inline double max(double a, double b)
{
    return a > b ? a : b;
}
```

The C++ compiler will invoke the function whose parameters match the actual parameters' types. If no match is found, an error will result. This particular example could have been more efficiently and flexibly implemented by using a C++ function template, but it illustrates the point of overloading function names. Overloaded functions still serve a useful purpose when the code within each variant of the function is quite different.

5.2.11 Embedded System Support

Moving from C to C++ for an embedded project may be fine, so long as the C++ compiler offers the same specialized functionality that you need:

- Inline assembly code
- `interrupt` functions
- Correct implementation of `const` and `volatile`
- `char` and `short` bit fields
- Packed structures and classes
- Efficient implementation of C++ constructs (which may require an optimized linker).

5.2.12 Change Involves Effort

Newcomers to C++ might hope that C++ is simply a superset of C and, therefore, by simply taking the existing C code, they can run it through a C++ compiler, and everything still works.

Unfortunately, that is not quite the case. It is very likely that some changes will be necessary. The language must be studied more carefully, and a migration strategy should be developed.

5.2.13 Massage C Code into C++

To cushion the blow, you may convert some C modules to C++, leaving others in C. In this case, each C++ module must have special declarations for all C `extern` functions, as well as for all its own C++ functions that are directly called from C. The special declarations are enclosed in the following construct:

```
extern "C" { ... }
```

This construct may enclose types and variables—even entire header files. For example:

```
extern "C"
{
    #include "externs.h"
    int forward_func(int i);
}
...
int forward_func(int i)
{
    ...
}
```

Note that functions with `extern "C"` linkage cannot be overloaded. This is not a problem if you are linking C code into a C++ application but could cause you problems when you're working the other way around.

5.2.14 The Hard Part: Designing Objects

Now that you are ready to take advantage of the features of C++, you should spend some time designing objects that will be useful to you and to other members of your organization. When designing objects, you should ask what fundamental operations define each object. It is important to exclude nonessential operations. For example, as shown in Ref. [2], the fundamental operations on complex numbers are converting from float, adding, multiplying, and so on. Extracting the components is not a fundamental operation:

```
class complex
{
    double re, im;
public:
    complex(double r=0.0, double i=0.0)
        { re = r; im = i; };
    friend complex operatorl(complex, complex);
    friend complex operator*(complex, complex);
    ...
    // do not do this:
    double first()
        { return re; }
    double second()
        { return im; }
};
```

The reason that first() and second() should not be included is that they allow users of complex objects to write programs that depend on the underlying representation of complex numbers to be Cartesian coordinates. This dependency can exist even though the users have no direct access to re and im. If the provider of complex later changes the underlying representation to the polar coordinates rho and theta, and first() and second() return rho and theta, respectively, programs that use complex would be invalidated.

5.2.15 If It Ain't Broke, Don't Fix it

Some C programs should just be left in C. If your program is fairly stable, and you don't expect to make significant changes to it, converting it to C++ may be an interesting exercise, but it is not justified from an engineering standpoint. However, if you are continually changing your program, you can make a gradual transition to C++ using the techniques described in this chapter.

5.3 Clearing the Path to C++

Over many years, engineers have considered the transition from C to C++ for embedded applications. I have given many seminars on the topic and, in collaboration with Lily Chang in 1993, wrote a piece for NewBits, upon which this article is based.

CW

5.3.1 A Strategy for Transition

Object-oriented programming techniques and the C++ language have rapidly gained popularity among software developers in all spheres of computing. For embedded systems development, especially of larger, more complex systems, the object-oriented programming approach is particularly apposite, since it lets you hide intricate parts of your program.

Compared to other object-oriented programming languages, C++ has its pros and cons, but that is not the subject of this article. C++ is successful because the C language is well established and represents (almost) a complete subset of C++. As a result, the learning process and the application of the C++ language may be approached incrementally. We will explore this approach to using the language in this article.

5.3.2 Evolutionary Steps

To make the move from programming purely in ANSI C to full-scale application of C++ requires a series of evolutionary steps. There are three "stepping stones":

1. Applying reusability (i.e., linking C and C++ modules): New modules written in C++ may be linked with older C code, or old C language modules may be reused in a new C++ project.
2. Writing Clean C: C language code should be written with additional care (and older code overhauled) to ensure its acceptability as a C++ subset. This permits older C code to be treated as if it were C++—i.e., processed by the C++ language build and debug tools.
3. Introducing C++: C++ language features can be gradually introduced as they are learned, and their suitability established for the application in hand.

5.3.3 Applying Reusability

The first stage is to simply link together C and C++ modules to add C++ to an existing C project, or reuse C modules in a new C++ project. However, this procedure is not totally straightforward. One of the very attractive features of C++ is *type-safe linkage*. This feature ensures that the correct correspondence between a function's definition and its use (in terms of parameter number and types and function return types) is monitored. Type-safe linkage is realized by the compiler. It is achieved by "mangling" the function name to incorporate the types of its parameters. In addition, the return type information is included for the name mangling of pointers to functions. Applying the same mangling to a definition and call should yield the same modified function name and, hence, no link errors. Having the compiler mangle function names has the advantage of the independence of the binary formats in use and provides built-in support for overloaded functions (which was really the original motivation for the technique).

Of course, a C function will not have had its identifier mangled if it is processed by the C compiler. However, a call to the function from C++ code will reference a mangled form of its name, resulting in link errors. To overcome this problem, you can define in C++ the external linkage of a function as of some "foreign" form. By declaring a function as `extern` `"C"`, its name is not mangled by the compiler.

Less known is the technique whereby conventional `extern` declarations can be enclosed by braces and preceded by `extern "C"` to encompass all of them in its effect. For example:

```
extern "C"
{
    extern void alpha(int a);
    extern int beta(char *p, int b);
}
```

This is particularly useful if conditional compilation (`#if ... #endif`) is used to activate the facility during the transitional phase from C to C++. This code is portable between ANSI C and C++ compilers:

```
#ifdef _cplusplus
extern "C"
{
#endif
    extern void alpha(int a);
    extern int beta(char *p, int b);
#ifdef _cplusplus
}
#endif
```

Ironically, this problem did not occur with early definitions of the C++ language (prior to V2.0) since function name mangling was originally used just to differentiate between overloaded functions. The first (and, maybe, only) instance of a function did not need to be mangled to have a unique name; only the second and subsequent instances needed new names. When it was realized that mangling all the names gave type-safe linkage and more secure function overloading, this method was incorporated into the language, and the `overload` keyword rendered redundant. With current versions of C++, this necessary complication is a major incentive to progress to the next step in the move to C++.

5.3.4 Writing Clean C

Since ANSI C is essentially a subset of C++, it would seem reasonable to use the C++ build and debug tools on the old C code and simply write all the new parts in C++. However, there are numerous minor exceptions that make the conversion process less than 100% straightforward. Even if you could not circumvent these exceptions, the fact that C++ is a superset of C is a major asset when learning and understanding the new language.

The solution to these incompatibilities is to write C (or modify it) to take them into account. We call the resulting dialect "Clean C," which is ANSI C written in such a way that it is acceptable as a C++ subset. The following sections describe the major issues to address when writing Clean C.

Type Checking

C++ enforces a higher degree of type checking on function calls; function prototypes are mandatory. In Clean C, make sure that all functions have valid prototypes, even when the only calls to a function are elsewhere in the same file.

It is easy to code an erroneous function call in C:

```
void q()
{
    char *p;
    f(p);
}
f(int i)
{
    ...
```

In Clean C an error is flagged:

```
void f(int);
void q()
{
    char *p;
    f(p); // compile-time error
}
f(int i)
{
    ...
```

Casting

C++ is more "careful" about loss of data from implicit casts of types from a higher precision to a lower one. In Clean C, always perform such casts explicitly. For example, in Clean C:

```
long big; char little;
little = (char) big;
```

Enumerated Types

ANSI C does not take enum variables very "seriously"; they are viewed as being a convenient alternate #define notation. A function parameter, for example, may be declared as an enumerated type, but a call is still permitted where an integer value is provided.

C++ is stricter in this respect. For example:

```
enum greek { ALPHA, BETA, GAMMA, DELTA };
void func(enum greek);

...
func(GAMMA); // not func(2)
```

A call in the form func(4) would result in an error from the C++ compiler.

Character Strings

In ANSI C a `char` array that is statically initialized to be a character string need only be large enough to hold the characters of the string; i.e., it is not essential to accommodate the null terminator. This is not the case in C++ and, hence, in Clean C.

In C, it is possible to allocate minimal space:

```
char str[3] = "xyz";
```

In Clean C, the `NULL` should be accommodated:

```
char str[4] = "xyz"; // room for NULL
```

or better, allow the compiler to allocate the space:

```
char str[] = "xyz";
```

Scope of Struct and Enum

In C++, the scope of `struct` and `enum` data types has limitations that are not present in ANSI C. In C, if you define a `struct` within a `struct`, you can use the inner structure outside of the outer one. Because you can't do this in C++, define a nested structure in Clean C only if its use is suitably confined.

Inner-nested structures can be used in C:

```
struct out
{
    struct in
    {
        int i;
    } m;
    int j;
};
struct in inner;
struct out outer;
```

In Clean C, the scope of structures must be clarified:

```
struct in
{
    int i;
};
struct out
{
    struct in m;
    int j;
};
struct in inner;
struct out outer;
```

Multiple Declarations

In ANSI C, it is permissible to include multiple declarations of a (single) variable in a file, even though the second and subsequent declarations are redundant and the practice can lead to confusion. This technique is not legal in C++, and thus should not be done in Clean C.

A variable may be declared several times in C:

```
int i;
f()
{
    ...
}
int i;
...
```

In Clean C, only one true definition can be included:

```
int i;
f()
{
    ...
}
extern int i;
...
```

Extra Keywords

C++ has additional keywords that are not reserved in ANSI C. These include:

```
asm         friend      private     try
catch       inline      protected   this
class       new         public      virtual
delete      operator    template    throw
```

In Clean C, these keywords must be avoided. Some keywords that can be particularly troublesome are class, new, template, and delete. You can, however, apply a naming convention that avoids all clashes with keywords, such as initial letter capitalization.

These extra keywords are not an issue in C:

```
struct class
{
    int private;
};
void copy (char *old, char *new);
```

In Clean C, capitalization solves the problem:

```
struct Class
{
    int Private;
};
void Copy (char *Old, char *New);
```

Compiler Assistance

Some of the requirements of Clean C are easier to apply than others. A number (mandatory prototyping, for example) may be available as options with an ANSI C compiler. Such facilities permit the transition to Clean C to be commenced at an early opportunity.

5.3.5 C+—Nearly C++

Having programmed in Clean C and having passed the code through the C++ build tools, you can now take advantage of C++ features. As you learn about and discover needs for these features, incorporate them into your code incrementally. The use of C++ features will increase at a gradual rate until the program is transformed from a C/C++ dialect that we call "C+" to code that is entirely written in C++.

Let's first look at some of the simpler, syntax-enhancing C++ features.

Comment Notation

C++ has an alternate end-of-line comment form, specifically the // notation. This notation is actually being reintroduced; it was lost many years ago in the transition from BCPL to C. This is particularly useful for commenting out code and adding annotation to variable declarations.

In C, comments must be terminated:

```
int *vp; /* pointer to store */
```

In C+, a comment can finish at the end of a line:

```
int *vp; // pointer to store
```

Reference Parameters

Reference parameters are useful for writing clearer code to hide the necessary use of pointers.

In C, using pointers is the only option:

```
void swap(int *to, int *from)
{
    int temp;
    temp = *from;
    *from = *to;
```

```
        *to = temp;
    }
```

Reference parameters make the code clearer in C+:

```
void swap(int &to, int &from)
{
    int temp;
    temp = from;
    from = to;
    to = temp;
}
```

References also avoid null pointers dereferenced at a call site and avoid infinite recursive calls.

Placement of Variable Definitions

In C++, local variable definitions can be placed almost anywhere, unlike C, where they must be placed before the executable code.

C requires variables to be defined at the top of the function:

```
w()
{
    int x, y;
    for (x=0; x<3; x++)
        f(x);
    ...
    for (y=5; y>0; y-)
        g(x,y);
}
```

In C+, variables may be defined at the point of use:

```
w()
{
    for (int x=0; x<3; x++)
        f(x);
    ...
    int y;
    for (y=5; y>0; y-)
        g(x,y);
}
```

As your C++ implementation progresses, you can incorporate more sophisticated C++ features such as the following:

Constructor Functions

The constructor function capability in C++ can facilitate the initialization of `struct` member variables (as, in C++, a `struct` is just a degenerate class).

In C, a structure needs to be explicitly initialized:

```
struct list
{
     int data;
     struct list *ptr;
};
void init(struct list *);
struct list links;
init(&links);
```

Initialization can be implicit in C+:

```
struct list
{
   int data;
   list *ptr;
   list(int d=0, list *p=NULL)
   {
       data = d; ptr = p;
   };
};
list links; // initialized automatically
```

Dynamic Memory

A common error in C is the allocation of dynamic memory without the corresponding deallocation. The use of a C++ class destructor provides a means to avoid this problem.

It is easy to forget to deallocate memory in C:

```
void f()
{
   char *sp;
   sp = (char *) malloc(1024);
}
g()
{
   f();
}
```

A destructor can handle deallocation in C+:

```
extern "C" { char* malloc(int);}
struct memory
{
   char *p;
   memory(int s=1024)
   {
       p = (char *) malloc(s);
   };
```

```
        ~memory() { free(p); };
    }
    g()
    {
        memory mob; // allocate memory
    }
        // clean up automatically
```

Error Handling

In C, the handling of error conditions (exceptions) is not specifically accommodated in the language, so it is difficult to implement securely. A useful feature of C++ is exception handling.

The definition of an exception, in this context, should be appreciated. An exception is an unintended synchronous event, not an interrupt. For an embedded system, the controlled handling of exceptions is particularly important. Simply terminating the program with an error message is rarely an option.

The typical use of exception handling in C++ is to facilitate communication of an error condition between reused code and the new application using it; e.g., between a library function and its caller.

The programming of exception handling is a two-stage process:

1. A function that has detected an error "throws" an exception, using the keyword throw and possibly an object of an appropriate type to convey details of the exception. This is in the hope that the exception is "caught" later by the caller or some other routine in the call chain.
2. The calling function encloses calls to the reused code in a try block, followed by a series of catch blocks (exception handlers) corresponding to one or more types of "thrown" objects. Of course, the opportunity still exists for there to be no catch block corresponding to a thrown object, and for an error to go unchecked. In that case, an unexpected exception is raised.

5.3.6 Conclusions—The Path Ahead

With C++ established as the most appropriate language for the next generation of embedded systems, now is the time to take the first steps toward its use. There are various strategies that may be adopted, but a cautious, incremental approach has many merits.

5.4 C++ Templates—Benefits and Pitfalls

In the 1990s, as C++ was beginning to gain popularity among embedded developers and certain features were getting "bad press," I wrote a piece for NewBits, in collaboration with Michael Eager, about the usefulness of templates; this article is based upon that work. Another article in this chapter, "Looking at Code Size and Performance with C++," expands upon some of the issues raised here.

CW

As C++ evolved, various new features were included in the language. Important examples include multiple inheritance, transparent function overloading, exception handling (EHS), and class/function templates. In the embedded context, the first two are widely understood; the latter two, less so. In this article, we will focus on function templates from the viewpoint of the embedded systems programmer: examining what they are, how to use them, how they are implemented, and pitfalls in their use.

5.4.1 What Are Templates?

In broad terms, templates are a means whereby the designer of a function or class parameterizes more of the function's or class' properties than was previously possible. The most obvious property that may be subjected to this treatment is the type of data contained within an object or processed by a function.

In this article we will concentrate on function templates, simply because they can be described with minimal preamble. All the same concepts and much of the syntax are identical for class templates.

Using Function Templates

Consider this code:

```
void swap(int& x, int& y)
{
    int temp;
    temp = x;
    x = y;
    y = temp;
}
```

This is a simple C++ function, which effects a swap of the contents of the two integer parameters x and y. It is generally unremarkable, except that it takes advantage of C++ reference variables to render the code more readable than the explicit pointer manipulation than would be necessary in a C implementation of the same function.

The problem with this swap() function is that it can only deal with a pair of ints. Of course, a similar function that operates upon, say, two floats, could easily be written. With C++ function overloading, it could even have the same name in order to enhance readability. However, this does not solve the problem of the function being limited to two ints or two floats. To accommodate all eventualities, an indefinite number of overloadings would be necessary.

Another solution to this kind of problem may well be to use a #define macro. A macro may be able to cope with calls of this form:

```
swap(v1, v2)
```

Since the expansion of this macro would most likely make more than one reference to the parameters, the fact that the expressions v1 and v2 have no side effects (because they are just variables) is significant. However, the following would still cause difficulties:

```
swap(v[i++],w[i++])
```

since i++ and j++ would be referenced twice and, hence, incremented twice. The solution is to use a function template, thus:

```
template <class X> void swap(X& x, X& y)
{
    X temp;
    temp = x;
    x = y;
    y = temp;
}
```

This example defines a generalized function that operates on data of type **x**. The actual type of x will not be determined until the template is created (i.e., *instantiated*). Using templates allows the designer to focus on implementing an algorithm that can be used for any type of data.

As another example of using a specific algorithm for different data types, consider the binary search algorithm. Although this algorithm is described in many books, you can still make a mistake in writing the function. But once the function is written and tested for one type, you can reuse it wherever needed.

Here is a template that implements the binary search function:

```
template <class T>
int bsrch(const T data[], int num_elem, const T val)
{
    int max = num_elem, min = 0, mid;
    while (max > min)
    {
        mid = (max + min) / 2;
        if (data[mid] == val)
            return mid;
        if (data[mid] < val)
            min = mid +1;
        else
            max = mid - 1;
    }
    return -1;
}
```

This function expects you to give it an array that is sorted in ascending order, the number of elements in the array, and a value to search for. It will return the index into the array if the value is found or −1 if it is not.

To use this template, all you have to do is call the function with the correct parameters. The compiler will automatically create a function that will work on the data type of the array.

Care must be taken when writing a function template. For example, the `bsrch` function template will work for any type that has the operators == and < defined. But if you use it to search an array of pointers to strings, you will be surprised. The function will compare the values of the pointers and not the strings.

In this case, the best solution is to create a string class that has the == and < operators defined as member functions. These functions would call `strcmp()` and return `TRUE` or `FALSE`.

5.4.2 Template Instantiation

The definition of a template does not, in itself, result in the generation of any code. It is simply a specification to the compiler for a function that should be instantiated when a corresponding (otherwise unresolved) function call is encountered. Typically, function templates are placed in header files; use `#include` to make them visible to the compiler.

This code:

```
int i, j;
float a, b;
...
swap(a, b);
swap(i, j);
```

would result in two instantiations of the `swap()` function: one for two ints (`i` and `j`) and one for two floats (`a` and `b`).

5.4.3 Problems with Templates

Although templates represent an elegant solution to a very real problem, their use may cause unexpected difficulties. Using a template with many different types of data can result in many copies of essentially identical code. Moreover, the implementation of templates can result in multiple copies of identical binary code, which may limit their usefulness for embedded systems development.

Program Overhead

As discussed previously, the attraction of function templates, compared with overloaded functions, is that you write the code just once. This single instance of source code is a real advantage in both productivity terms (i.e. writing the code in the first place) and in ongoing efficiency (i.e. later maintenance of the code; there is only one function to modify).

However, this efficiency may lead to a false sense of security. It must be remembered that, each time a template is applied to a new situation (i.e. instantiated for a new parameter data type), a new version of the (binary) code is generated. In other words, the use of templates maximizes programmer efficiency (which is good) but does not have any impact upon the space/time efficiency of the resulting program. If the development tools do not perform well, a very significant program overhead may unexpectedly develop.

Duplicate Instantiations

To reiterate: when the compiler "sees" a function template, it accepts it as a specification for functions that it can generate when corresponding calls are encountered. The resulting functions (template instantiations) are effectively static functions; they are local to the current module. This is necessary because the responsibilities of the compiler itself are limited to the module. The implication of this seemingly innocuous implementation detail is that identical code may be present in many modules, where identical template instantiations have occurred.

For host computers, with very large memory capacities, this redundant code duplication may be a reasonable price to pay for the programmer efficiency gained by the use of templates. For embedded systems, where resources like memory are precious, the cost can be too high.

What is the solution? With many C++ development toolkits, the only solution is to limit or prohibit the use of function templates. A solution may be effected by implementing a "smart" linker. At link time, identical functions (which can be template instantiations or "out-of-line" inline functions) can be identified and merged, so that only one copy of the code exists in the executable image. The memory requirements of many programs built with tools employing this facility may be reduced very substantially.

5.4.4 Multiple Template Parameters

There are limitations in the function templates presented so far, since only a single data type has been parameterized. This need not be the case. For example, this function template:

```
template <class Y> int bigger(Y a, Y b)
{
    return (a>b);
}
```

yields a function, bigger(), which returns a logical value indicating whether or not the first parameter is larger than the second. As implemented, this function will work satisfactorily for, say, two ints or two floats. However, it will not work for two parameters of differing type, even if this would be logically acceptable. This capability can be imparted by changing the first line of the template to:

```
template <class Y, class Z> int bigger(Y a, Z b)
```

The type of each parameter is now considered individually.

5.4.5 Other Template Applications

Although function parameter and internal object types are the most common template parameters, the facility is not specifically limited to this application. The template parameters may, in principle, specify anything that may be determined at compile time.

A useful example is shown here:

```
template <class X, int S> void f(X n)
{
    X buff[S];
    buff[0] = n;
    ...
}
```

where an array size is parameterized. This example is particularly pertinent to embedded systems because it eliminates possible heap usage, which is a common source of runtime difficulties.

One limitation is specific to function templates (and does not apply to class templates): at least one of the template parameters must be a function parameter. This is because the C++ function overloading mechanism relies on functions being differentiated by the type of the parameters in the call. Without this difference existing between template instantiations, ambiguities would result.

5.4.6 Conclusions

Templates are a useful addition to the C++ language. Their use can improve programmer productivity and code readability. However, their use must be considered carefully because the code generated may be unexpectedly large. Also, careful selection of development tools will be beneficial in avoiding redundant code overheads.

5.4.7 Postscript

Since the article was written, C++ templates have developed significantly, and much of this development was not anticipated. The C++ standard library consists of a large number of class templates, as well as a few function templates. Newer class libraries, such as Boost, extend the Standard Template Library with powerful components such as reference-counting pointers.

Templates, in many ways, have extended C++ to be a much richer and more robust language, although at the expense of ease of understanding and debugging. Error messages that involve templates remain less than clear and understandable, even in the best of compilers.

ME

5.5 Exception Handling in C++

While C++ is widely used for embedded systems programming today, it was in the early stages of acceptance in 1995 when I, in collaboration with Michael Eager, wrote a piece for NewBits, upon which this article is based.

CW

This article is an introduction to a less-understood facility in the C++ language—the exception-handling system (EHS).

5.5.1 Error Handling in C

In C, there is no built-in facility to deal with error conditions that may occur during the execution of the code. For example, if you call a library function and it detects an error, how is the error dealt with?

The convention is to utilize a global variable, errno, which is set to a specific nonzero value by a library function if it detects an error condition. This convention has several problems:

- It is awkward. Writing code that checks errno for each library function call is impractical.
- It may be impossible. If a library call is made by code generated by the compiler, the programmer may be unaware that a check is required.
- Reentrancy can be a problem. What happens if an interrupt (or task swap) occurs between the library call and the errno check? This situation may result in another library call that corrupts errno. Some modern implementations do circumvent this problem.

Exception handling in C++ attempts to solve these difficulties by incorporating an error-handling facility into the language.

5.5.2 Does Not Involve Interrupts

A common misconception associated with C++ exception handling is that it has something to do with interrupts, which are confusingly called "exceptions" on some devices. Exception handling in C++ does not involve asynchronous events; it is a method to deal with synchronous error conditions.

5.5.3 C++ Exception Handling

The conceptual model for exception handling in C++ is very simple:

- A library routine that detects an error can `throw` an exception.
- An exception handler can `catch` the exception.
- Blocks of code that activate exception handling `try` to catch exceptions.

These actions, which are keywords, are described in the next section.

Exceptions are characterized by type. The type may be a built-in one (`int`, `float`, `char`, etc.) or user defined (a class). User-defined types are most common. An exception can also have a value that can be interrogated by the exception handler.

Keywords

Exception handling is implemented using three keywords: `throw`, `try`, and `catch`. The syntax for them is shown in this section. Code to activate EHS should be enclosed in a `try` block, thus:

```
try
{
    // code that calls library functions
}
```

This is followed by one or more `catch` blocks:

```
catch (int n)
{
    // exception handling code
}
```

where, in this case, exceptions of type `int` are caught.

A library function that has detected an error can throw an exception with the following:

```
throw 99;
```

resulting in an exception of type `int`, with the value 99 being thrown. Figure 5.1 illustrates the modularity of the EHS.

An Example of Exception Handling

This code illustrates a more complete example of the use of exception handling:

```
char store[100];
void scopy(char* str)
{
    if (sizeof(store)+1 < strlen(str))
        throw -1;
    strcpy(store, str);
}
```

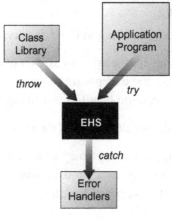

Figure 5.1
EHS modularity

```
void get_string()
{
    char buff[100];
    cin >> buff;
    scopy(buff);
}
main()
{
    try
    {
        ...
        get_string();
        ...
    }
    catch (int err)
    {
        cout << "String too long!";
    }
}
```

The function `scopy()` takes a string parameter and copies it to an array, `store`, using the standard `strcpy()` library function. Before attempting the copy, the function checks the size of the string and, if it is too large, throws an integer exception with the value −1.

The function `get_string()` accepts some text from the console into a buffer and passes it as a parameter to `scopy()`. Notice the C++ input notation usage.

A `try` block exists in `main()`, which includes a call to `get_string()`. Following the `try` block is a single `catch` block, which catches `int` exceptions. This block prints a useful message. The value of the exception is collected in `err` but not used.

Multiple Catch Blocks

In real applications using EHS, multiple `catch` blocks accommodate exception types that may be thrown by library functions. For example:

```
catch (int n)          // catches ints, including value
{
     // handler code
}
catch (float)          // catches floats, discarding value
{
     // handler code
}
catch (complex c)      // user-defined data type
{
     // handler code
}
```

5.5.4 Special Cases

Situations can occur that cause unusual exception-handling requirements.

No Matching Catch Block

What if an exception is thrown where no `catch` block exists that corresponds to the exception type? If this occurs, the `terminate()` library function is called. In practice, the version of this function in the library is most likely a stub and does not attempt to handle the situation—it just terminates the program. You must supply your own `terminate()` function and incorporate it during linking.

Catch Everything

A series of `catch` blocks are normally provided, which includes entries for all possible exception types. Circumstances can occur that require a simpler processing model. You also may be unaware of all possible exception types that can be thrown by a library. Both situations can be addressed using a special `catch` block construct, as follows:

```
catch (...)
{
   // handler code
}
```

This block is a handler definition for all exceptions, regardless of type. It is impossible to ascertain the exception value or type in the handler.

This construct also means "catch all *other* exception types." It may be used after more specific exception handlers to catch any exceptions that were not explicitly addressed.

Exception Objects

A common technique is to create a hierarchy of exception classes that are all derived from a single base class. This allows us to identify the nature of an exception more precisely than by just throwing an int or a float. The following code illustrates this approach:

```
class general_exception
{
public:
     general_exception() {}
};
class memory_error : public general_exception
{
public:
     memory_error() {}
};
class timeout : public general_exception
{
public:
     timeout() {}
};
class io_error : public general_exception
{
public:
     io_error() {}
};
void func1()
{
    try
   {
        func2();
   }
    catch (timeout)
    {
        cout << "Timeout\n";
    }
}
void func2()
{
    static int x = 0;
    if (++x % 2)
        throw timeout();
    else
        throw memory_error();
}
int main()
{
    for (int loop=0; loop<10; loop++)
    {
        try
```

```
    {
        func1 ();
    }
    catch (general_exception)
    {
        cout << "General Exception\n";
    }
    }
}
```

First, the base class for our hierarchy, general_exception, is defined, from which three further classes are derived: memory_error, timeout, and io_error. Note that none of these classes contain any data. We could add data members to each of them to pass details about the exception from the place where the error is encountered to the place it is caught.

Our main() routine, at the bottom, consists of a simple loop in which we call func1() 10 times. This is within a try block, so if func1(), or any function that it calls, throws a general_ exception, we can catch it in the succeeding catch block.

func1() also has a try block. In this try block, we call func2(). If a timeout exception is thrown, we will catch it in the local catch block and print a message.

func2() will alternately throw a timeout exception or a memory_error exception each time it is called. In a real application, this might be a routine that encounters a variety of different situations and throws different exceptions.

When func2() throws a timeout exception, it is caught by the lowest level catch block that can accept a timeout exception. This is in func1(), which will issue a message and then return to where it was called in main().

When func2() throws a memory_error exception, the EHS will search for a matching catch block. A catch block matches the thrown exception if the types match or if they are convertible. As the runtime routine searches for a matching catch block, it checks the catch specification in func1() and sees that it does not match. The catch block in main() specifies general_exception, which is the base class for memory_error. A derived class can be converted to a base class (losing whatever data is added in the derived class), so the catch block in main() will catch memory_ error or any of the other exceptions derived from general_exception.

As this example shows, we can nest exception handlers so that the higher-level handlers take care of general (and often more extreme) actions such as restarting a system. The lower-level handlers take more specific actions when they encounter an error, such as retrying an I/O operation.

Using a class as an exception object allows us to group similar exceptions together by creating derived classes. All of these can be caught by the same exception handler. This object-oriented approach to exceptions offers great flexibility in organizing the exception structure of a program and allows us to add exception types to the program as they are needed.

5.5.5 EHS and Embedded Systems

Like the C++ language, exception handling was not specifically designed for use with embedded systems. Embedded system environments demand special consideration.

Overheads

Embedded systems developers are concerned about the efficient use of limited resources. Hence, a common question is, "Does EHS usage result in a runtime or memory overhead?" Of course it does. In software development, a *Law of Conservation of Effort* exists: you cannot get something for nothing. C++ exception handling provides an example in which the compiler performs operations, which would normally be the programmer's responsibility. It eliminates the need to write code to check for error conditions.

A very important factor concerning overhead is that compilers supporting EHS tend to generate additional code, regardless of EHS usage. This could result in a serious problem of unexpected overhead. Some compilers include a switch to control EHS support, with the switch "off" by default. You need to explicitly activate EHS if you are planning to use EHS (thus implicitly accepting the overhead).

5.5.6 Conclusions

The exception-handling facility in C++ may be useful in an embedded systems application. It offers a clear and consistent method of handling unusual conditions within a program.

A Case for Simple Exception Handling

Since overheads are so important in many embedded systems, care needs to be taken in the application of EHS. Keeping it simple is the best approach. To appreciate the point here, consider a nonembedded example.

You are running a Microsoft Windows program, and a runtime error occurs; maybe you have run out of memory. A dialog box pops up explaining the situation. Clicking OK results in shutting down the application. Although the result may ultimately be the same, a number of possible causes for this problem are detailed in the dialog box. C++ EHS provides a straightforward means of programming such an application, if the overhead is acceptable.

With an embedded system, the situation may be very different; let's consider another example. A microprocessor-based heart pacemaker, programmed in C++, finds a problem. What action can it take? Popping up a dialog box is not an option. Clearly the priority is to return to a stable state so that the operation of the pacemaker may continue. The best option is to simply reset the software (and maybe the hardware) so that it starts up afresh. The fact that an error has occurred possibly could be logged away for future reference.

Although the life-or-death nature of a heart pacemaker is dramatic, the need for a very simple response to runtime errors is common among typical embedded systems: telephone switches, instruments, engine management systems, and so forth.

The obvious way to handle exceptions, if the requirements are simple, is to make use of the catch (...) construct. It turns out that the code generated by the compiler for this construct is substantially simpler (and faster) than more selective exception handling. An alternative is to allow exceptions to be thrown but provide no catch blocks at all. Then the `terminate()` function can be called (and effect the reset or whatever).

5.6 Looking at Code Size and Performance with C++

Although C is still widely used, the popularity of C++ for embedded programming has steadily increased. However, there are still frequent objections to the language on the grounds of excess overhead and inefficiency. This article is based upon one written for NewBits in 1995 by Nick Lethaby (now with Texas Instruments), in which he addressed these objections.

CW

The C++ language has gained rapidly in popularity among embedded developers in recent years and has clearly established itself as the language of choice for developers of larger systems. C is still in widespread use among embedded and real-time developers, but many of these developers are turning to C++ for new projects.

As its popularity has increased over the years, the C++ language has also undergone significant evolution, adding new features such as templates (parameterized types), runtime type identification, and structured exceptions. The language is now fairly stable, with the establishment of an ANSI standard.

5.6.1 How Efficient Is C++ Compared to C?

While the benefits of C++ and object-oriented programming have been well publicized, very little has been said about the efficiency of C++. Unlike host computer systems that have abundant memory and CPU resources, embedded systems must typically minimize such resources to minimize production costs and/or power consumption. Many embedded programmers are rightly concerned about program size and performance degradation when applications are developed in C++.

C++ is essentially a superset of C and may be used exactly like C. Under these circumstances, applications developed using C++ will incur no more overhead than those written in C. However, although C++ does provide improved type checking over C, along with some other niceties, most developers are attracted to C++ for the promised benefits of object-oriented programming techniques, including improved reusability and maintainability. These benefits require using the many new language features of C++, which may have program size implications, depending upon how they are implemented by the compilation system.

Using object-oriented programming methods with C++ can lead to performance problems unless the underlying implementation of the objects used in the program is well understood. For example, it is important to understand how long it takes to construct or destruct an object. Such an understanding is critical if a programmer is to avoid arbitrarily defining objects with long creation times in performance-critical sections. Proper programming practice is by far the best defense against the adverse effect on performance of poorly designed constructors and destructors.

5.6.2 How C++ Affects Application Memory Requirements

Several C++ language features, including templates, inline functions, virtual functions, and inheritance, can affect application size. All of these features are widely used by C++ programmers as well as by the ANSI C++ libraries and other commercial C++ class libraries.

Templates

Templates provide an excellent mechanism for sharing source code because they allow you to write just a single implementation of a class or function to handle multiple data types. However, the C++ language definition says nothing about whether templates should share target code. This convention is left to the compilation system and is therefore highly implementation-dependent. A poorly implemented solution can result in significant increases in code size, which would be unimportant in host-based software but could be a disaster for an embedded application.

In some compilation systems, the compiler will simply instantiate a copy of the template function (or class) for each data type invoked within a particular source file (or module). In Figure 5.2, for example, the compilation of each source file will result in its own instantiations of the max() function. The consequences of this approach are obvious: if an application has 500 source files that use the max() function, 1,500 copies of the max() template function are generated. If the linker simply links all of these into the final object, the application will be significantly larger than necessary. In reality, only three copies (actually, variants—they are all composed of different code) of the max() template function are needed: one for int, one for char, and one for float, as shown in Figure 5.3.

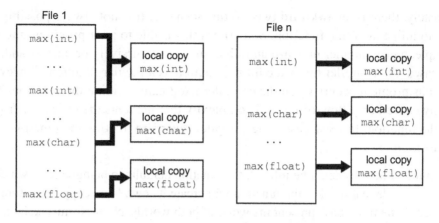

Figure 5.2
Inefficient template implementation

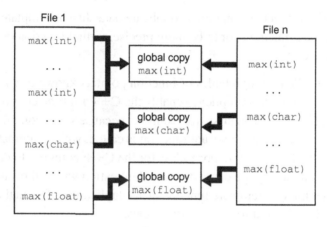

Figure 5.3
Efficient template implementation

Inline Functions

Inline functions can also result in unwanted duplicate code, although the problem is less obvious than with templates. When a programmer defines a function with the `inline` keyword (or defines it inside a class definition, thereby making it inline by default), the expected result is for the function to be inlined everywhere the function is called. This approach improves performance by eliminating the overhead of a function call. Whether or not code size increases depends on the size of the function. If the function body is smaller than the code required to set up and return from a function call, as many inline functions are, code size will decrease. If not, code size will increase.

Unfortunately, there is an awkward twist to this scenario. It is not always possible for the compiler to inline an inline function. Sometimes this is due to poor programming style. For example, a programmer may use an inline function recursively or take its address. In these cases, the compiler treats the inline function as a regular function. However, a more serious problem occurs when the compiler itself cannot inline the function. This is typically caused by a flow-of-control complexity that prevents the optimizer from inlining the function. In these cases, the compiler reverts to treating the function as a regular function.

When an inline function cannot be inlined, the function defaults to being static, which means that each file that invokes the function will create its own local copy. Therefore, these "non-inlined inline functions" present the worst of both worlds. No performance increase is realized because the function call is still being made, and code size is increased unnecessarily because a copy of the function exists in every source file that invokes it.

Virtual Functions
The other side of the inefficiency equation is duplicate data. Just as template classes can result in duplicate data, virtual functions, or to be more precise, virtual function tables, can also result in duplicate data.

Virtual functions are called through tables of function pointers known as *virtual function tables*, which are created by the compiler to enable the C++ runtime environment to call the correct function. Inefficiencies can arise, for example, because some compilers create virtual function tables in every file where a virtual function is called, which automatically results in duplicate function tables. A better approach is for the C++ compiler to create virtual function tables only where they are defined. Some duplication will still occur, however, if the virtual function is an inline or template function. This duplication of virtual function tables further compounds the potential for size inefficiencies.

Inheritance
The layout of inherited objects also contributes to space requirements. Space can be saved by optimizing the layout of objects and virtual function tables. Some C++ toolkits are inefficient in the way that they lay out certain types of inherited objects. Consider the following code fragment for deriving a class called MDerived from several classes that are in turn derived from a virtual base class:

```
class BaseClass { ... };
class DerivedA : virtual public BaseClass { ... };
class DerivedB : virtual public BaseClass { ... };
class DerivedC : virtual public BaseClass { ... };
class DerivedD : virtual public BaseClass { ... };
class MDerived : DerivedA, DerivedB, DerivedC, DerivedD { ... };
```

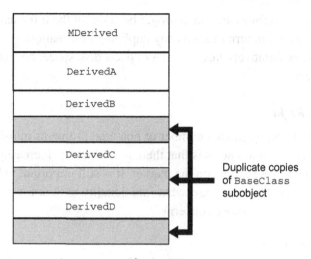

Figure 5.4
Internal layout of object of type MDerived

In inefficient toolkits, if you derive a class from *n* classes that in turn are derived from the same virtual base class, the derived class will have at best $n-1$ copies (and often *n* copies) of the virtual base class members, as illustrated in Figure 5.4.

An efficient implementation requires only one copy of virtual base class members. Since instances of such a class are created at runtime, this inefficiency has the potential to consume valuable RAM if an application creates thousands of objects containing duplicate virtual base class members.

Duplicate Debug Information Impacts Productivity
While minimizing the final application size is important in helping to reduce the cost of the finished product, the problem of duplicate debug information also affects productivity during the development cycle. Although this problem existed with C programming tools, the changes in programming style introduced with C++ have greatly magnified its effect in that C++ applications tend to include every header file in each source file. The complex interrelationships of classes in an application make it difficult to decide exactly which header files are required to successfully compile a certain source file.

In the past, compilers have tended to generate debug information for all the types declared or defined in the header files, regardless of whether the source file actually used them. Obviously, including every header file in each source file can result in vast amounts of duplicated debug information. This extraneous debug information can affect productivity in several ways:

- Link times are much longer because the linker must process all the debug information in each file.

- Debugger initialization times are much longer because all the information must be read by the debugger, which in turn discards any duplicate information.
- Files can become prohibitively large, which impacts disk space, network traffic, and backup procedures.

5.6.3 Doing C++ Right

Although this article has highlighted a number of potential problems in using C++ for embedded applications, the good news is that there are solutions. Increasingly, these issues are being addressed by C++ toolkits on the market. It is still important to be able to identify the problems, if they persist, and to know what functionality to look for when selecting tools. We will now take a look at just those concerns.

Faster Compile and Link Times

The old way of implementing C++ was to translate it to standard C first. Although this approach was instrumental in the success of the language (compared with other object-oriented languages based upon C), major performance and functional benefits are to be gained from modern, C++ to assembly, compilers.

It is important for a compiler to be discriminating about the debug information that it outputs. The compiler should output only debug information for functions and variables used in the source file rather than for every program construct declared or defined in the header files included by the source file. Typically, this optimization results in a 50% reduction in the final size of an application (absolute file) built with debugging enabled.

The main benefits of this optimization are seen downstream from the compiler. Link times are shortened dramatically because the linker is now processing smaller files. Since linkers tend to be I/O-bound applications, any reduction in size of the files processed automatically translates into shorter link times.

Reduced Program Size

The problems of duplicate code and data must be addressed in the linker as well as in the compiler. This is because the linker is the most practical place to optimize across file boundaries.

Compiler optimizations may address virtual function tables and object layout. Virtual function tables should be created only in files where the function is defined. This practice significantly reduces the problems of duplicate virtual function tables. Improvements to the layout of objects can eliminate the problem of derived objects containing duplicate copies of a virtual base class (see Figure 5.5). This improvement can save significant amounts of RAM in applications with large numbers of objects derived via multiple inheritance and virtual base classes.

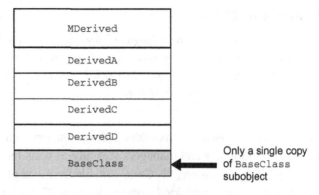

Figure 5.5
New layout of object of type MDerived

The linker may be enhanced to address the issue of duplicate template functions and inline functions, as well as duplicate virtual function tables generated by inline, template, or virtual functions. Solutions involving the compiler tend to lead to considerable increases in compile times.

Linker optimizations that reduce final application memory requirements are performed by special algorithms that merge duplicate copies of code and data. The compiler tags legitimate duplicate copies, such as functions defined as static by the user, so the linker knows not to merge them. Algorithms can also be used to merge duplicate debug information.

Linkers tend to be I/O-bound programs. As a result, performing these optimizations does not greatly increase link times because the new linker simply throws away extraneous code and data rather than putting it into a file. In contrast, a compiler-based approach would increase compile times if the compiler had to perform complex cross-file optimizations to remove duplicate code and data.

5.6.4 Conclusions

As C++ is the standard language used for the implementation of large complex embedded systems, it is important that C++ compilation technology provide compact code and data size to reduce memory requirements and improve compile-link-debug cycle times. Because conventional C optimization technology does not address potential code and data size inefficiencies introduced by some of the C++ object-oriented programming constructs, new optimization techniques need to be implemented to eliminate the unnecessary overhead often associated with C++.

5.7 Write-Only Ports in C++

The debate whether C++ really is a good language for embedded applications has continued for many years. I have given many seminar and conference presentations and conducted numerous training classes on the topic. In 1995, I captured a key example in a piece for NewBits, which was the basis for this article.

CW

C++ is widely, but not exclusively, used for embedded software programming. C also continues to be popular. Many engineers and managers seek advice upon the choice of development language. Typical questions are, "Is C++ suitable for my application?" and "What real benefit does C++ provide for the embedded systems developer?" This article provides guidelines to enable you to answer the first question yourself. In addition, it will offer at least one answer to the second. To work toward this answer, we will define a specific embedded systems programming challenge, consider a solution programmed in C, and investigate how it may be solved in C++. Then we will develop the idea to provide a more general solution. First, however, we need to consider the fundamental philosophy of using C++ in this context.

5.7.1 Encapsulating Expertise

If you ask any programmers what the benefits of an object-oriented approach are, their responses will most likely focus on the reusability of the code. This is a true benefit because the avoidance of "reinventing the wheel" is important in any computer programming context. However, reusability has no *specific* benefit to embedded systems programming.

The difference between "normal" programming and embedded systems work may be hard to define; it has as much to do with the composition of the development team as the techniques employed. Writing software for a large embedded system ideally requires a good knowledge of software techniques, a reasonable understanding of the hardware, and a thorough appreciation of the real application. It is rare to find an individual who can actually bring all of this expertise to a project team. In general, a team consists of various specialists: software engineers, hardware engineers, and applications experts. It is quite common, for example, for a software engineer to have no understanding at all of the hardware used to implement the system. This leads to difficulties in implementing low-level drivers and other elements.

An advantage of using object-oriented techniques is that the difficult-to-understand driver code may be contained in an object, effectively encapsulating the expertise of the hardware engineer. The software engineer can then use the object quite safely because their access to the internals of the driving code is strictly controlled.

5.7.2 Defining the Problem

To illustrate this point, we need an example of an embedded system feature that may present difficulties to a software engineer—e.g., write-only ports. A write-only port is an output

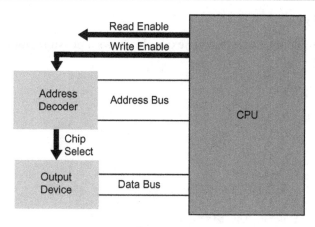

Figure 5.6
Write-only port implementation

device register (usually memory-mapped), to which data can be written, but from which data cannot be read. An attempt to read from the device can, at best, yield useless data; at worst it might result in a processor exception. Figure 5.6 illustrates the implementation of a write-only port.

Note that the Read Enable signal is simply not connected to the port electronics. Write-only ports are not the work of a deranged hardware designer with a strange sense of humor or who harbors a grudge against software engineers (even though we may all be acquainted with individuals who fit this description). Write-only ports are an inevitable result of rational design; the ability to read from the device requires additional and otherwise useless electronics.

The problem for the software designer is exacerbated by a common feature of write-only ports—they are frequently collections of unrelated output bits. This means that different parts of the software may be concerned with the setting or clearing of different bits on the port. Without care, one routine will set a bit, for example, then another will unwittingly change its state.

The usual solution is to maintain a "snapshot" of the last value written to the port. This "snapshot" is commonly referred to as the *shadow*. It is then an apparently simple matter of updating the shadow and outputting it to the port each time a bit needs to be changed. However, this approach may be beset by a number of difficulties:

• Initialization
• Control
• Distribution of code and data—not encapsulated
• Reentrancy.

5.7.3 A Solution in C

To illustrate these problems, the following example shows a possible implementation of write-only ports in C:

```
extern int ports[10];
int shadows[10];
void wop_set(int port, int bit)
{
    int val;
    val = 1 << bit;
    shadows[port] |= val;
    ports[port] = shadows[port];
}
void wop_clear(int port, int bit)
{
    int val;
    val = ~(1 << bit);
    shadows[port] &= val;
    ports[port] = shadows[port];
}
```

In the program, `ports` is an array that has been mapped onto a series of output ports at consecutive addresses, and `shadows` is an array that contains shadow copies of the ports. Two functions have been provided: `wop_set()` and `wop_clear()`. This code is rather inflexible and does not address the issues of initialization or reentrancy. Along with these shortcomings, the programmer is provided with a very limited means of setting and clearing bits.

5.7.4 A First Attempt in C++

The first objective in attempting to implement a write-only port object is to simply encapsulate the necessary functionality into a C++ object. The code shown in the following example is a first shot:

```
class wop
{
    int shadow;         // not accessible -
    int* address;       // to user
public:
    wop(long);          // constructor
    ~wop();             // destructor
    void or(int);       // "operator" -
    void and(int);      // functions
};
```

The class `wop` has two private member variables, `shadow` and `address`, which hold the stored port value and the address of the port, respectively. Along with the constructor and destructor, two member functions provide OR and AND functionality, which is quite sufficient to set and clear bits on the port.

The constructor:

```
wop::wop(long port)
{
    address = (int*) port;
    shadow = 0;
    *address = 0;
}
```

which is called automatically when the object comes into scope, simply initializes the `address` variable, and writes 0 (an arbitrary value) to the port and the shadow copy.

The destructor:

```
wop::~wop()
{
    *address = 0;
}
```

which is called when the object goes out of scope, just resets the port to 0 again. The `or()` and `and()` member functions:

```
void wop::or(int val)
{
    shadow | = val;        // set bit(s) in copy
    *address = shadow;     // update port
}
void wop::and(int val)
{
    shadow &= val;         // clear bit(s) in copy
    *address = shadow;     // update port
}
```

just take the provided parameter, operate on the shadow data, and write the new value out to the port. This `main()` function illustrates the creation and use of a `wop` object:

```
main()
{
    wop out(0x10000);
    out.or(0x30); // set bits 4 and 5
    out.and(~7); // clear bits 0, 1 and 2
}
```

5.7.5 Using Overloaded Operators

In C++, it is possible to redefine operators to enable them to work with new data types. This concept sounds confusing, but it can lead to a very natural way of using objects if the usual behavior of the operators is carefully retained. The example is modified as follows:

```
class wop
{
    int shadow;
    int* address;
```

```
public:
      wop(long);
      ~wop();
      void operator I =(int);     // overloaded
      void operator&=(int);       // operators
};
```

The member variables and the constructor and destructor functions are unchanged. The `or()` and `and()` member functions have been replaced by two operator functions that overload the `|=` and `&=` operators:

```
void wop::operator|=(int val)
{
      shadow |= val;        // set bit(s) in copy
      *address = shadow;    // update port
}
void wop::operator&=(int val)
{
      shadow &= val;        // clear bit(s) in copy
      *address = shadow;    // update port
}
```

It turns out that the code for the two functions is actually unchanged.

A modified `main()` function shows the natural use of the new operators on a `wop` object:

```
main()
{
      wop out(0×10000);
      out |= 0×30;     // set bits 4 and 5
      out &= ~7;       // clear bits 0, 1 and 2
}
```

At this point, it would be entirely reasonable to hand over such an object to the applications programmers, who could use it in any way they wished, without having to worry about the write-only nature of the port.

5.7.6 *Enhancing the* wop *Class*

Although the `wop` class defined in the previous section would be useful, it is quite straightforward to add other functionality. For example, the initialization of the port to the value 0 was purely arbitrary; this matter could be put under the control of the programmer. This code shows the necessary modifications:

```
class wop
{
      int shadow;
      int* address;
      int initval; // stored initial value
```

```
public:
    wop(long, int);
    ~wop();
    void operator |=(int);
    void operator&=(int);
};
```

A second parameter has been added to the constructor to provide the means of setting the initial port value. The parameter is optional; if the programmer omits it, the value 0 is utilized. This retains compatibility with the previous version of the class definition:

```
wop::wop(long port, int init=0)
{
    address = (int*) port;
    initval = init;
    shadow = initval;
    *address = initval;
}
```

Another private member variable, `initval`, has been added to store the initial value of the port. This value needs to be retained because it will be required by the destructor:

```
wop::~wop()
{
    *address = initval;
}
```

A small adjustment to the `main()` function illustrates the use of this enhancement:

```
main()
{
    wop out(0x10000, 0x0f);
    out |= 0x30;
    out &= ~7;
}
```

5.7.7 Addressing Reentrancy

In the C++ code presented so far, most of the defined problems with the implementation of a write-only port have been addressed: initialization, control, and the encapsulation of code and data. The remaining issue is reentrancy. In many systems, reentrancy would not be a concern. It is a problem only if an interrupt may occur between the updating of the shadow data and the writing to the actual port. This, in turn, is a problem only if the interrupting code makes use of the port.

In general terms, we can make a resource reentrant by providing a locking mechanism on that resource; i.e., code that uses the resource locks it, uses it, and unlocks it again. As a first step, we will modify the `wop` class to include dummy `lock()` and `unlock()` functions:

```
class wop
{
    int shadow;
    int* address;
    int initval;
    void lock() { };               // dummy
    void unlock() { };             // functions
public:
    wop(long, int);
    ~wop();
    void operator|=(int);
    void operator&=(int);
};
```

These are called by the operator functions to utilize the resource:

```
void wop::operator| =(int val)
{
    lock();
    shadow |= val;
    *address = shadow;
    unlock();
}
void wop::operator&=(int val)
{
    lock();
    shadow &= val;
    *address = shadow;
    unlock();
}
```

Before proceeding, we should consider how useful it would be to make the `wop` object reentrant. This would be an unnecessary overhead in a system where reentrancy is not a problem. So we really need some flexibility. The answer lies in object-oriented programming techniques. We can define a new object, `rwop`, derived from the already existing `wop` object, which imparts reentrancy. In turn, to make this workable, we need to make the dummy `lock()` and `unlock()` functions replaceable; i.e., we need them to be virtual functions:

```
class wop
{
    int shadow;
    int* address;
    int initval;
    virtual void lock();           // replaceable
    virtual void unlock();         // functions
public:
    wop(long);
```

```
        ~wop();
        void operator|=(int);
        void operator&=(int);
};
```

The implementation of the rwop object is as shown:

```
class rwop : public wop
{
    int flag;
    void lock() // replacement functions
    {
        while (flag)
        ;
        flag = 1;
    };
    void unlock()
    {
        flag = 0;
    };
public:
    rwop(long, int); // constructor
};
```

The locking mechanism needs to be initialized by the constructor:

```
rwop::rwop(long port, init init=0) : wop(port, init)
{
    unlock();
}
```

The new lock() and unlock() functions use a simple token variable, flag, to effect the locking. This is a simplistic example and not strictly accurate, but it serves to illustrate the point.

Once again, the applications programmer can use this object with no knowledge at all of the underlying mechanisms. Just using rwop instead of wop:

```
main()
{
    rwop out(0×10000, 0×0f);
    out |= 0×30;
    out &= ~7;
}
```

5.7.8 Using an RTOS

Most commonly, reentrancy problems arise as a result of the use of a real-time operating system (RTOS). In this case, it is only logical to make use of the RTOS facilities to control access to the resource. One approach is to use a mailbox to contain a token, which imparts ownership of the resource. In the following example, a new type of object, vwop, is defined, which uses RTOS calls to maintain a mailbox and create a reentrant write-only port:

```
class vwop : public wop
{
    int* mbox;
    int err;
    void lock()
    {
        pend(&mbox, 0, &err);
    };
    void unlock()
    {
        post(&mbox, (char*)1, &err);
    };
public:
    vwop(long, int);
};
```

No change is necessary to the wop base class. In addition to the new lock() and unlock() functions, a constructor is required for the vwop class to initialize the mailbox:

```
vwop::vwop(long port, init init=0) : wop(port, init)
{
    mbox = (int*)0;
    unlock();
}
```

This last example is a very powerful illustration of the use of C++ for embedded systems development. It uses fairly simple object-oriented techniques, and yet the applications programmer can utilize vwop objects very straightforwardly, without any knowledge of write-only ports or the use of an RTOS.

If, at some future point, a different RTOS is selected, the vwop class will most likely need modification. However, the wop base class and the applications code will not need changing.

5.7.9 Expertise Encapsulated

We started out with the idea of the encapsulation of expertise. In working through this example, we successfully encapsulated the expertise of the hardware specialist, by hiding the write-only functionality. Then we encapsulated the expertise of the RTOS specialist by the transparent addition of reentrancy protection. At every stage, the applications developer could concentrate on their expertise—the application.

Interestingly, this approach would have been useful, even if the hardware designers had taken a benevolent attitude to the software engineers and implemented a read/write port. Such a device might still need careful handling to avoid reentrancy problems.

5.7.10 Other Possibilities

Many of the ideas discussed here can be extended further: an XOR operator, external access to the shadow, separate initialization and shut-down values, and so on.

Also, the same principles employed in this article could easily apply to other embedded systems programming problems. For example, many serial I/O chips have multiple internal registers that need to be selected before they can be accessed. A suitable class could be designed to hide this detail from the applications programmer and also to protect against errors and reentrancy problems.

5.7.11 The Way Forward

C++ is very commonly the language of choice for embedded systems development. But many developers are still concerned about whether the benefits outweigh any possible downsides and take an "if it ain't broke, don't fix it" approach. Hopefully this article has given them some useful ideas.

The best guideline in the use of the language is the simple suggestion that you make conservative use of the object-oriented techniques. Do not get buried in deep inheritance swamps; concentrate on encapsulating vital expertise.

5.8 Using Nonvolatile RAM with C++

I wrote a piece in NewBits in late 1997 to illustrate a practical and useful application of C++ and object-oriented programming for embedded applications. This article is based on that piece.

CW

Nonvolatile RAM (NVRAM) is memory that retains data after the power has been removed. It is increasingly common for embedded systems to require storage of data (often configuration parameters) after they have powered down. This necessitates some careful coding to accommodate integrity checking (such memory can be corrupted) and initialization. Various NVRAM technologies are available, each of which may need different handling.

To use NVRAM, a means of reading and writing data in a transparent, secure manner is required. The actual mechanism for accessing the memory and effecting the security should be hidden from the applications programmer. C++ facilitates this by permitting the definition of a class that describes the functionality of an NVRAM variable.

5.8.1 Requirements of Using Nonvolatile RAM with C++

The problem of using NVRAM with C++ will be addressed from the perspective of the applications programmer. How would someone be comfortable using NVRAM?

This code illustrates the way an applications programmer might like to make use of NVRAM:

```
main()
{
    nvram x(0×10000,10,0×100);
    nvram y(0×10000,11,0×100);
    char a;
    x = 10;
    a = x;
    y = a + x;
}
```

Two NVRAM variables have been defined, x and y. In both cases, the declaration includes three parameters (which are passed to the constructor): the base address of the NVRAM area, the offset into the area of this particular variable, and the total size of the area. After the definitions, the variables are treated as if they are regular char variables.

The preceding code example indicates what is required by the applications programmer from the class definition:

- A constructor that accepts the specified parameters.
- An implementation of the = operator to enable values to be assigned to the nvram variable.
- An implementation of an int conversion function that permits the value of the nvram variable to be extracted.

5.8.2 NVRAM Implementation

Various NVRAM technologies are available. Some have special access requirements—e.g., timing. In all cases, integrity checking (checksum, CRC, etc.) is necessary to ensure that the memory was not corrupted during the power-down period. The memory may be affected by power spikes, or some types are prone to "forget" after a time. It is also a requirement to track whether the memory has been initialized or just contains garbage values, because it will be the first time that it is powered up.

In the case study, some assumptions and arbitrary decisions have been made (as illustrated in Figure 5.7):

- The NVRAM is a simple technology, not requiring any special timing.
- The last 6 bytes of the NVRAM area contain a null-terminated signature string: "NVRAM".
- A single-byte checksum is placed just before the signature. This is a simple exclusive-OR of the data bytes.

5.8.3 A C++ nvram Class

Here is the code for an implementation of a class to handle nonvolatile memory, nvram:

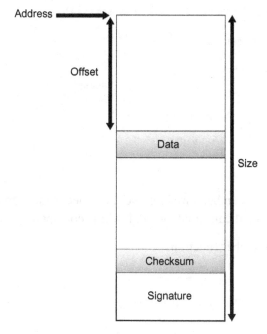

Figure 5.7
NVRAM architecture

```
class nvram
{
     int    check_sig();
     int    eval_checksum();
     char   *address;
     int    size;
     char   *data;
public:
     nvram(long, int, int);
     void operator=(int);
     int operator int();
};
```

The nvram class includes two private member functions that check the integrity of the signature string (check_sig()) and evaluate the checksum (eval_checksum()). There are three private member variables: a pointer to the base of the NVRAM area (address), the total size of the area (size), and a pointer to the specific location in the NVRAM that is allocated to this object (data). There are three public member functions: the constructor, an overloading of the = operator, and an int conversion function.

Here is the code for the constructor:

```
#define CHECKSUM (address + size - 7)
#define SIGNATURE (address + size - 6)
```

```
nvram::nvram(long memaddr, int offset, int memsize)
{
    address = (char*)memaddr;
    size = memsize;
    data = address + offset;
    if (!check_sig() || *CHECKSUM != eval_checksum())
    {
        for (int i=0; i<size-6; i++)
            *(address + i) = 0;
        strcpy(SIGNATURE, "NVRAM");
    }
}
```

This code sets up the three member variables and then checks the signature and checksum integrity. If either of them indicates that the NVRAM is corrupt, it is initialized to all zeros.

Here are the two utility member functions:

```
int nvram::check_sig()
{
    return !strcmp(SIGNATURE, "NVRAM");
}
int nvram::eval_checksum()
{
    char cs = 0;
    for (int i=0; i<size-7; i++)
        cs ^= *(address + i);
    return cs;
}
```

The check_sig() function simply compares the last 6 bytes with the string "NVRAM," returning TRUE if there is a match. The eval_checksum() function calculates and returns the checksum value. This is the exclusive-OR of all the bytes of the data area of the NVRAM (i.e., all but the last 7 bytes).

Here is the code for the operator functions:

```
void nvram::operator=(int val)
{
    *data = val;
    *CHECKSUM = eval_checksum();
}
int nvram::operator int()
{
    return *data;
}
```

The = operator overloading function permits a value to be assigned into the nvram object. It stores the value into the location in the NVRAM allocated to the object and updates the checksum. This functionality could have been implemented by means of an alternate constructor, with a single `int` parameter. However, doing so would have given the applications programmer the opportunity to attempt to define an `nvram` object erroneously using a constructor call with a single parameter.

The `int` conversion function obtains the data from the `nvram` object and returns it as an `int`. This permits the `nvram` object to be used anywhere in an expression where an `int` would be valid. Why `int` and not `char`? When a `char` is used in an expression, the rules of C/C++ demand that it be converted to an `int`. So conversion to the `int` type makes more sense here.

Using the previous definition of the `nvram` class, the original application code may be employed.

5.8.4 Enhancing the `nvram` Class

Various enhancements and variations upon the `nvram` class implementation may be considered:

- Each element of an NVRAM could have a flag that indicates its validity. This would permit each object's constructor to initialize the element to a specific value, if the flag indicated that the NVRAM had been corrupted.
- A popular NVRAM implementation is flash RAM. This type of memory has specific timing and access requirements, which may be readily incorporated into a C++ class definition.
- An additional public member function could be implemented that enabled the entire NVRAM to be initialized to zeros.
- The `nvram` member functions could be made reentrant by including appropriate locking mechanisms into the operator functions. This would permit their straightforward use with an RTOS.
- With the current implementation, every object definition requires the base address and size of the total NVRAM area, even if only one such area is in existence. It would be more convenient if this information could be provided just once, the relevant data being saved in static member variables. A single object that accepted the base address and size as constructor parameters could be defined initially.

5.8.5 Conclusions

In C++, objects may be created that encapsulate specific expertise. These objects may be used by application programmers to exploit that expertise without needing to understand it. Further objects may be derived from existing ones that include additional, mutually exclusive expertise. Using object-oriented techniques, the knowledge and experience of a large project team may be segmented, encapsulated, and distributed.

Classes are designed, implemented, and maintained by an "expert." The access and usage by an applications programmer may be restricted to that which is strictly necessary.

Careful design of classes permits straightforward definition and subsequent manipulation of quite complex objects. In the particular case of nonvolatile RAM, the applications programmer may be provided with an entirely intuitive interface, whereas the systems integrator has the freedom to change the underlying nonvolatile RAM technology, as required.

Further Reading

[1] N. Wybolt, Experiences with C++ and object-oriented software development, ACM Software Engineering Notes 15 (1990) 2.

[2] S.C. Dewhurst, K.T. Stark, Programming in C++, Prentice Hall, Englewood Cliffs, NJ, 1989.

Real Time

Chapter Outline

In my original plan for this book, I had a single chapter allocated to all real-time related articles, whether this included real-time operating system (RTOS) topics or not. I then concluded that I should emphasize the fact that there's more to real-time than RTOSes. So these articles got a chapter to themselves.

6.1 Real-Time Systems

Back in 1992, when I wrote the original NewBits article, upon which this piece is based, there was an increasing feeling among embedded developers that "real-time" meant "real-time operating system" (RTOS). This was an attitude only encouraged by vendors of commercial RTOS products. My intention was to set the record straight.

CW

What exactly is a "Real-Time System"? There are those who would say that an embedded system is a real-time system (RTS). I have a little trouble with that statement, since a definition of the term "real-time" is rather hard to come by. That seems to be the way with computer jargon. There are many terms, which we use quite freely (and accurately), but for which we cannot offer a precise definition.

Looking up "real-time system" in a rather old computer dictionary yields: "Any system in which the processing of data input to the system to obtain a result occurs virtually simultaneously with the event generating that data." It cites airline booking systems as being an example. Clearly this is not an adequate definition of what most embedded systems do. If you consider typical embedded systems, they are generally performing supervisory, control, or data acquisition functions. Such processing is considered "real-time" because it happens as and when required, not when the system "gets around to it." It also has very little to do with being fast, just being on time. A system that gathers some data once an hour is just as much "real time" as one which reads sensors every 10 ms.

Embedded Software: The Works. DOI: 10.1016/B978-0-12-415822-1.00006-4

6.1.1 Implementing an RTS

Having arrived at some idea, if not a complete definition, of what an RTS is, we can now look at the options available when implementing one. There are, very broadly, four ways to implement an RTS:

1. A simple processing loop
2. A background processing loop, with interrupt service code
3. A multitasking system using a scheduler
4. A multitasking system using an RTOS.

6.1.2 A Processing Loop

For a small system of low complexity, the simplest (and, therefore, by definition, the best) approach is a closed processing loop. Each part of the code is processed in sequence endlessly. The loop needs to be infinite, of course, since, in general, an embedded system does not have a logical state which corresponds to the code having stopped.

Although simple, this approach is not without challenges. Typically, the problem is ensuring that particular parts of the code are run often enough which, in turn, depends upon the execution time of other parts of the code. Without care, this method can lead to serious debugging problems especially with timing, which are only located when the code is run on the final hardware. Facilities such as the cycles count of instruction set simulator debuggers may be invaluable here. The worst-case execution time of sections of code can be determined long before real hardware is available.

A more sophisticated approach is needed when an application becomes so complex that timing requirements result in certain code sequences being called several times in different parts of the loop. This immediately suggests that other implementation techniques such as those described next are necessary.

6.1.3 Interrupts

Using interrupts is the next most complex approach to designing an RTS. Typically, there are two circumstances where the use of interrupts may be appropriate: if the servicing of a device is not particularly regular, but must be timely (e.g., a communications application), or when regular processing is required under the control of a real-time clock.

Such a system usually still has a background continuous processing loop, which is interrupted to perform the more urgent work. With a number of microcontrollers, it may be better (from a power consumption viewpoint) to halt the CPU most of the time and perform all the processing in interrupt service routines.

Since high-level languages such as C are normally used to implement modern embedded systems, the ability to write interrupt service routines straightforwardly in C is clearly a requirement. Most modern C compilers, intended for embedded applications, support an additional keyword, `interrupt`, to accommodate this need. (Another article in this book, "Interrupt Functions and ANSI Keywords," covers this topic in more detail.)

An application which uses a number of interrupts (particularly those from a real-time clock) to trigger sizeable sequences of code can be implemented better using a multitasking approach.

6.1.4 Multitasking

The concept of multitasking is really quite straightforward: a number of programs (we will call them "tasks") appear to be running on the same CPU simultaneously. This is achieved by allowing each task to run for a short while and then interrupting it, saving its machine registers' contents, loading the registers with the data for the next task, and continuing processing. The stimulus for the task swap is possibly a real-time clock interrupt; this is called *time-slicing*.

Writing a scheduler to implement time-sliced multitasking is not, in general, very difficult and may be well worthwhile if it provides a simple means of dividing up the CPU's power in a predictable manner. A further consideration is the use of software team members. Since each task can be quite independent from the rest, assigning different tasks to each team member can be an efficient way to divide up the work.

If the multitasking application becomes too complex (e.g., multiple task priorities and mailbox communication), you should consider the purchase of a commercial real-time operating system (RTOS). (Another article in this book, "Bring in the Pros—When to Consider a Commercial RTOS," looks at this selection process in more detail.)

Implementing an application using multitasking has distinct challenges associated with it. A compiler must be capable of generating reentrant (i.e., sharable) code, as it is very common for functions to be shared between many tasks. Unlike "real" computer operating systems, it is uneconomic for each task to have a complete, private copy of all the code. The programmer is also responsible for generating reentrant codes. Only automatic (and register) variables will be replicated for each instance of the code; the static storage of data must be avoided. Also, any library functions that are called by shared code must themselves be reentrant. Lastly, to keep track of the current task, a machine register may need to be reserved for use by the scheduler.

Reentrant code generation, reentrant libraries (within the bounds of ANSI compliance), and register reserving are all supported by most embedded C compilers.

The debugging of a system using an in-house implemented task scheduler has its own set of challenges. Primarily, these difficulties relate to task interaction and the tracking of shared code. Since a conventional debugger has no "knowledge" of the scheduler, it cannot assist with debugging any aspect of task control or interaction. Furthermore, even something as simple as an execution breakpoint is a problem, if it is set on some shared code. When the breakpoint is hit, the code could have been executing in the context of any task and may not be the task currently being debugged. (Further articles in this book, "Debugging Techniques with an RTOS" and "A Debugging Solution for a Custom Real-Time Operating System" address this topic in more detail.)

6.1.5 Using an RTOS

For a more demanding application, where task scheduling and interaction are very complex, a commercial real-time operating system (RTOS) is needed to marshal the power of a modern high-end microprocessor. Such systems typically provide multipriority task scheduling, sophisticated intertask communications, dynamic task creation, and deletion and memory allocation. All of this is performed in a manner which is appropriately fast and predictable for a real-time application.

An RTOS package is likely to include a number of optional components beyond the kernel, such as protocol stacks, file systems, and graphics. Also, a debugger, tailored to the needs of the RTOS, is essential.

6.2 Visualizing Program Models of Embedded Systems

This article, based upon one by Richard Vlamynck in NewBits in 1996, takes another look at the core features of real-time systems. The use of the term "model" is interesting, as this term gets used in a variety of ways. Here it refers to an approach to system implementation. Richard uses a novel "question and answer" format, which I have left unchanged.

CW

6.2.1 Which Programming Model Is Best for Building Real-Time Systems?

The answer depends on the kind of system being built. There are two basic programming models for building real-time systems, each with its advantages and disadvantages. The first model views the real-time application as a single thread of execution, while the second model views the application as multiple threads of execution.

6.2.2 What Purpose Do Models Serve for a Real-Time System?

A model is a preliminary work or construction that serves as a plan from which an end product is drawn; models aid in testing and perfecting the final product. More importantly, the model describes the system, accounts for its known properties, and can be used for further study of its characteristics. The real-time engineer uses program models to develop software and hardware tuned to the real-time system under consideration. A model allows the engineer to predict and meet performance constraints and functionality objectives.

6.2.3 What Differences Exist Between Models, and What Gains Require What Sacrifices?

Some programming models allow the engineer to write programs more easily, but such programs are harder to debug. In other cases, the program can be harder to write, but the system is easier to debug. Some models allow the programs to run faster, but at the cost of consuming more memory resources. Still others are more robust and require less maintenance.

6.2.4 What Is the Single-Threaded Programming Model?

The single-threaded programming model has many uses for building small real-time systems. Write a main() function that calls other functions and does work. The program executes instructions one at a time—a "single thread of execution" that can be traced through the program.

6.2.5 What Are the Advantages and Disadvantages of the Single-Threaded Programming Model?

Single-threaded models are usually fast and simple to program and reprogram. New functions plug in easily, changing the nature of the machine's response. The disadvantages of a single-threaded program lie in its restricted usefulness in limited application domains; it is not easily reprogrammed "on the fly." This makes it difficult to adapt a running system to a different behavior or a different environment.

6.2.6 Is a Polling Loop a Single-Threaded Program?

Yes (see Figure 6.1). Visualize the function main() coded as an endless loop, running forever in an embedded system. This programming model applies to certain classes of real-time systems. Consider a simple petrochemical blender that opens or closes valves in response to some other control program. In such a case, a simple polling loop sufficiently implements the system. A polling loop requires little effort to write, but demands resources to maintain

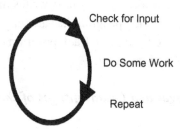

Check for Input

Do Some Work

Repeat

Figure 6.1
Polling loop visualization

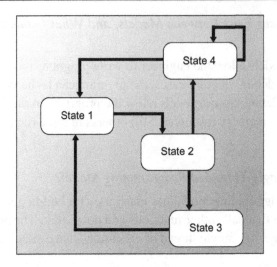

Figure 6.2
State machine visualization

if the system gets too complicated. Accurate timing predictions based on the model of a polling loop are difficult for embedded systems.

6.2.7 Is a State Machine a Single-Threaded Program?

Yes (see Figure 6.2). A state machine, though single threaded, has a greater complexity. A state machine is generally a "Moore machine" or a "Mealy machine"; either is programmable in software. This programming model yields a real-time system that is more difficult to write than a polling loop but that is usually easier to maintain as the system gets more complex.

6.2.8 What Is a Multi-Threaded System?

Visualize a multi-threaded system as a multiple number of polling loops or state machines, all potentially running at the same time (see Figure 6.3). A multi-threaded system advantageously lets the embedded systems engineer break up a system in different ways than would be possible with a single-threaded programming model. An engineer can conceptualize parallel rather than linear flows of work. The engineer writes multiple programs that can all potentially run at the same time. Each program has an entry point similar to the one that `main()` possesses in a single-threaded program. Each program can be coded as an endless loop or state machine or merely work and then exit.

6.2.9 What Are the Advantages and Disadvantages of the Multi-Threaded Programming Model?

The multi-threaded model allows the engineer to divide the system's work into logical sections, and then write an independent program to process that work. All programs

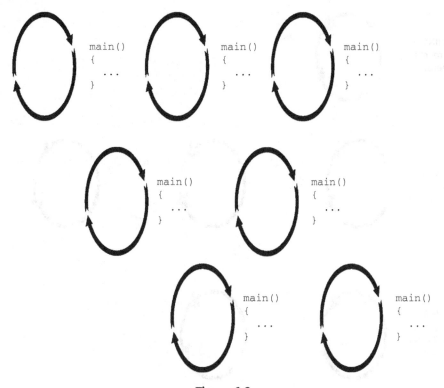

Figure 6.3
Multitasking visualization #1: many tasks, all with the potential to run at once

potentially make progress in parallel. The engineer can introduce new models of communication and cooperation among the program tasks due to higher throughput. This must be done with care, as race conditions might be introduced.

6.2.10 Can Multiple Threads of Execution Really Run Simultaneously on One CPU?

No. All threads appear to execute simultaneously, but at the microsecond level, each program runs for an instant only, alternating with some other program which runs for an instant, and so on (see Figure 6.4). All programs make some progress through their code.

6.2.11 How Can a Multi-Threaded Environment for a Real-Time System Be Acquired?

Broadly, you have a choice between implementing it yourself and obtaining a commercial RTOS. Take a look at the articles in Chapter 7 of this book where you will find plenty of guidance in making this choice.

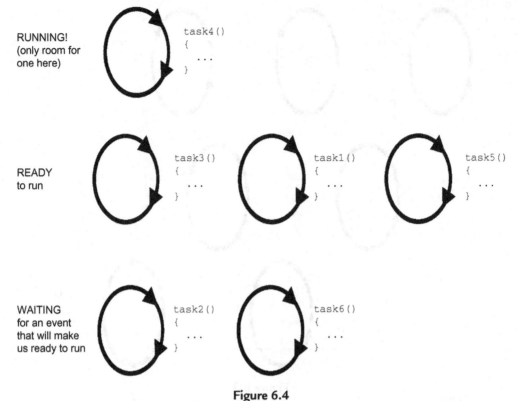

RUNNING!
(only room for
one here)

```
task4()
{
    . . .
}
```

READY
to run

```
task3()
{
    . . .
}
```

```
task1()
{
    . . .
}
```

```
task5()
{
    . . .
}
```

WAITING
for an event
that will make
us ready to run

```
task2()
{
    . . .
}
```

```
task6()
{
    . . .
}
```

Figure 6.4

Multitasking visualization #2: one task running and others waiting for an event

6.3 Event Handling in Embedded Systems

In this article, based upon a piece by Richard Vlamynck in NewBits in 1996, we will look at the basics of a real-time embedded system. Again, we see Richard's interesting "question and answer" style.

CW

6.3.1 What Is an Event?

An event is anything that needs the attention of the embedded system. Embedded systems exist to service events. When an event occurs in the real world, an embedded system must recognize the event and take some action in a timely fashion. Embedded systems are often called *event-driven* systems.

Consider if a thermocouple needs to input an updated temperature to the embedded system, or perhaps a transducer must interrupt and let the embedded system know that some pressure vessel has reached an upper or lower limit. At a higher level, an event may signal the fact that some software function has made an error, and an exception needs to be raised.

6.3.2 Is a Signal the Same Thing as an Event?

In a general sense, the words "event," "interrupt," "signal," and "exception" all refer to the same thing: something important happened and we have to take care of it. At a high level of abstraction, a UNIX software signal is the same thing as a hardware interrupt: something happened "out there," and our system needs to respond to that time-critical event.

6.3.3 What Events Are the Most Time Critical?

Usually, the most time-critical events in real-time embedded systems are associated with exceptions. In this case, we will be less abstract and more specific and take the meaning of the word "exception" to mean only those events that are handled directly by the CPU (microprocessor) hardware.

6.3.4 What Does the Microprocessor Do When it Detects an Exception?

When an exception occurs, the microprocessor hardware itself stops the program thread that was running and starts running a different program thread. Generally, the microprocessor hardware selects the new program thread from an array of pointers to "exception handlers." This exception handler table is located by a pointer in a special register in the microprocessor. Because the response to an exception is mostly hardwired by the microprocessor itself, that response can occur very quickly, usually on the order of a few microseconds. This timely response is exactly what the embedded systems engineer wants.

6.3.5 Are All Exceptions the Same?

Exceptions are broadly classed as synchronous or asynchronous exceptions. From the point of view of the embedded control program, an exception that is synchronous occurs as a result of something that it did, and an exception that is asynchronous occurs as a result of something that someone else did.

6.3.6 What Is a Synchronous Exception?

A synchronous exception occurs when the program thread that is running either causes some error to occur or deliberately causes a trap. The two most classic program errors are when a thread tries to calculate the results of a bad arithmetic expression, causing a divide by zero exception, and when a thread de-references a bad pointer, causing a bus error exception. In each case, the microprocessor stops that thread and vectors to a new thread that will service the exception. Some software systems provide the means to trap to system services; in this case, a program thread will execute a machine instruction that deliberately causes a synchronous exception associated with an exception handler which will invoke some system service.

6.3.7 What Is an Asynchronous Exception?

An asynchronous exception occurs when the program thread that is running gets interrupted by some external event. That is why an asynchronous exception is usually referred to as an *interrupt*. A classic example of an interrupt is when data arrives on some input/output (I/O) channel. The program thread has no idea when data may be received or transmitted from any number of I/O channels, so the embedded control program (program thread) just runs along doing its work and depends upon the interrupt mechanism to service events that happen on an unpredictable basis.

6.3.8 How Do Interrupts Get Generated and Serviced?

Most microprocessors have a special input pin or pins which are used by external hardware to signal the microprocessor that some external event needs attention. These are the interrupt pins. For example, when a data packet arrives on a network device such as an Ethernet chip, the device usually does some preprocessing of the packet and then applies a signal to an interrupt pin on the microprocessor to get its attention and interrupt whatever software is currently running. The microprocessor then saves the program thread that was running and vectors to the interrupt handler which is associated with the interrupt that was generated by the Ethernet chip. The Ethernet interrupt service routine runs to completion, doing whatever work needs to be done to take care of the arrival of the packet from the network. When the interrupt service routine is finished, it executes a special instruction which signals the microprocessor that it wants to return from exception. The processor-dependent return instruction will perform some hardwired machine instructions, which will return control flow back to some other (usually non-interrupt handler) program thread.

6.3.9 What State Gets Saved by the CPU?

When we say that the microprocessor saves the program thread that was running, we mean that it saves only enough machine state in order to start running that thread again later. Typically this means that it pushes the values of registers and maybe other data onto the stack.

6.3.10 Is the Machine State the Same as the Thread State?

The machine state that is associated with a thread could be much less than the full state of that thread. For instance, the machine state saved by the microprocessor interrupt mechanism may be only the general CPU registers and the program counter, but the full state of a thread (in a multi-threaded environment) may also include things like the thread-ID number, its priority, and the ID numbers of the system resources owned by that thread. This extra thread state may or may not have to be saved during exception handling, depending upon how well the kernel or operating system was optimized for a particular microprocessor.

We should now understand basically what happens when an exception fires: the hardware saves some thread state and vectors us to an exception handler. We now need to address the two related issues of how to implement the exception handler and how much work it should do. Today, an embedded systems engineer should have the choice of low-, intermediate-, and high-level languages, which are assembly, C, and C++, respectively.

6.3.11 Should I Write Exception Handlers in Assembly Language or in C?

Most exception handlers are implemented in either assembly or C language, for reasons of expediency. This brings up the related topic of how much work should be done in the exception handler. Interrupts need to be handled very quickly, and you never know how soon you will get the next interrupt, so the less work you do in the interrupt service routine, the better off you are.

6.3.12 How Do I Avoid Doing Work in the Exception Handler?

One way of avoiding work in the interrupt service routine (or the synchronous exception handler) is to acknowledge the event that caused the exception by servicing the interrupting hardware device so that it squelches the interrupt signal and then defers all the rest of the work associated with that interrupt to a "higher" level of software. In the Ethernet example above, this could mean sending the freshly input data to a task for processing.

What this means in the world of an RTOS is that an interrupt or synchronous exception handler can post an event to a system object and wake up a high-priority task. Thus, a system object (such as a mailbox, queue, or event flag) can be used to associate an exception with a particular task. An interrupt handler can get in and out very fast because of the services provided by an RTOS.

6.4 Programming for Interrupts

Interrupts are an aspect of programming which may not be familiar to every software engineer, but embedded developers need at least a basic understanding of the concepts. This article is based upon one of the "Byte Site" pieces that I did for New Electronics magazine in 1996.

CW

Most embedded systems need to work in real time; they need to respond to external events in a timely fashion. There are a number of ways that such a system may be implemented:

1. The code may be written as a large loop, which continuously goes around checking inputs and setting outputs. For a simple system, this is by far the best approach. For a more complex requirement, it may be less appropriate, as the code could become less readable and maintainable and the servicing time for inputs may be extended.
2. The program may be implemented as two loops, with control being exchanged between them (co-routines). This yields more readable, more responsive code at the expense of initial programming complexity. This approach is a first step toward the implementation of a multitasking scheme.

3. A multitasking operating system may be used, which permits the code to be divided into a number of discrete programs that appear to run concurrently. This model is advantageous for more complex systems, where the functionality may be readily divided up into distinct processes. This structure can also ease the division of programming labor across a team of engineers.
4. Non-time-critical code may be placed in a loop, with the input/output response managed by interrupts.

In the remainder of this article, we will look at the last of these options and consider the programming requirements for handling interrupts.

For some devices, interrupts are referred to as "exceptions." In general, these terms may be considered to be synonymous.

6.4.1 Setting up Interrupts

Given that the hardware has been correctly configured to trigger an appropriate interrupt upon detection of an external event, three aspects of interrupt handling software must be considered:

*   Interrupt service routines
*   The interrupt vector
*   Initialization and enabling of interrupts.

6.4.2 Interrupt Service Routines

The code executed as a result of an interrupt being serviced is called an "interrupt service routine" (ISR). This routine is similar to a normal subroutine (or C function), except for two special requirements:

*   An ISR must save the "context" (contents of CPU registers, etc.) of the code which was running when the interrupt occurred and restore it when it has finished.
*   At the end of an ISR, instead of the usual "return from subroutine" instruction, a special "return from interrupt" instruction is commonly required.

These special requirements, together with the need to maximize execution performance of an ISR, have resulted in it being common to write the code in assembler. Nowadays, however, the quality of code generated by compilers is perfectly adequate for use in most ISRs. Additionally, compilers often implement the `interrupt` modifier (as an extension to ANSI C), which enables a function to be compiled as an ISR.

Here is a simple example:

```
interrupt void alpha(void)
{
    indata = device_data;
```

```
        device_control = 0×80;
        ei();
    }
```

The function `alpha()` is implemented as a very simple ISR, which just reads from a data register, writes to a control register, and returns (reenabling interrupts). The "function" `ei()` may actually be a #define macro, which expands into an assembler insert; this minimizes runtime and stack overhead. Note that `alpha()` is a `void` function, as it has no means to return a value and has a null parameter list, as it cannot receive parameters from anywhere.

6.4.3 Interrupt Vector

Most microprocessors support multiple interrupts. The usual way that these are managed is to have a table of addresses of the ISRs, the interrupt (or exception) vector.

Traditionally, all coding that appertains to interrupts has been performed in assembler. However, this is rarely necessary and a high-level language (C or C++) may generally be used. The interrupt vector may be coded in C as an array of pointers to functions, thus:

```
void (*interrupt_vector[])()=
{
    alpha,
    beta,
    gamma
};
```

This interrupt vector has three entries, which point to three interrupt service routines, implemented as C functions: `alpha()`, `beta()`, and `gamma()`. All that is necessary is to ensure that the array `interrupt_vector` is located at the appropriate address, using the linker. Although the C syntax for pointers to functions is arcane, this code is easier to read and maintain than its assembly language equivalent.

6.4.4 Initialization

It is frequently necessary to initialize I/O devices to function in "interrupt mode"; the precise requirements vary from one device to another. This process is normally just a matter of writing a few values into control registers, which may be readily accomplished in C.

Most microprocessors also require interrupts to be enabled by executing an "enable interrupts" instruction. In C, this may be achieved by calling a library function, or, better, by including some inline assembler code (e.g., `asm("EI");`), if this facility is supported by the compiler.

6.4.5 Conclusions

Interrupts are an important facet of a real-time program. Programming to accommodate them need not be a complex task, being readily accomplished in C using modern development tools.

Real-Time Operating Systems

Chapter Outline

There is nothing mandatory about using an RTOS, but most larger embedded systems tend to include a kernel or OS of some kind. I draw a distinction between real time operating systems, which are covered here, and desktop-derived operating systems, like Linux, which are covered in a later chapter. The articles in this chapter look at various aspects of using, choosing, or implementing an RTOS and take glimpses at the internal workings.

7.1 Debugging Techniques with an RTOS

Debugging is a very broad subject. Typically, embedded software engineers appreciate only a small subset of the gamut of possibilities. In particular, debugging with a real-time operating system presents a wide range of options. I have made presentations on this topic at numerous seminars and conferences and authored many papers and articles. A piece that I did in NewBits in early 2003 provided the basis for this expanded article.

CW

7.1.1 Introduction

Most modern embedded system designs make use of an RTOS. This permits the software designer to employ the multitask paradigm to distribute the available processor resources across the required functionality. At the same time, the opportunity arises to distribute the detailed design, programming, and debugging effort across a team.

Embedded Software: The Works. DOI: 10.1016/B978-0-12-415822-1.00007-6
243

An RTOS may be developed in-house or it may be a commercial product, licensed for use in the design at hand. In either case, the multitasking environment introduces some new challenges. Not the least of these is debugging.

This article aims to provide a complete overview of the issues, scope, and limitations of debugging when an RTOS is in use.

The term *task* is used throughout. This may variously be translated into *thread*, *process*, or *program*, depending upon the specific RTOS in use. The precise semantics are, for the most part, not relevant in this context.

7.1.2 Multiprocess Concept

The key function of an RTOS is to allocate time to each task in the system in a rational way. There are a variety of mechanisms whereby this is achieved. The simplest is a round-robin, where each task simply passes control on to the next. The next possibility is some kind of time slicing, where the RTOS allocates an equal-sized period of time to each task. Commonly, an RTOS has a much more complex facility, where each task has a priority and is allocated resources accordingly.

The perception of this allocation of processor time is important in the use and design of debugging solutions.

Apparent Simultaneity

Superficially, it would seem logical to keep in mind the mechanism by which the RTOS works: quickly swapping between tasks, so that they give the appearance of executing simultaneously (see Figure 7.1). After all, this is a reflection of what is really happening. However, this model does not prove to be very useful. With the possible exception of simplistic round-robin schedulers, it is never really possible to predict which task will run next. So a programmer cannot profit from this model.

A debugging solution, built with this concept in mind, is likely to be "task-aware"—primarily focused on the current task and the need to isolate it from the rest of the system. This is rather inflexible.

True Simultaneity

Even though it is not a true reflection of reality, it proves much more useful to conceive of a model where all the tasks are running simultaneously (see Figure 7.2). This is what most programmers practice because it is the only way to write secure code with proper management of shared resources.

A debugging solution, designed with this mind-set, lends itself to multitask debugging—viewing a number of tasks and their interactions—which is much more flexible. Furthermore, the model is scalable upward to a system that contains multiple processors, which really do execute concurrently.

Figure 7.1
Apparent simultaneity

Figure 7.2
True simultaneity

7.1.3 Execution Environment

The conventional, and only really viable, configuration for debugging an embedded system with an RTOS is to connect a host computer (generally a PC, running the debugger) to the target device. The target may be real hardware or perhaps a host-based simulation.

Simulation
A simulation of the target device, running on the host computer, may be useful in a number of circumstances:

- Before the hardware is ready
- Before the hardware is reliable
- Before the hardware is available in volume.

Two types of simulation are available when debugging with an RTOS:

- **Instruction set simulation:** Where the simulator accepts the real code, built for a real target, and simulates its execution instruction by instruction. The execution speed is relatively slow, but it is a very precise environment with great opportunities for nonintrusive "instrumentation" of the code (i.e., monitoring the detailed execution behavior of the code, without actually changing it). Although slower than real time, the temporal model may be accurate and has the unique possibility for "stopping real time." Since the entire functionality of the processor is simulated (perhaps along with some of the surrounding hardware), there is no reason why a complete system—RTOS, drivers, and application code—cannot be run.

- **Native system execution:** Where the code is built to actually run on the host computer. This results in a very good execution speed—probably similar to or faster than the real target hardware would be—but with a different real-time profile. A couple of alternative approaches may be taken to facilitate host execution of the target code: a special version of the RTOS runs as a host process; alternatively, the RTOS calls may be mapped onto calls to the host OS.

Hardware Available

At some point, real hardware becomes available and/or limitations of simulation are reached. The hardware may be:

- Prototype hardware
- Production hardware
- Evaluation boards (similar architecture to final target).

Although apparently better than simulation, using actual hardware is not without problems— hardware glitches and faults and visibility of execution are among them. So, even when hardware is at hand, simulation may remain useful to address certain types of bugs.

If target hardware is to be used, a suitable connection to the host computer must be established.

7.1.4 Target Connection

In broad terms, there are two types of host/target connection that may be employed for debugging:

- **Dedicated debugging connection:** The link has no additional function, other than debugging.
- **Communications link:** May be used for other purposes as well as debugging.

In some cases, both may be useful.

Dedicated Debugging Connection

The use of a host–target connection, which is dedicated to debugging, is becoming quite common. Typically, this is based upon the JTAG standard, which, while never intended for this purpose, has been found to be ideal for many such applications. Prior to JTAG becoming common, Motorola (now Freescale) introduced Background Debug Mode (BDM), which worked in a similar way. Almost all new processor designs incorporate a JTAG interface for debugging.

A JTAG connection is quite cheap and easy to incorporate into a design. There are only a small number of signals—it is a synchronous serial protocol—and low-cost connectors are used. Some kind of adapter is required to interface to the host computer. This may be a simple "intelligent cable," costing $100–1,000; it may be an interface box, which includes other functionality (e.g., trace), costing $1,000–10,000.

A major advantage of this type of debugging connection is that it will even function with a target that is barely working. So long as the processor has power and a valid clock, a JTAG connection may normally be established. Of course, little progress may be made until the memory system is also functional.

Typically, a target processor is "frozen" (i.e., not executing code) while it is communicating with the host over a JTAG connection.

Communications Link

Historically, the use of a conventional communications link for debugging was very common. Generally this would be a serial line (RS232, RS422) or Ethernet. However, there are many other viable possibilities: USB, PCI Bus, Bluetooth, IRDA, and 802.11.

A key, but obvious, requirement is a working target. The hardware must be fully functional—the processor must be capable of executing code and the communications interfaces working. The software must, of course, be working to the extent that the communications protocols can be supported. This leads to an interesting "chicken and egg" issue: if this is the debugging interface, how can the drivers and protocol stacks be debugged? The answer is a combination of simulation and the use of a dedicated debugging connection, which is ideal for such "low-level" debugging.

This type of RTOS-aware debugging is necessarily more complex. This need not be a great concern to the user of a commercial RTOS product but would probably discourage the user of an in-house kernel.

If the communications link is required for the application, sharing it for debugging purposes may be very attractive. Otherwise, the overhead of the additional hardware and software may be disproportionate to the benefits gained.

The big advantage of RTOS-aware debugging over such a link is flexibility. The target may continue to run while the debugger communicates with it. This is called *run-mode debugging*.

7.1.5 Debugging Modes

There are broadly two modes of RTOS-aware debugging: stop mode and run mode. Stop mode is almost always an option. The availability of run mode (as well) may be advantageous.

Stop Mode

This mode, which is sometimes also called *freeze mode*, occurs when the entire system is stopped—when no code executes at all—when a breakpoint is hit or the operator intervenes.

Stop mode is a totally host-resident debugging implementation, with no overheads on the target at all. Typically, the target connection is JTAG.

Stop-mode debugging is quite satisfactory for most situations and is only problematic when:

- The bug being sought is closely tied in with the dynamic interaction between tasks.
- Halting the system, as a whole, would cause disadvantageous side effects (that may hamper locating the bug).

This debugging mode is ideal for use with an in-house RTOS because no target support is required, and the implementation may be quite straightforward.

With simulation, stop-mode debugging is sufficient for any conceivable multitasking debugging scenario.

Run Mode

Run mode is a more sophisticated RTOS-aware debugging setup because it allows just parts of a system to be stopped. For example:

- Just the current task stops (others continue execution).
- A defined group of tasks stop (others continue execution).
- All tasks stop (but interrupts are still serviced).

This facilitates some subtle debugging, which may not be possible with stop mode.

Run mode needs sophisticated software support on the target—a debugging monitor. This is most likely to preclude its use with an in-house RTOS. A memory overhead is also incurred.

Run mode requires a working communications link to the host computer. This may be a further overhead and a debugging challenge in itself (to get the communications working). Use of JTAG is an outside possibility, but the host to target communication interrupts code execution (in a less controlled way that with the use of interrupts by, say, an Ethernet driver), which may compromise real-time integrity.

7.1.6 RTOS-Aware Debugging Functionality

RTOS awareness may mean different things to different people and is very dependent upon the tools and RTOS in use. The two key areas of functionality are the ability to view data, in an RTOS-aware fashion, and the control of RTOS functions and objects.

Viewing

All of the information concerning the status of a running system is contained in its memory. Meaningful viewing of this data is simply a matter of knowing where it is located and how it is structured and formatted.

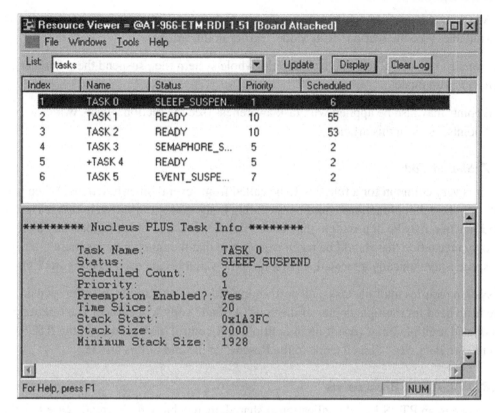

Figure 7.3
Task information display

For tasks, it is necessary to be able to view a list of tasks in the system, with their running state (typically "current," "ready," "blocked," or "suspended") and priority (see Figure 7.3). For each individual task, it is desirable to be able to view local variables, registers (including the program counter), and stack.

Lists and information about other RTOS objects (e.g., mailboxes, queues, semaphores, mutexes, and messages) are also required.

Controlling

A task-aware debugger may also exercise some control over the target system. There are various possibilities for controlling tasks:

- A task may be suspended—not available to be scheduled.
- A new instance of a task may be spawned (in an RTOS that supports dynamic task creation).
- A task's priority may be adjusted.

- A task may be paused, while others continue to execute (if run-mode debugging is available—see the section "Run Mode" earlier in this article).

Additionally, it is useful to be able to stop the whole system (i.e., suspend the scheduler), when using run mode.

Breakpoints may also be applied with task awareness (see the section "Task-Aware Breakpoints," later in this article).

7.1.7 Shared Code

In C, it is very common for a function to be called from several other functions. When using an RTOS, the calling functions could be executing in the context of different tasks. Thus, the called function may be in use more than once "simultaneously," which is not really a problem. The only precaution that should be taken is to ensure that the function is reentrant—i.e., all its data space is dynamically allocated, so that no static variables are used (see Figure 7.4).

If a system requires multiple tasks, all performing the same operations on different data, there is no need for multiple copies of the code. A single copy will suffice. The task may be instantiated multiple times (resulting in multiple task-control block entries in the RTOS), each with its own data. See "Shared Code to the Rescue" at the end of this article.

7.1.8 Task-Aware Breakpoints

If any code in an RTOS-based application is shared, then debugging is greatly eased by the availability of task-aware breakpoints. Such a breakpoint is similar to any other, except that it is qualified by the identifier of the task (or maybe tasks) in which it is required to be active.

In fact, a breakpoint can (almost) never be truly "task aware." The apparent awareness is the result of the action taken by the debugger when a breakpoint is hit. A decision is made as to whether execution should continue (the current task is not the one in which the breakpoint is required) or stop.

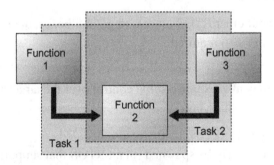

Figure 7.4
Shared code

This task-context decision always represents a (usually small) time overhead, which has various levels of impact, depending upon its implementation. This overhead is called *breakpoint latency*. Broadly, the task-context decision may be made on the target (by a monitor program running there) or on the host (by the debugger itself).

Target
If the task context is evaluated on the target, the breakpoint latency is reduced. The delay may actually (only) occur at the time a breakpoint is hit, or it may impact the task-context switch each time the task is scheduled. This is an implementation detail.

In either case, a monitor program is required in the target, which may be quite complex to implement. However, the resulting facility is most flexible.

Host
The alternative is for every breakpoint to halt the target and for the task-context decision to be made by the debugger. Even if this decision is made very quickly, the breakpoint latency will tend to be higher because of the host–target communications.

With this approach, no support is required at all on the target, and the coupling between the debugger and the RTOS is much looser. Since the implementation is quite simple, it is an ideal choice for an in-house RTOS.

Breakpoint Scope/Action
It is important to differentiate between the scope of a breakpoint and the action that it will take.

A breakpoint's scope is the task or tasks in the context of which it will be activated. This is only relevant if code is shared between tasks. Typically, a breakpoint may have the scope of a single task, several instances, or all instantiations of the shared code.

The action of the breakpoint is what happens when an active breakpoint is hit. Possibilities include:

- Stop the system (this is the only option if stop-mode debugging is used).
- Stop just the scoped task.
- Stop a number of tasks (normally including the scoped one).
- Stop all the tasks (but interrupts continue to be serviced).

7.1.9 Dependent Tasks

It is very common for task interdependencies to occur in an RTOS-based application.

A simple case occurs when one task is generating data that is sent to another for processing. These tasks are interdependent (see Figure 7.5). The Receiver depends on the Sender for data to process; the Sender is dependent on the Receiver to accept the data that it has

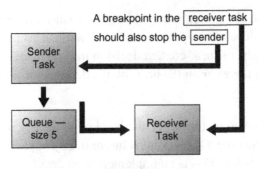

Figure 7.5
Dependent tasks

generated (otherwise, its buffer will fill up). This second case is very relevant to debugging. If a breakpoint results in (only) the Receiver task being stopped, the Sender task will rapidly reach an unstable state.

What is required is a means by which a breakpoint on the Receiver task will also result in the Sender task being stopped. This is accommodated by a synchronized breakpoint.

Synchronized Breakpoints
The idea is simple: when a task-aware breakpoint is set in one task, a list of other tasks is provided; these other tasks should also be halted when the breakpoint is hit. The additional tasks may be other instances of the scoped task or they may be different code entirely (or both).

Synchronized breakpoints can be implemented only when run-mode debugging is in use because the remainder of the system continues to execute.

It is possible to implement this facility on the host or on the target. A host implementation is relatively simple but results in a small delay (latency) while the synchronized task list is processed; this results in each of the additional tasks over-running slightly, which is rarely a problem. A target implementation may be much more complex, but latency is reduced.

7.1.10 Memory Management Units
The use of a memory management unit (MMU), in some form, is common with many modern microprocessors. The necessity of using an MMU may be to implement a simple inter-task memory protection or for the full implementation of a "process" model. The advantages and disadvantages from the design and implementation perspective are widely documented, but the debugging implications are rarely considered.

Simple Memory Protection
An MMU may be used in a relatively simple way to afford memory protection between tasks. Each task is permitted to access only specific areas of memory. Thus, each task is protected from interference by others and its own erroneous behavior (e.g., illegal memory references, stack violations, etc.).

From the debugging perspective, this use of an MMU is not a significant problem. The debugger simply needs to deactivate the MMU to access all of the memory which has no impact upon the integrity of the software itself.

The Process Model

An MMU may be used to remap the memory address space for each task so that it appears to have control of a complete processor. Such a task is commonly referred to as a *process*. It offers greater flexibility and protection, but complicates debugging. The debugger needs access to the remapping information so that it can use the MMU to access the memory space of each task.

It is possible to have a "hybrid" memory model, where each process itself contains multiple threads of execution (as in Windows and Linux). This, in turn, provides the opportunity for a hybrid debugging capability: a debugger, which is associated with a single process in the system and provides thread-aware debugging of its code.

For process model operating systems, it is common for debugging tools to be available for the system and separately for the applications. These may actually be individual tools or a debugger may have system and application modes.

7.1.11 Multiple Processors

The deployment of multiple processors in embedded systems is increasingly common. This may be by means of multiple boards in the system, multiple devices on a board, multiple processor cores on a chip, or any combination. From the debugging perspective, architectural details are relatively unimportant.

The processors may be all the same type—an array of processors may be a useful solution to certain types of problem. Alternatively, there may be a mixture of architectures, where different processor capacities (8-, 16-, or 32-bit) or capabilities (e.g., DSP) are useful.

Debugging, with multiple processors, may be approached in various ways. Multiple debuggers could be employed—one for each processor—but this approach could be complex because each is likely to be a different tool. There may also be connection capacity issues. A single debugger is possible, if the debugging architecture of choice accommodates the "true simultaneity" paradigm. This results in a less-steep learning curve (only one tool) and opens the possibility for advanced debugging functionality, like synchronized breakpoints.

Of course, multiple processor systems are just as likely to use an RTOS and, hence, have particular debugging needs. This can get more "interesting" because multiple RTOS products may actually be deployed in a single system. This would be a challenge for any debugging technology—a challenge but not an insurmountable one. The situation gets even more interesting if symmetrical multiprocessing (SMP) is used, where a single instance of an operating system is run over a number of CPUs. This topic is considered further in a later chapter on multicore embedded systems.

7.1.12 Conclusions

It is clear that debugging, when an RTOS is in use, may be complex business, and the development of suitable tools can be nontrivial. As systems become increasingly complex, the capabilities of the debugging tools need to track the needs of embedded software developers.

Understanding RTOS debugging is important for both developers and managers. Even if the functionality is transparent (because it has be addressed by the RTOS/tools vendor), the developer can use this knowledge to appreciate the possibilities and limitations of the technology. A manager can efficiently deploy resources and make informed purchasing decisions.

Shared Code to the Rescue

An embedded developer contacted their RTOS/tools supplier with a problem: his design had a fixed amount of code memory, which was all used, and further enhancements were required to the application.

The support engineer looked at what was being done. He tried improving compiler optimization, which was helpful, but nowhere near enough extra memory was freed. Then he took a closer look at the application, made a fairly small change to the structure of the code, and reduced the code memory requirements by *nearly 90%*!

The application was some kind of data router that handled 10 channels of data. Each channel was handled identically by its own task. Originally, each task had its own copy of the code. The modification that saved so much memory was to share one copy of the code between all 10 tasks.

This introduced a new problem, however. Previously, a breakpoint, placed in the code of one task, would only take effect in the context of that task. The code was not shared and, hence, only executed in the single-task context. Now the breakpoint would trip regardless of which task utilized the line of code.

The solution was a debugger with task-aware breakpoints, which the RTOS/tools vendor could readily supply.

So the story had a happy ending for all concerned.

7.2 A Debugging Solution for a Custom Real-Time Operating System

Users of in-house real-time operating systems are challenged when it comes to debugging. In a piece written for NewBits in 1998, Robin Smith (who is now with Oracle) and I described how to effectively utilize a debugger with a custom RTOS. I have adapted the material for this article.

CW

A discussion of task-aware debugging is worthwhile when a commercially supported real-time operating system (RTOS) is in use. It is an entirely reasonable expectation that the RTOS vendor is able to provide a complete debugging solution with a choice of host- or target-based technology. If an in-house designed ("custom" or "home brew") RTOS is being employed, it is unlikely that the development of a custom debugger is also feasible.

The only real possibility is to select a debugging technology that allows significant user-customization. This necessitates a host-based approach, but this is generally quite satisfactory and a great improvement on a total lack of task-aware debugging.

The remainder of this article will describe the basic details of such an implementation.

7.2.1 Implementing Task-Aware Debugging

To illustrate an example of the implementation of task-aware debugging, we need to select a target microprocessor, a debugging technology, and define an RTOS.

Example Target Device
No specific processor is considered because this is not necessary for the discussion. The assumption is simply made that the device has a 32-bit architecture. None of the ideas presented are at all specific to any particular device.

Example Debugger
The specific debugging technology employed is also unimportant. For the purposes of describing RTOS-awareness implementation, we have assumed that the debugger simply has a scripting facility that looks a lot like C language.

Example RTOS
Obviously, every custom RTOS is different, and it would not be useful to describe a particular example here. However, it turns out that there is very little that we need to know in order to present the ideas.

For the purposes of this article, assume the following about the RTOS:

- Each task has a task-control block (TCB), which describes characteristics of that task and holds its current status.
- The format of the TCB is illustrated in Figure 7.6.
- The TCBs are assembled into a table that is a doubly linked list with NULL pointer terminators.
- The names of pointers to the start of the TCB table and to the TCB table entry for the currently running task are TCBList and TCBCurrent, respectively. And these symbols are accessible to the debugger.

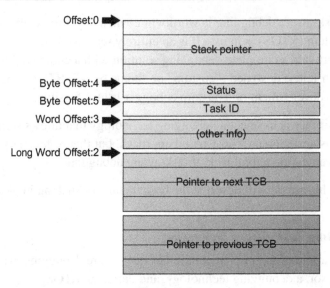

Figure 7.6
TCB layout

7.2.2 Task-Awareness Facilities

In this article, the implementation of task awareness will be limited to two specific facilities:

- Displaying status of all the tasks in the system
- Task-aware breakpoints.

Given an understanding of these two facilities, you will be well equipped to design and implement similar and additional facilities for your custom RTOS.

Task Information Display

To implement a task information display, the first requirement is a set of utility scripts that extract the necessary information from TCB entries:

```
void *GetNextTCB(TCB)
int *TCB;
{
    return(*(TCB + 032));
}
```

This returns a pointer to the next TCB in the chain, which enables the list to be "walked." It returns a NULL pointer when the end of the list is reached. A similar script could be implemented to traverse the list backward, if that were required.

```
int *GetStackPointer(TCB)
int *TCB;
{
    return(*(TCB + 0×0));
}
```

This returns the value of the task's stack pointer. This information can be used to gain access to the context information (registers, etc.) of the task, which were pushed onto the stack, noting that it is assumed that stack entries for this processor are aligned on integer boundaries (32 bits in this case).

`GetID()` and `GetStatus()` each return other useful information about the task:

```
char GetID(TCB)
unsigned char *TCB;
{
      return(*(TCB + 0x5));
}

      char GetStatus(TCB)
      unsigned char *TCB;
{
      return(*(TCB + 0x4));
}
```

The task information display command is implemented using a further script:

```
int GetTaskInfo()
{
      unsigned char id;
      unsigned char status;
      char *pp;
      unsigned int sp;
      void *tcb;
      tcb = TCBList;
      printf("ID TCB Status StackPointer\n");
      printf("-- --- ------ ------------\n");
      while(tcb != 0)
      {
          id = GetID(tcb);
          status = GetStatus(tcb);
          if (status == 0)
              pp = "Ready";
          else
              pp = "Wait";
          sp = GetStackPointer(tcb);
          printf ("%02u %08X %s %08X\n",id,tcb,pp,sp);
          tcb = GetNextTCB(tcb);
      }
}
```

This makes use of the utility scripts to traverse the TCB table and display information about each of the tasks in the system.

A useful enhancement would be the addition of some code to detect when `tcb` has the value of `TCBCurrent` and maybe add a "*" to the listing line. This would highlight the currently running task.

Task-Aware Breakpoints

With many debuggers, when a breakpoint is set, a script may optionally be associated with it. When the breakpoint is hit, the script is executed. If the script returns a value that may be interpreted as TRUE (nonzero), execution continues when the script has completed. The user may be quite unaware that a breakpoint has been hit. If the script returns FALSE, execution is suspended in the usual way.

This script may be used to implement task-aware breakpoints:

```
int fortask(id)
int id;
{
    unsigned char taskid;
    void *tcb;
    tcb = TCBCurrent;
    taskid = GetID(tcb);
    return(!(id == taskid));
}
```

Every task has a unique identifier (ID). The task ID is passed as a parameter to the script. Using the current task TCB pointer, TCBCurrent, the ID of the current task is compared with that specified; execution is suspended only when they match.

This script could be enhanced to issue an appropriate explanatory message when it stops the execution. Optionally, a facility to display instances where the breakpoint was hit, but execution was not stopped (i.e., a different task was active), could be added for diagnostic purposes.

7.2.3 Conclusions

When using any RTOS, it is desirable to have an RTOS-aware debugging solution. If a commercial RTOS technology is being employed, a range of debugging tools should be available from the RTOS vendor. If a custom or noncommercial RTOS is used, a good debugging solution may be implemented using the techniques described in this article.

7.3 Debugging—Stack Overflows

This article is based on a short piece, which appeared in the Winter 2003 issue of NewBits, by Larry Hardin, customer support manager at Accelerated Technology. The ideas tie in well with my article, "Self-Testing in Embedded Systems" in Chapter 3.

CW

Let's talk about one of the types of problems we see in support regarding debugging—stack overflows. We'll discuss a few causes and then propose suggestions, based upon our experiences, to avoid the problem.

First, using `printf()` as a debugging aid. Although common in a Windows environment, this practice may present problems in an embedded system. Be aware that using library calls may bring in other modules to support that call, and `printf()` is no exception. Math subroutines may also be needed. And although it's not as common, the I/O functions provided with some toolkits may still not be reentrant. Keep these two factors at the back of your mind.

Monitor recursion carefully. Sometimes the level of recursion isn't factored correctly and each function call is pushing data onto your stack. Be sure to use stack-checking during development, if your design includes recursion! If your kernel doesn't have a stack-checking capability, write your own. Fill your stack with a preset pattern and monitor your stack fence. Your stack should always have a bit of "extra" space factored in. So, if you see the pattern cleared up to the stack fence, you may well have just blown that stack. Blowing a stack can either be easy to find or a bear to uncover—depending upon what data structure (and what part of that structure) your application just overwrote.

Some kernels incorporate a history-saving capability. Functions make an entry into a buffer at function entry, so a record of execution is formed. This can be a useful tool to show where you've been at crash time. If you aren't using an analysis tool that shows this information, peeking at the memory of the history buffer may help you locate the faulty function/code.

7.3.1 Conclusions

Although the use of multiple stacks with an RTOS can introduce difficult-to-locate errors, some care and forethought can prevent the majority of such problems.

7.4 Bring in the Pros—When to Consider a Commercial RTOS

Selling any product can be a challenge in the modern world. You need to get the right price, features, and performance for your customers, and you need to understand how to defeat your competition. In the commercial real-time operating system business, most of this is still true, but with one twist: sometimes—actually quite often—our customers are also our competitors! I am often asked why one should buy an RTOS, when writing one in-house is a possibility. I have made many seminar presentations and conference keynotes on this topic, and I wrote about my ideas in NewBits in 2003. That piece was the basis for this article.

CW

Although any self-respecting commercial real-time operating system supplier might wish to deny the possibility, writing your own RTOS is an option. It is an option that is exercised by many developers, even though a wide selection of RTOS products is on offer. In this article, we will review the wisdom of the "do-it-yourself" approach and consider how this compares with a commercial RTOS product.

7.4.1 Commercial and Custom RTOSes

Surveys suggest that a significant proportion of current embedded designs are implemented using an in-house, custom RTOS. The reasons for making this choice, despite a great many commercial products being available, are varied. To help understand how such decisions are made, we need to investigate the pros and cons of each approach.

7.4.2 Advantages of a Commercial RTOS

For a commercial RTOS product to be attractive, it must be well established and reliable. Clearly this is largely down to the company that is supplying it. If they are a secure, well-known firm of reasonable size, confidence is increased. Of course, the product should be mature, with a sizeable user base. A "one-man shop" can hardly engender such confidence.

Both the purchaser and supplier of an RTOS are expressing a long-term commitment. The user is committing expertise and investment going forward. The vendor is committing to support new processor architectures and the latest technology, as it comes along.

Documentation is important. A vendor should be quite prepared to provide copies to prospective customers. In past years, there were excuses (e.g., shipping costs, lead times); now a PDF by return email is a reasonable expectation. If the source code is provided, or at least available, it may be regarded as additional documentation (but not a replacement for it!), so long as it is readable and is well commented.

You need RTOS-aware debugging. Stop mode is essential; run mode may be useful too. A commercial product should have an adequate provision for these capabilities.

A commercial RTOS is not just a kernel. All the surrounding technology—the middleware—is likely to be available off the shelf. This includes networking protocols, file systems, graphics, and Java. Of course, these must be options. If a file system, for example, is not required for an application, there is no reason why it should be purchased or be included in the RTOS memory footprint.

A difficult area of any embedded software design is the low-level interface to hardware—the drivers. A commercial RTOS will be provided with a selection of drivers for standard and integrated devices. Although this provision cannot encompass custom hardware, the RTOS vendor will most likely provide tools or templates to make driver development as straightforward as possible.

7.4.3 Commercial RTOS Downsides

The decision to not use a commercial RTOS is influenced by a number of possible downsides. In some cases, they may be valid, but others are misconceptions.

An initial, and very sensible, consideration is often cost. Specifically, for many applications, there are concerns about ongoing license costs. For a high-volume, low-price embedded product, per-unit production costs are critical. It is unlikely that paying a royalty on an RTOS would be acceptable, and a royalty-free RTOS makes sense. However, the business model for a low-volume, high-ticket product is quite different, and a "pay-as-you-go" royalty fee may be quite appropriate. For other applications, other business models (e.g., subscription plans) may be relevant.

The key to success is flexibility. Every embedded system is different, and the RTOS vendor needs to accommodate those differences in their product offerings.

Some engineers are nervous about their lack of internal knowledge of a commercial RTOS. This really should not be a concern. You use a PC without needing to understand the intricacies of Windows; you can drive a car without ever lifting the hood. Why do you need to know exactly how an RTOS works, so long as it performs as specified? It may even be argued that employing staff in possession of this internal knowledge is an unnecessary expense. Of course, the availability of source code may alleviate these insecurities.

Suppliers of any product are keen to get vendor lock-in from their customers, and RTOS providers are no exception. On the other hand, their customers want to maintain flexibility and may feel that such a lock-in is a drawback. Although there is some validity in this concern, migration from one RTOS to another need not be a very big issue. Also, selecting a product that adheres to a standard (e.g., POSIX or OSEK) gives the flexibility to switch suppliers if the need arises.

The most fallacious argument against a commercial RTOS is that it will intrinsically carry the burden of excess functionality. Although it may be true that an RTOS possesses many capabilities that are not required in a particular design, just about every modern commercial RTOS product is scalable—i.e., only the required functionality is included in the final memory image. Take a look at Figure 7.7 to see how this works. Of course, the implementation of scalability may be more efficient in some products than others, and this should be verified. There is the additional argument that this excess functionality is being paid for, which is true but irrelevant because the costs are factored over a great many designs (from multiple customers).

7.4.4 Why Write a Custom RTOS?

The primary motivation for creating an in-house RTOS is to retain control. If it is developed in-house, obviously all the code is owned and internal knowledge is retained. Although this is true, it presupposes that there is a stable team who understands and can maintain the RTOS. Since this requires a specialized skill set, this may be an expensive proposition.

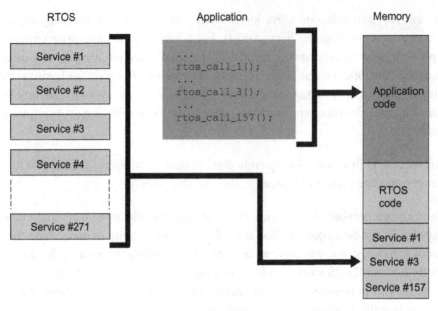

Figure 7.7
RTOS scalability

Of course, there is the argument that there are no ongoing license costs. This is true, but as we have seen, this may also be the case with a commercial, royalty-free RTOS.

Engineers often argue that an in-house RTOS is an exact match to their needs. This argument is hard to sustain because a properly scalable commercial RTOS will scale to precisely meet their requirements. Most often, the functionality of a custom RTOS is limited by what can be achieved within the available time/budget.

7.4.5 Reasons Not to Create a Custom RTOS

There are a great many factors that make the development of an in-house RTOS questionable.

As usual, the analysis starts with money. An RTOS has a development cost associated with it. This cost may be very substantial, but not visible, as it gets "lost" in a project's budget. This is bad management and it is illogical. It is not normal practice to pay for capital equipment and other reusable resources (like an RTOS) on the back of one project, when it may benefit many others (now and in the future).

The costs continue, if you take into account long-term support of the RTOS. This cost may escalate if there is a need to retain specific skilled staff for the purpose.

Given that the costs of developing and supporting an RTOS are understood and accepted, there is another requirement: debugging. Embedded software developers spend a lot of their time debugging. The debugger really needs to have some kind of RTOS awareness. It has been suggested that the cost of developing a debugger from scratch may by 10 times that of developing an RTOS. There may be an option to adapt a commercial debugger to work with a custom RTOS—this should certainly be considered.

Although there may be a clear intention to develop a single RTOS, there is a distinct danger that, in a less strictly controlled development environment, multiple implementations will result. (See the "Once Upon a Time" sidebar on the next page.)

If the requirement is for just a kernel, then writing it may be a contained and manageable task, but will it stop there? It is rare for modern embedded systems to be that simple. There is generally some requirement for additional middleware—protocols, graphics, and so forth. This could be accommodated in three ways:

- Write this too, but that may be a lot of work.
- Persuade a commercial vendor to port to your RTOS, but that would be difficult and expensive.
- Port an open-source product to your RTOS, which could be as hard as the first option.

Whatever device is being used today, it is certain that a different processor architecture will be used in the future. Such a change brings with it many challenges. Since an in-house RTOS is likely to be written using a significant amount of assembly language, its portability is going to be limited. Additionally, such code may include endianity dependencies and will also rely on the interrupt structure of the processor.

A final concern about developing an in-house RTOS is to consider core competencies. If a company's business is developing networking hardware, for example, their goal is to address that business as effectively as possible. They aim to produce the fastest, cheapest, most functional (or whatever) routers on the market. Getting distracted by spending resources developing software, which could be readily obtained off the shelf, is not playing to the strengths—the core competencies—of the company. (See the "Painting the Car Red" sidebar later in this article.)

7.4.6 Conclusions

It is fair to say that, as I am employed by an RTOS and embedded tools vendor, my view is biased. However, it is certainly clear that, in the vast majority of situations, there is very limited justification for deploying a custom RTOS in a new project.

Once Upon a Time

There was a young engineer working on embedded software (although this was so long ago that the word *embedded* had yet to be coined). He was working on small, 8-bit systems built around the Z80.

One day, he was starting a new project and concluded that a small kernel would be useful—just a time-sliced task swapper really. He spent a couple of weeks working on this project, resulting in about 2 K of (assembler) code, which worked well. The project was completed with no problems.

For the next project, he decided to use the kernel again, but he spent a few days incorporating new ideas for enhancements to the kernel into his code. The updated kernel was used in the project, and again, it was completed without undue difficulties.

By the time the third project came along, this engineer felt he knew a lot about kernels and didn't hesitate before including yet more ideas before utilizing his kernel again. Once again, all went well with the project.

At this point, it is reasonable to feel that it was high time this young engineer had a "reality check"; things have been going just too well for him. That's just what happened.

The hardware from the first of these three projects was enhanced, and this meant that revisions to the software were necessary. So our hero got out the listings and considered what was to be done. But he had a shock: what was this strange kernel that someone had used?

Of course, he had been the victim of "creeping elegance." He set out to create a kernel but ended up with three. Sure, they were related, but the urge to "do a bit better" meant they were not particularly compatible.

At around this time, the engineer (who will continue to remain nameless) left the company to join an embedded tools and RTOS vendor, leaving his three kernels behind him. Since he had put some effort into documentation, he did not leave his employers with too big a maintenance problem, but that was more luck than judgment.

Painting the Car Red

In the modern business world, successful companies are the ones that focus on what they do best and excel at it. The embedded systems business is no exception, but we can look elsewhere to find everyday examples. Take the automotive business, for instance.

Henry Ford famously promoted his company's flexibility by offering the Model T in any color, "so long as it's black." This would not work in today's car market, where the choice of color and the quality of the paint finish are key attributes in a car-buying decision.

But how many car makers manufacture their own paint? Answer: none (essentially—there may be the odd maverick). Car makers always outsource it. They are very careful where they get their paint from and specify it in very great detail, but they leave paint making to the paint manufacturers and get on with making cars.

This approach is applied to many aspects of an automobile: the upholstery, the electronics, the suspension, even the engine and gearbox. Some manufacturers license an entire car design from elsewhere if their core competency is finishing and production. We can learn from the car makers.

7.5 On the Move

It is frequently assumed that, having utilized a particular real-time operating system, you are locked in and that changing the RTOS will be problematic. I wrote a piece for NewBits in late 2004, upon which this article is based, to dispel some of the myths.

CW

7.5.1 Migrating from One RTOS to Another

It is inevitable that, at some time or another, an embedded software team will face the prospect of moving from using one real-time operating system (RTOS) to another. This migration may be brought about by many factors, including:

- RTOS availability for new chip architecture
- Customer demands
- Adherence to standards
- Change in technical requirements
- Change in commercial business model.

In this article, we will look at a number of facets of such a migration and its impact on embedded software development. For example:

- How can migration of code be managed?
- What about migration of skills?
- What preparations can be made for this inevitable migration?

In general, these matters are independent of whether the migration is between commercial RTOSes or from an in-house kernel to a bought-in product.

7.5.2 Code Migration

An initial challenge, when adopting a new RTOS, is the body of existing code, which is written against a specific application programming interface (API). Clearly this code has to be modified or adapted in some way to accommodate the API of the new RTOS.

Generally, this is done by hand, which may be quite satisfactory if the application is small or the code that interacts with the RTOS is localized. The difficulty of this task is also strongly affected by the nature of the old and new RTOSes. If they have similar APIs, the job will be easier. It may also be simplified if the "target" RTOS has a rich, orthogonal API because this is intrinsically more expressive.

Some tools are available to assist with this endeavor, but their availability is limited to specific RTOS migration paths. A creative way of addressing this problem is the use of a "wrapper."

7.5.3 Wrappers

A wrapper is simply a layer of software between the RTOS and the application code. It exposes an API and maps calls to this interface onto the actual API of the underlying RTOS. This process is illustrated in Figure 7.8. Here, call 1 to an old RTOS is mapped onto call a of the new one; similarly, call 157 is mapped onto call x. Call 3 in the old RTOS does not have an exact equivalent in the new RTOS but results in two calls to services c and d. In some cases, additional processing of data may be required, and "straight" mapping may not be possible.

Strategies

A number of different strategies may be employed in the use of wrappers, depending upon particular circumstances:

- **Map an old RTOS API onto a new one:** This obvious approach may be very successful because it directly addresses the problem at hand. It is somewhat inflexible, but, for a one-off migration, the inflexibility may not be important. Once an application has been migrated to a new RTOS in this way, new code can make use of the native API of the new RTOS. References to the old RTOS gradually will "fade away," and eventually, the wrapper may be discarded (see the section "API Availability" later in this article).

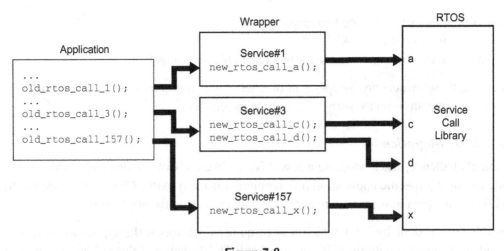

Figure 7.8
API wrapper implementation

- **Design a "neutral" API and implement a wrapper for each selected RTOS:** This preemptive strategy is used by some developers to facilitate relatively easy changes of the RTOS. The intention is to avoid locking in the application code to a particular vendor's API.
- **Use a wrapper to implement a standard API on a proprietary RTOS:** This strategy is similar to the preceding one, but it leverages accepted standards. Some commercial implementations of standard APIs are implemented in this way. See the sidebar "RTOS Standards" later in this article.

Implementation

The obvious way to implement a wrapper was illustrated previously in Figure 7.8. The wrapper is simply a library of functions that answer the required API and, in turn, make calls to the underlying RTOS. Done properly, such a wrapper is scalable in a way similar to most modern RTOS products. It may be possible to optimize this implementation by having the wrapper make jumps, instead of calls, to the RTOS service calls. The return from the actual service call will go back to the calling applications code.

If the wrapper is implementing an API that is not very different from the existing one, it may be possible to achieve a satisfactory result using C language #define macros. This is likely to introduce little or no overhead.

C++ offers a couple of different approaches to implementing an RTOS wrapper. A series of classes, representing various RTOS objects, may be created with all the API references hidden inside the class member functions. So, for example, creating a task may simply be a matter of instantiating a task object—the constructor would make the actual RTOS call to create a task. Alternatively, application-oriented classes may be employed, where the RTOS specifics are again contained, localized, and hidden within member functions.

Overhead

What does it cost? Nothing comes free in this world. So it is reasonable to expect that a wrapper will have a memory and/or performance hit. Unless a pure macro-based approach is viable, that indeed will be the case.

Taking the example of the call-based approach, every RTOS service call will incur an additional call/return time overhead. As previously commented, this overhead may be reduced if a jump/return sequence can be implemented. The memory overhead may be minimized by implementing the wrapper as a library (so that it scales automatically and no excess code is included).

In most cases, a modest overhead is entirely acceptable. In some cases, applications actually run faster using a wrapper on top of a modern RTOS than they did with the old RTOS.

Challenges

The level of difficulty in creating an RTOS wrapper is broadly governed by how well the architecture of one RTOS maps onto the other. Very large conceptual differences would render a wrapper implementation impossible or, at best, very inefficient. In most cases, the challenges are associated with specific details of the mapping that affect the wrapper. For example:

- **What kind of task identifier is used?** This may be a name, a number, or a pointer to a data structure. It may even double as the task priority.
- **How many priorities does the RTOS support?** If the new RTOS has less than the old one, this challenge could lead to difficulties. If the RTOS supports more priorities, should the translation spread the old ones over the new, wider range? There is also the issue of how priority is represented; do priorities start from 0 or 1? Is 0 the highest or lowest priority?
- **How are other facilities implemented?** Even if both RTOSes have the same facilities, there will be implementation differences. For example, if the old RTOS has bigger mailbox data structures than the new RTOS, the wrapper may need to manage the splitting of data over two mailboxes. This kind of mapping is messy but manageable.

API Availability

It is important to appreciate that using a wrapper for moving code from an old RTOS to a new one should normally be regarded as a temporary measure. As illustrated in Figure 7.9, there are three stages:

1. At the beginning, there is just the application code and the old RTOS. The wrapper is implemented to facilitate the use of the new RTOS; the application code is unchanged.
2. Over time, the application code is enhanced, making use of the native API of the new RTOS. The calls using the old API, via the wrapper, may be selectively replaced to optimize the code.
3. Eventually, no references to the old RTOS will remain, and the wrapper may be discarded.

The visibility of the native API of the underlying RTOS can also be useful when a wrapper is being used to implement a standard API. It may happen that the standard API lacks a particularly useful feature, which may be found in the RTOS API. Using a wrapper means that such a feature may be employed.

It should be remembered that an API is not necessarily confined to the kernel—it extends to all the "middleware" (like networking, file system, and graphics). So the use of a wrapper on top of a commercial RTOS with a very wide range of middleware permits unfettered access to all of this additional functionality.

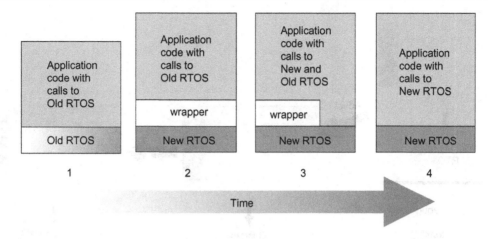

Figure 7.9
API availability

7.5.4 Drivers and More

One of the toughest issues with migrating between RTOSes is the "low-level" code—start-up, drivers, and so forth. There is little that can be automated in this area. The good news is that, if you are moving to a commercial RTOS product, the RTOS vendor will certainly be able to help. Standard devices should have drivers available off the shelf. And there certainly should be tools or templates to help with custom hardware.

7.5.5 Debugging Issues

An alternative, somewhat creative, approach to RTOS migration is illustrated in Figure 7.10.

The idea is to accept the fact that debugging matters and make it a priority, which can be viewed as a three-stage process:

1. A custom RTOS is in use, and there are no debugging tools.
2. Adapt a commercially available debugger that already has kernel-awareness capabilities to work with the custom RTOS. Work with this setup for a while to become proficient with the sophisticated debugger.
3. Move to the new RTOS and reconfigure the debugger. It is at this point that subtle bugs may be introduced (in the migration process). It is, therefore, very fortuitous that a good debugger and the skills to use it are on hand.

7.5.6 Conclusions

Migration from one RTOS to another, for various reasons, is an inevitable event for embedded developers from time to time. It need not be overly traumatic if the right strategies are adopted and a forward planning approach adopted. Use of a standards-based RTOS may be a preemptive action, which renders migration unnecessary.

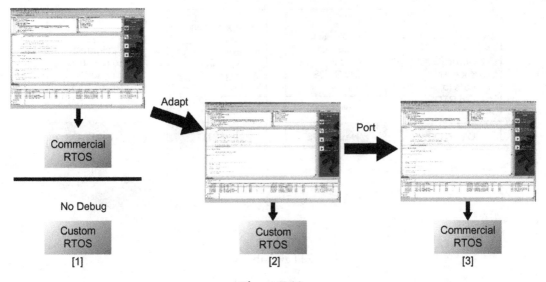

Figure 7.10
Debugging driving RTOS migration

RTOS Standards

Standards are a good thing—such a good thing that, in the RTOS world, we have a bunch of them. A variation on an old joke, but an illustration of the clear fact that, for embedded systems, one size does not fit all.

A big attraction, from the user's perspective, of a standards-based RTOS is a high degree of vendor independence. Suppliers are forced to compete on support, customer service, and product quality, instead of simply locking in their customer base.

Of course, code portability is cited as important, but it is not as significant as portability (and, hence, availability) of skills. Code portability is most useful when considering the licensing or reuse of code from outside of the development team. The topic of standards-based RTOSes could fill a lengthy article, but here we will limit ourselves to a brief discussion of key examples:

POSIX

This is the standard UNIX API and is, hence, understood by many programmers who are used to working on the desktop. A real-time variant of the POSIX standard was defined to accommodate the requirements of embedded applications.

micro-ITRON

In Japan, more than half of embedded software developers work with TRON standard operating systems. The micro-ITRON standard is, therefore, well established and also of interest to companies outside of Japan who collaborate with Japanese development teams.

OSEK/VDX

OSEK was developed by the main automotive companies in Germany, along with the University of Karlsruhr. Later this standard was merged with VDX, which had been developed by their counterparts in France. OSEK/VDX is now very widely used in the automotive industry in Europe and increasingly popular elsewhere in the world.

The standard addresses much more than just the API. All of the key features of an RTOS are covered:

- The kernel architecture (OS)
- Interprocess and interprocessor communications (COM)
- Network management (NM)
- System configuration (OIL)
- Debug interface (ORTI).

It is interesting to note that nothing about OSEK/VDX is specific to automotive applications. It has useful capabilities for many embedded applications, particularly when they have the following characteristics:

- Safety critical (or sensitive)
- Distributed (multiprocessor)
- Hard real time
- Static configuration (no dynamic object creation/destruction).

A good possible example would be medical electronics systems.

It is not possible to build an OSEK-compliant RTOS using a wrapper on top of an existing (dynamic) RTOS product. The architecture must be intrinsically static to be deemed compliant.

Java

The Java language was originally designed with embedded systems in mind. Although it has found many other applications, it is still applicable to embedded applications where dynamic changes to code functionality are required.

But it is a language, so why is it being mentioned under the heading "RTOS Standards"? The answer is that, unlike C and C++, Java "knows" about multitasking. Multi-threading is incorporated into the language itself (as it was previously in the Ada language). There is, obviously, some underlying scheduler/RTOS, but it is quite invisible to the Java programmer and can be implemented in a variety of ways. So, it is reasonable to regard Java as an RTOS API.

7.6 Introduction to RTOS Driver Development

It is hard to write about device drivers without being very specific about a particular real-time operating system. In a piece for Nucleus Reactor in 1997, Neil Henderson aimed to dispel some of the mystique around writing device drivers; this article is loosely based upon that work.

CW

Writing device drivers is generally thought of as a difficult job requiring special skills and knowledge. This may indeed be the case for "bigger" operating systems, which take a UNIX-like approach because they are large complex pieces of software. But for a conventional real-time operating system, which tends to be leaner and simpler, this need not be true. The key requirement is an understanding of the basics, which is the theme of this article.

Device drivers generally consist of two pieces: one for task-level processing, the other for interrupt processing. The manipulation of data and the coordination between both pieces are most important and will be addressed in more detail.

7.6.1 The Two Sides of a Device Driver

Most device drivers are written to support interrupts. True, device drivers can be more efficiently written without interrupts; however, the reality is that most are written to support them. Since it is easier to write non-interrupt-driven device drivers, if you learn how to write one that uses them, all other implementations are trivial.

Let's quickly review the basics of a device driver: initialize, send, receive, control, and close. There, that was rather painless! When you get right down to it, there is not much to a device driver in most cases. Sure, some extra error processing might need to be done, or you may want to provide an `ioctl()` function, but what it all boils down to is that you are sending and receiving data. Not too complicated! So why do we get all tied up in knots when we have to write a device driver? Well, it's generally because, unless you have some sophisticated tools, you cannot see what the device is doing until it works! However, with the right framework, making a device work quickly is quite straightforward.

So what are the two sides? One side interfaces with the tasking environment, and the other side is an interrupt service routine (ISR) and its related processing. What are the potential problems with coordinating the two sides? One is data corruption, and the other is task scheduling. If the ISR is manipulating pointers into the receive buffer, and those pointers are not protected from dual access, then your buffer data or control structures can be corrupted. Likewise, if you do not properly handle scheduling issues, a task waiting for a device driver service can bring a system to its knees.

7.6.2 Data Corruption

An RTOS is likely to provide interrupt control and protection mechanisms to address this matter. The one you use depends upon circumstances. If you are in an ISR, where you are permitted to call upon the RTOS services, then the protection mechanism is best because it minimizes the time when interrupts are disabled.

7.6.3 Thread Control

Why is thread control needed in device drivers? Well, it doesn't have to be, but to develop a driver efficiently, you should treat the driver itself as a part of the requesting thread. This means the thread will need to suspend execution until buffers are empty (for sending) or data is available (for receiving). Furthermore, the thread control has to be coordinated between the task thread and the interrupt thread. For example, you want to receive some characters that are coming from a serial port. You really cannot do any processing in the task thread until those characters are available. So, you invoke the driver to request a character. However, none are currently available so you want to suspend. You need a mechanism to do this. To go one step further, when the character is available, you need the task to be scheduled. You need a mechanism to do this also. The same scenario applies to sending characters when the transmit buffer is full. You may want your task to suspend until there is room in the buffer. Then, when room becomes available, you want to be scheduled so the rest of the characters can be sent.

7.6.4 Program Logic

To see how the parts of a driver fit together, let's step through a simplistic example. This example is described in pseudo-C code to illustrate the main logic and does not relate to any specific RTOS.

In the applications program, we want to write a string of text to an output device, so we call a driver's OutputCharacter() function repeatedly:

```
char str[] = "Hello World!";
char *sptr = str;
while (*sptr)
    OutputCharacter(*sptr++);
```

The driver OutputCharacter() function is run as part of the task and looks something like this:

```
void OutputCharacter(char c)
{
    if (PortReadyFlag != TRUE)
    {
        AddToPortWaitList(ThisTaskID);
        SuspendTask();
    }
    PORTOUTPUT = c;
}
```

The code checks whether it can output the character now. If it cannot, it sets up a mechanism so that the ISR can alert the task when the port is ready and goes to sleep until the port is available and then does the output.

The ISR for this write port is the other part of the driver:

```
interrupt void WritePort(void)
{
    TASKID tid;
    PortReadyFlag = TRUE;
    tid = GetFromPortWaitList();
```

```
       if (tid != NULL)
       {
           PortReadyFlag = FALSE;
           ResumeTask(tid);
           RunScheduler();
       }
   }
```

The ISR checks whether any task is waiting for the port to become available. If there is one, it wakes up the task.

The design of this simplistic example driver assumes that the device can accept a character from any task at any time—there is no control of "ownership" of the port, just a list of tasks pending upon it. This alternate logic would be straightforward to implement.

7.6.5 Conclusions

Historically, writing device drivers has been an arduous task. This stems from the fact that many real-time operating system environments are closed and there's an inclination to use UNIX-like facilities to manage device drivers. In an RTOS designed for deeply embedded applications, a much more open approach is necessary. No nasty tables should have to be maintained or special hoops that have to be jumped through—just simple templates that make creating custom device drivers rather less of a chore.

7.7 Scheduling Algorithms and Priority Inversion

For many embedded applications, even those that are truly real time, a deep understanding of how an operating system functions is not required. When it is, the topic can quickly become very complex. This article, based upon a piece by James Ready (now with Montavista) in NewBits in 1996, introduces some of the concepts.

CW

7.7.1 Introduction

A simple way to define a real-time system is one where the right results must be delivered on time. The objective of a software engineer writing a real-time application is to design the software so that the right results are delivered on time. In this article, we will explore the issues surrounding the development of a real-time application and discuss the available theory and technology that can facilitate the software engineer in building such a program successfully.

7.7.2 Real-Time Requirements

The development of a real-time application is essentially an exercise in resource allocation, and in the reduced case, the resource to be allocated is CPU time. For a given application, with a given workload, on a given processor, the trick is to find some ordering of execution such that all the activities that the software must perform are completed within the deadlines associated with each activity. The set of rules with which the activities are ordered for execution is called a *scheduling algorithm*.

Historically, the development of these rules has been done heuristically and validated by (hopefully) exhaustive testing. Thus, the development of real-time systems was often more an "art" and less of an engineering discipline. However, the United States Department of Defense (DOD) and other organizations who used real-time systems had a considerable interest in seeing that the development of real-time systems become far more of a science and that some scientific reasoning could be applied to their development. Consequently, over the past 25 years, considerable funding of research has been invested in the development of real-time scheduling algorithms.

7.7.3 Scheduling Algorithms

A number of scheduling algorithms have been developed that allow precise statements to be made about the timing correctness of the real-time application. Typically, an analysis can be made that will indicate that for a given type of application, characterized by its workload, on a CPU of a known capacity, the deadlines will be met completely, or a deadline or deadlines will be missed at some time. As is typical in an emerging technology, only the simplest kinds of applications can be scheduled completely, and scheduling algorithms for more complex and dynamic applications remain research studies. What follows is a brief summary of two well-known scheduling algorithms and the type of workload that they can handle successfully.

Rate Monotonic
In applications where the workload consists of a set of periodic tasks each with fixed-length execution times, the Rate Monotonic Scheduling (RMS) algorithm can guarantee schedulability. The RMS algorithm simply says that the more frequently a task runs (the higher its frequency), the higher its priority should be. Obviously, the most frequent task has the highest priority in the system. If an application is naturally completely periodic or can be made so, then the developer is in luck because the RMS algorithm can be employed and a provably correct assertion can be made about the application meeting its deadlines.

Deadline Driven
For applications with a mix of periodic and aperiodic tasks, or where the execution length of a task may vary with time, the Deadline Drive algorithm can be used. Here the criterion for scheduling the next task to run is based on finding the task with the earliest deadline. Upon completion, the next earliest deadline task is selected and run.

7.7.4 Implications for Operating Systems and Applications

The discussion of scheduling algorithms so far has centered upon a fairly high-level view of the application software, and no particular details of the runtime environment have been assumed. For example, we have not indicated whether an operating system supports the application.

But, just as the time behavior of the application must be understood to select the right scheduling algorithm, if we do employ an operating system, its own time-dependent behavior also enters into the equation. For example, if the underlying operations of the operating system are unpredictable because of some memory allocation/deallocation strategy, it may well be impossible to reliably schedule the application. In other words, the operating system behavior violates the predictability required by the scheduling algorithm we hoped to use.

It turns out that the problem of operating system interference is quite real, even with an operating system that claims to be real time. This interference can occur within the operating system itself, or be induced at the application level by an inadequate capability in the system call set of the operating system.

There is a phenomenon that can occur within an application, called *priority inversion*, which serves as a good example of how a lack of care in developing an application may compromise the use of the powerful scheduling algorithms we discussed previously.

The Priority Inversion Problem

The conditions under which priority inversion can occur are quite commonly found in both real-time applications and within operating systems. We will discuss here the occurrence of priority inversion at the application level. The conditions are:

- Concurrent tasks in the system share a resource that is protected by a blocking synchronization primitive such as a semaphore or exchange.
- At least one intermediate priority task exists between any two tasks that share a resource.

The following scenario illustrates how priority inversion occurs (see Figure 7.11):

1. A low-priority task makes a memory allocation call, which in turn uses a semaphore to protect a shared data structure (see part A in Figure 7.11). It takes the semaphore and enters the critical section (illustrated in part B of Figure 7.11).
2. An interrupt occurs that enables a high-priority task (part C). The kernel switches control to the higher-priority task (part D).
3. The high-priority task now makes a memory allocation call (part E) and attempts to enter the critical section. Since the low-priority task currently holds the semaphore, the higher-priority task is suspended (blocked) on the semaphore, and the low-priority task runs again (part F).
4. An interrupt occurs (part G), and a medium-priority task becomes ready to run. The medium-priority task runs (part H) because it is higher in priority than the lower-priority task holding the semaphore and because the high-priority task is suspended on the semaphore.

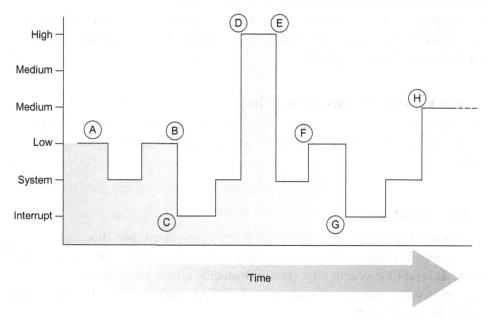

Figure 7.11
Priority inversion

5. At this point, *all* tasks of a priority higher than the low-priority task that become ready to run will do so at the expense of the low-priority task (which is how it should be), but they also run at the expense of the high-priority task, which is held up as long as the low-priority task is held up.

In effect, the high-priority task's priority has become lowered or "inverted" to match that of the low-priority task—hence the term *priority inversion*.

If the operating system does not offer any specific facilities to address this problem (priority inheritance), there are three possible solutions:

- Avoid sharing resources between tasks of differing priorities.
- If the operating system allows, turn off preemption of the low-priority task for the critical section during which it holds the semaphore. (This, of course, inhibits multi-threading, which is normally undesirable.)
- The low-priority task could raise its own priority while in the critical section (priority inheritance done "by hand").

7.7.5 Conclusions

The understanding of scheduling algorithms is of fundamental significance to developers of hard real-time applications. In some cases, application developers are required by certification agencies to employ the RMS algorithm because of the mission-critical nature of the applications. Even if the use of these scheduling algorithms is not mandated, they provide a sound basis upon which to build real-time applications such as those found in communications and control systems.

7.8 Time Versus Priority Scheduling

This article is based upon a white paper by John Schneider, which he wrote in response to some of the "FUD" (fear, uncertainty, and doubt) marketing of various RTOS vendors.

CW

7.8.1 RTOS Scheduling

With time-domain bounding, every domain is given a guaranteed time allocation no matter what the priority. If a higher-priority domain runs out of time, it must wait for its next time slice. In a hard real-time system, time-domain bounding also requires the user to plan CPU allocation for the worst-case scenario, with no opportunity to take advantage of the less-than-worst-case scenario.

Most true real-time operating systems use a priority-based, preemptive scheduler instead of time-domain bounding. This guarantees that higher-priority tasks run when needed. The worst-case example for a priority-based scheduler is when a higher-priority thread consumes a large, contiguous block of CPU time. This scenario could essentially lock out lower priority threads causing starvation. In this case, two approaches can be used. The first option is to have the threads at the same priority as the time-consuming thread and allow the scheduler to preempt the time-consuming task on regular intervals to service other threads based on round-robin scheduling. A better option is to have the time-consuming task relinquish the processor to lower-priority threads. This can be accomplished by using a *sleep* command at appropriate, noncritical intervals in the code. The *sleep* command will allow lower-priority threads to run for a specified amount of time. This implementation allows the user to decide when and where the lower-priority threads get access.

The following example of a network connected data acquisition system shows the differences between the two scheduling schemes.

7.8.2 Perfect World

In a perfect world, a network connected data acquisition system would operate in a manner such that none of the threads interfere with each other and the system is always available when a thread is needed. A sample would never be missed, network access would only happen after the computation is complete, and calculations would always be the same length as CPU cycles. This is what it would look like in a simple timeline:

```
|-Acquisition-|      |-Acquisition-|      |-Acquisition-|      |-Acquisition-|
| Available  |      | Available  |      | Available  |      | Available  |
   As---Ac              As---Ac              As---Ac              As---Ac
       Cs----Cc             Cs----Cc             Cs----Cc             Cs----
           Ns---Nc              Ns---Nc          Ns---Nc
```

Key

As: Data acquisition start
Ac: Data acquisition complete
Cs: Data computation start
Cc: Data computation complete
Ns: Network access start
Nc: Network access complete

7.8.3 Real World with Priority Scheduling

In the real world, network access is often asynchronous to the rest of the system and can use up variable amounts of CPU cycles (this is largely due to the nondeterministic nature of TCP/IP rebroadcast schemes and variations in data packet size being sent or received), computations are never identical, and so on. These deviations from an ideal case can be overcome by the use of a priority-based scheduler. In the following case, the data computation has strategically placed explicit sleep calls (noted by XS) in noncritical places in the code. This allows the lowest-priority thread to run until the highest-priority thread preempts. When the highest-priority task is complete, the system reschedules based on priority and returns to the data computation thread.

```
|-Acquisition-|      |-Acquisition-|      |-Acquisition-|      |-Acquisition-|
| Available  |      | Available  |      | Available  |      | Available  |
As--Ac               As--Ac               As--Ac               As--Ac
    Cs--XS                                    -------XS             --Cc
        Ns--Nc           -------                                Ns ----Nc
```

Program Priority

A typical distributions of task priorities is:

Highest priority: Data acquisition
Medium priority: Data computation
Low priority: Network access
xs: Explicit sleep
I: Idle

7.8.4 Time Domain Bounded Without Relinquish

The timeline that follows shows the same time sequence but split up in a time-domain-bounded system without the ability to relinquish unused portions of the allocated time slice. In this example, we've allocated 50% of the time for the most important task, data acquisition; 30% of the time to the data computation task; and 20% to network access. You can see that the computation never gets completed, which is due to the large block of idle time at the end of the acquisition stage. In our example, we had to allocate this extra block of time due to the worst-case response time of the hardware. This leaves the user having to allocate for the worst-case scenario and never being able to take advantage of the extra time that results in the lower response times from our system.

```
|-Acquisition-|      |-Acquisition-|      |-Acquisition-|      |-Acquisition-|
| Available   |      | Available   |      | Available   |      | Available   |
As--Ac               As--Ac               As--Ac               As--Ac
       Cs----                ------               ------               ---
            Ns--                    --Nc                         Ns--
       III                III                  III                  III
    AAAAAAAAAACCCCCCNNNNAAAAAAAAAAACCCCCCNNNNAAAAAAAAAAACCCCCCNNNNAAAAAAAAAAACCC
```

A: Data acquisition domain—50%
C: Data computation domain—30%
N: Network access domain—20%
I: System idle time

7.8.5 Time Domain Bounded with Relinquish

Even in a time-domain-bounded system that has the ability to relinquish CPU resources when finished, the lag time shifts to the front of the acquisition while the acquisition task is waiting for the acquisition to be ready. Since the other tasks have used their allotment, they can't run. This leads to wasted time. The timeline shows the identical timeline as the one shown previously but with the ability to relinquish.

```
|-Acquisition-|      |-Acquisition-|      |-Acquisition-|      |-Acquisition-|
| Available   |      | Available   |      | Available   |      | Available   |
As--Ac               As--Ac               As--Ac               As--Ac
       Cs----                ------             --------               ------
            Ns--                    --Nc              Ns--                  -Nc
         I                        II
    AAAAAAACCCCCCNNNNAAAAAAAACCCCCCNNNNAAAAAAAAAACCCCCCNNNNAAAAAAACCCCCCNNN
```

7.8.6 Conclusions

On first sight, a scheme that allocates time in a very ordered fashion may appear to be the most deterministic and maybe the most applicable to a hard real-time system. However, as discussed previously, even in a simple system, time-domain scheduling is unable to cope with changing demands upon CPU resources.

7.9 An Embedded File System

Software engineers, who are new to embedded systems are often rather disconcerted by the lack of a "supporting" environment." Even if there is an operating system (and there may not be one at all), typically it does not have all the capabilities of Windows or Linux. One of the obvious facilities that are lacking is a file system—normally code is run out of some kind of read-only memory and the only data store is RAM. A seasoned embedded developer would take a different view and be shocked if he were presented with an RTOS that required a file system—he would consider it unbearable overhead. As with many things in embedded systems, different users have different needs, and a file system is a commonly requested option. This article is based upon an Accelerated Technology white paper.

CW

Embedded systems continue to grow and mature at an astounding rate. Rarely will an embedded system remain stagnant. As time goes on, systems that were simply a CPU with a few peripherals add more peripherals, more powerful processors, and eventually features such as offline storage and networking. In many ways, the evolution of embedded systems has caused a convergence of desktop technology with real-time technology. However, due to the price performance advantages of non-desktop-oriented CPUs and the lack of standardization in the embedded industry, the base software for embedded systems has not kept pace with that of desktop systems. While there are millions of prospective users for desktop software packages, the numbers pale when it comes to embedded software.

One area in which this disparity is particularly evident is offline storage. Desktop systems are widely known for their capabilities in this area. Even though you may be able to embed MS-DOS or Linux in your embedded system, they may be too large and cumbersome, or they do not support the CPU that you choose. So, to get the file system advantages of a commercial operating system such as one of these, you must either write the capabilities that you require on your own, or you can purchase an off-the-shelf product from an embedded software provider, usually an embedded real-time operating system provider.

In two situations, an embedded system may use a file system:

* To provide a static store of information so that the embedded system could quickly reconfigure at start-up or after a downtime.
* To provide a way to transfer data from an embedded system to a desktop system for data reduction and analysis.

The first requirement can basically be satisfied with a simple file storage facility. In this way, data can be written to an offline device in a flat file structure. However, this type of solution generally does not have the capacity to improve easily with the growing needs of an embedded system.

The second requirement—to transfer data to another system—requires a more sophisticated file structure that is compatible with common desktop platforms. An MS-DOS compatible solution is generally considered most appropriate. It easily handles the first requirement and by its nature provides a complete solution for the second requirement.

7.9.1 Requirements of an Embedded File System

Aside from meeting the two requirements just mentioned, an embedded file system must permit a real-time system to operate in real time, it must have minimal impact on data and processing requirements, and it must allow multiple tasks to access file information.

Critical activities must be completed in a timely and reliable manner in real-time embedded systems. If they are not, catastrophic results can occur. In a somewhat ludicrous yet illustrative example, if a machine that is responsible for pumping blood through a patient and saving statistical information to a file causes the blood to stop pumping, the result could be fatal. Therefore, the file system must be integrated into the real-time embedded system in such a way that file system activity can be relegated to a low priority when necessary.

Due to the fact that every kilobyte of memory and every instruction of processing time in an embedded system can have a drastic effect on cost, memory and processing allocation is critical. An embedded file system must use memory wisely and process information quickly.

Most embedded systems, by their very nature, must be multitasking. The requirement to allocate CPU time to individual processing requirements based on priority can be very important; in many cases, for an embedded system to operate correctly, it is essential. An embedded file system must be accessible by multiple tasks simultaneously. This means, since a file system is required to have global data structures, they must be protected from reentrancy problems.

7.9.2 MS-DOS File System Overview

The MS-DOS file system is often referred to as a FAT file system. A FAT is the File Allocation Table, which coordinates access to data on the disk. In addition to the FAT, the file system also includes a boot sector. The boot sector contains information about the disk including its size, the size of the FAT entries (12-, 16-, or 32-bit), and other useful information. FAT12 entries are for small disk media, such as floppy diskettes. FAT16 entries are for medium-sized disks, up to around 2 G. FAT32 entries were generated to accommodate today's large multi-gigabyte disks.

The FAT is made up of a series of pointers that are associated with each cluster on the disk. A cluster is a physical sector or group of sectors where data is stored. The clusters are linked together to make up a file. By reading the clusters associated with a set of FAT entries, you can construct the data in any file.

An embedded file system maintains the FAT on an ongoing basis. Each time a file is written or appended, the FAT is updated in memory. When a file is closed or flushed, the FAT is written to the disk. The FAT can be buffered completely or cached depending on your memory requirements. If it is cached, only portions of the FAT are kept in memory.

An embedded file system would generally include facilities to enable you to indicate how large the in-memory FAT will be for any of the disks in your system.

7.9.3 Long Filenames

With the larger media becoming more widespread in the embedded industry, true MS-DOS compatible file systems must be able to support any new MS-DOS features. Among the most prominent is the need to recognize filenames beyond the "eight-dot-three" format. MS-DOS now allows filenames to contain up to 255 characters, to provide more descriptive file naming. Long filenames are also stored and displayed in a truncated format, to accommodate disk media with earlier MS-DOS versions that do not properly display them in long format. In the truncated format, all characters after the first six characters are truncated, a tilde (~) is added, and a number is assigned preceding the extension dot. For example, the filename, `Encyclopedia Data.doc` could be displayed in its entirety or in the truncated form `Encycl~1.doc`.

7.9.4 Formatting

When you format a disk, or disk partition, you are writing a boot block and an empty FAT. A file system will probably provide standard functions for formatting floppy (360 K, 720 K, 1.2 M, and 1.44 M) as well as hard disks. To format a disk, you fill in a structure with the disk's characteristics and issue a service call to the embedded file system, which then writes the appropriate information to the correct location on the disk.

7.9.5 Partitioning

Disk partitioning is not the same as disk formatting. A partition is a physical subdivision of the directory database stored on and managed by a specific directory server. Partitioning is common on large disk media, changing the appearance of one large physical disk into several smaller logical disks. Disk partitioning capabilities are likely to be provided via the IDE driver.

7.9.6 Devices

Standard device drivers for an embedded file system are likely to include floppy, fixed, flash, and RAM disk drivers. Device drivers require four functions: open, close, perform input/ output, and control input/output.

7.10 OSEK—An RTOS Standard

Although the OSEK standard had its origins in the European automotive industry, it is now gaining much wider acceptance. It is finding worldwide application in various transportation applications. There is also significant interest from developers in other safety-critical areas, such as medical electronics. The standard is well worth considering wherever the static architecture and simplicity of OSEK is a benefit. This article provides a brief introduction and is based upon an Accelerated Technology white paper by Zeeshan Altaf.

CW

7.10.1 About OSEK

OSEK/VDX is a joint project initiated by some major players in the automotive industry. It aims for an industry standard of an open-ended architecture for distributed units in vehicles. An RTOS, software interfaces, and functions for communication and network management tasks are thus jointly specified.

The term "OSEK" means "**O**ffene **S**ysteme und deren Schnittstellen für die **E**lektronik im **K**raftfahrzeug" (open systems and the corresponding interfaces for automotive electronics). This particular standard originated in Germany and included BMW, Bosch, Daimler-Benz, Opel, Siemens, VW, and the University of Karlsruhe. The term "VDX" means "**V**ehicle **D**istributed e**X**ecutive." This standard originated in France and included PSA and Renault. The functionality of the OSEK operating system was harmonized with VDX. For simplicity, the truncated term "OSEK" is generally used instead of OSEK/VDX.

The key motivations behind the development of the OSEK specification were high recurrent expenses in development, irregular management of non-application-related aspects of control unit software, and the incompatibility of control units made by different manufacturers due to different software interfaces and protocols.

The OSEK architecture comprises three primary modules:

- **Operating system (OS):** A real-time executive for electronic control unit (ECU) software and the basis for the other two OSEK modules.
- **Communication (COM):** The data exchange within and between control units.
- **Network Management (NM):** Configuration determination and monitoring.

Figure 7.12 shows a graphical depiction of the OSEK architecture.

7.10.2 OSEK Requirements

The specification for the OSEK operating system includes strict real-time requirements. Memory resource usage must be minimized, and the operating system must be scalable, reliable, ROMable, and cost-sensitive. The software must be developed in a manner to promote portability. It must serve as the basis for software integration from various manufacturers. Finally, the configuration, including scalability and scheduling policies, must be static.

The goals of the OSEK project are to support the portability and reusability of the application software by providing specifications for interfaces that are abstract and as application-independent as possible. Additionally, the specifications of a user interface should be independent of supplied hardware and network applications. The architecture will also be more efficiently designed because of the guidelines. The functionality shall be configurable and scalable to enable optimal adjustment of the architecture to the application in question. Finally, verification of the functionality and implementation of prototypes in selected pilot projects shall be available.

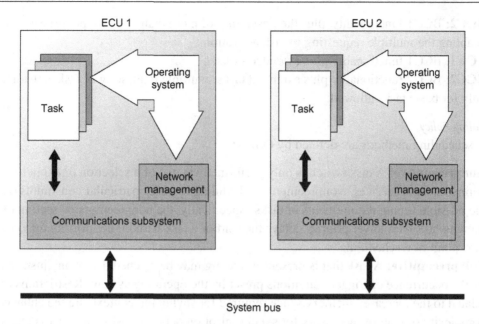

Figure 7.12
OSEK architecture

7.10.3 OSEK Tasks

Task Types

Two task types are defined in OSEK:

- **Basic** tasks release the processor only if:
  - they are being terminated;
  - the system is executing higher-priority tasks;
  - an interrupt occurs, and the processor switches to an ISR.
- **Extended** tasks can additionally have a waiting state when they are using the event mechanism for synchronization.

Conformance Classes

The conformance classes provide convenient groups of operating system features for easier understanding and discussion of OSEK. The conformance classes allow partial implementations, which may be certified as OSEK-compliant, along predefined lines. They also create an upgrade path from classes of less functionality to classes of higher functionality with no changes to the application using OSEK-related features. Four conformance classes are defined for OSEK functionality:

- **BCC1:** Only basic tasks, limited to one request per task and one task per priority (all tasks have different priorities).

- **BCC2:** BCC1 functionality, plus the possibility of more than one task per priority and enabling the multiple requesting of task activation.
- **ECC1:** BCC1 functionality plus extended tasks.
- **ECC2:** BCC2 functionality plus extended tasks: multiple requesting of task activation only for basic tasks allowed.

Scheduling Policy

Three scheduling methods are defined by OSEK:

- **Nonpreemptive:** A task switch is only performed via one of a selection of explicitly defined system services. Nonpreemptive scheduling imposes particular constraints on the possible timing requirements of tasks. Specifically, the nonpreemptable section of a running task with lower-priority delays the start of a task with higher priority up to the next point of rescheduling.
- **Full preemptive:** A task that is presently running may be rescheduled at any instruction by the occurrence of trigger conditions preset by the operating system. Restrictions are related to the increased memory space required for saving the context and the enhanced complexity of features necessary for synchronization between tasks. Access to data, which is used jointly with other tasks, must be synchronized.
- **Mixed preemptive:** A mixture of nonpreemptively and full preemptively scheduled tasks.

Synchronization Mechanisms

Two synchronization mechanisms are defined by OSEK:

- **Resource management:** Controls access to jointly used logic resources or devices.
- **Event management:** Event management for task synchronization—allowed only for extended tasks.

7.10.4 Alarms

Two types of alarms are defined by OSEK:

- **Relative:** Count values are defined relative to the actual counter value.
- **Absolute:** Count values are defined as absolute values. The count values can be defined dynamically at runtime (variant alarms) or statically at compile time (nonvariant alarms).

7.10.5 Error Treatment

OSEK implementations normally utilize user-definable hook routines for error handling and debugging. Scalability of error checking includes an extended status for the development phase and a standard status for the production phase, which is generally implemented by answering the OSEK API with macros, which may be expanded in a variety of ways.

Networking

Chapter Outline

There are a variety of add-on products for real-time operating systems. This is commonly called "middleware." Given that an increasing number of embedded systems are connected to the Internet or to private networks, it is unsurprising that networking products dominate the middleware market. In this chapter, the articles address a wide range of topics all under the heading of Networking, which range from broad introductions to detailed discussion of protocols.

8.1 What's Wi-Fi?

Wireless networking in homes and offices has become very popular in just a few years. Such networks are constructed from a number of embedded systems. So, of course, interest in incorporating Wi-Fi into embedded devices has soared. In early 2004, I wrote a piece for NewBits to broadly explain what 802.11 was all about. That was the basis for this article.

CW

When I was a little kid, back in the early 1960s, we had a "trannie"—a transistor radio. This was the cool gadget to have back then—the MP3 player of its day. After all, the transistor was the latest technology, having only been invented some 15 years before. We don't seem to have such a long "lab-to-living room" cycle nowadays! My grandmother didn't have a trannie, but in the corner of her kitchen, she had something much more mysterious: a big wooden box with dials and displays that lit up when you switched it on. Nothing else happened when you applied the power because it needed to "warm up" (I guess we would say "boot up" nowadays—although that term was understood then, it wasn't used in polite company). It was a vacuum tube (in England we say "valve") radio. But my granny called it her "wireless."

So my perception of this term was that it was a logical—receiving a signal with no wires—but rather old-fashioned—dare I say it—antiquated term. It never occurred to me as odd that this word, while logical, did not signify the successor to some kind of "wired radio." Neither did I suspect that this term would take on a whole new lease in life in my adulthood.

Today, we use the word "wireless" to refer to a whole slew of technologies, where one requiring no wires has replaced a wired approach. We are not limited to radio. Many other parts of the electromagnetic spectrum are used: light, infrared, microwave, and so forth. Maybe we are referring to voice or multimedia transmission (see the sidebar "Why Were Phones First?" later in this article), but often the terminology is used in relation to the transfer of data. It is this latter area, wireless data communications, that I will concentrate upon here.

8.1.1 Wireless Datacom

There is not a single technology that can be labeled "wireless datacom." There are a whole bunch of them. They can be categorized and characterized by their operating range.

Wireless Personal Area Network (WPAN)

This kind of network would typically encompass just a few meters around an individual. Infrared communications (IRDA) and the increasingly popular Bluetooth are common WPAN implementation technologies. ZigBee is emerging as another WPAN option.

Wireless Local Area Network (WLAN)

In this case, the range of the network is something like a building. It could be just part of a building or could encompass a small group of buildings. A shop or restaurant may be covered using a WLAN. Typically such an area is termed a "wireless hot spot." The most likely implementation technology is one of the IEEE 820.11 families, which are described in the sections that follow.

Wireless Metropolitan Area Network (WMAN)

This kind of network is similar in functionality to a WLAN but covers an area the size of a city.

Cellular Network

The voice-oriented cellular networks are gradually becoming more capable in terms of handling data traffic. GSM was very widely used, for example, with GPRS and EDGE available for high-speed data, but has largely been replaced by 3G, with 4G technologies coming along.

These four families can, for the most part, coexist. Having said this, the WMAN domain is being encroached upon from both sides—bigger WLANs and cost-effective cellular services. Some further comments on the comparison between Wi-Fi and Bluetooth follow later.

8.1.2 IEEE 802.11

The 802.11 family of standards addresses WLANs. The term "Wi-Fi" (wireless fidelity) was defined by the Wireless Ethernet Compatibility Alliance (WECA) to describe them. This term has since been widely used and misused. The 802.11 family includes many standards, and, because of its success, more standards are continuously being added. There is a real alphabet soup of 802.11 specifications, but some stand out as worthy of particular attention. Three of them, 802.11a, 802.11b, and 802.11g, are physical layer standards.

802.11b

This standard was the first generation, offering modest speed (11 Mbps). Equipment employing this standard is very widely used. The first widely deployed Wi-Fi standard was 802.11b. It utilizes the 2.4 GHz band of the radio spectrum, which is unregulated in the United States. As a result, it shares this bandwidth and is susceptible to interference by devices that include, but are not limited to, cordless phones, microwave ovens, and Bluetooth-enabled devices. The maximum throughput is 11 Mbps, but 6 Mbps is a better estimate of what can be achieved in a real-world deployment.

802.11a

Confusingly this is the second-generation standard, offering a higher data rate (54 Mbps) on a different frequency (5.4 GHz). The maximum throughput using 802.11a is 54 Mbps, but as with 802.11b, the real throughput achieved will be significantly less than this, around 33 Mbps.

802.11g

The third-generation standard, which is now most widely used, aims to combine the good things of the previous ones. It provides the speed of 802.11a on the same frequency band as 802.11b. It is backward compatible with 802.11b, so devices that support this standard will be able to communicate with both 802.11b- and 802.11g-enabled devices. At the time of writing and even faster/better standard is just beginning to take hold: 802.11n.

802.11i

This standard provides an alternative to Wired Equivalent Privacy (WEP), which has security holes. Applying the a, b, and g standards, it will provide authentication and encryption procedures.

8.1.3 802.11 Basics

The place of 802.11 in the protocol scheme is best illustrated by looking at the mapping onto the ISO/OSI seven-layer model (see Table 8.1). As can be seen from this table, the 802.11 standards address the bottom two layers, Data Link and Physical. The higher layers are satisfied by the standard Internet protocols. In a real system, there may, of course, be multiple Data Link and Physical layers, representing multiple 802.11 variants, other 802 standards, or other (nonwireless) networking technologies.

Table 8.1: ISO/OSI Seven-Layer Model

OSI Layer	Protocol
Application	HTTP, FTP, POP3/SMTP
Presentation	DNS, LDAP
Session	
Transport	UDP, TCP, ICMP, RSVP
Network	
Data link	Logical link control (LLC)
Medium Access Control (MAC)	
Physical	Physical (PHY)

In a Wi-Fi network there are broadly two types of devices: station devices (STA), which are users of the network, and access points (AP), which are wireless network hubs. Clearly each of these will require different software support.

Wi-Fi devices can function in one of two modes:

- **Ad hoc mode:** Where two station devices are directly linked.
- **Infrastructure mode:** The usual way of working, where a station device is linked to an access point.

8.1.4 Wi-Fi and Bluetooth

There is some confusion between the functionality offered by Wi-Fi and Bluetooth. In reality, they are quite complementary and offer different functionality—each has strengths and weaknesses in different situations.

Range
Bluetooth is intended to be used over about a few meters; Wi-Fi is intended for areas covering tens of meters.

Topology
Bluetooth is strictly point-to-point/master-slave and, as such, is very analogous to USB. Although Wi-Fi can be used in ad hoc mode, its real strength is when it is used in infrastructure mode, which is much more analogous to Ethernet, where multiple devices may be connected together.

Power Consumption
In designing embedded devices, which are wireless enabled, power consumption (=battery drain) is frequently an issue. This is a downside of Wi-Fi, where frequent data exchanges are needed to maintain network integrity. Wireless transmission is power hungry. Bluetooth transmits only when there is really data to send.

8.1.5 Where Next?

Today we are only seeing the very start of what wireless networking will provide in the future. Connecting together boxes that are recognizable as computers is useful but hardly scratches the surface of the potential. There has been a trend for embedded devices to become "connected"—the next trend is for them to be Wi-Fi enabled. So what about:

- A car alarm system that seamlessly integrates into the security system of the house, office, or parking lot?
- A digital camera that is always visible to nearby computers and immediately uploads and archives its images? (SD cards with this capability are already available.)
- Printers that you can locate anywhere you want and that can service numerous PC users?
- A network in your house that distributes multimedia content so that the TV can go anywhere there is a power outlet?

With rapidly decreasing prices on Wi-Fi chip sets and the ready availability of cost-effective, off-the-shelf software solutions, all of these ideas and many more will soon become reality. The rest is up to you.

Why Were Phones First?

In the early 1980s, several things came together. Parts of the radio spectrum became available at the same time as the possibility for cheap electronics to exploit these frequencies. All that was needed to make this commercially interesting was a pervasive, ubiquitous technology to replace or augment.

That technology was the telephone. Having been around for about a century, phones were everywhere in the Western world—every home, every office, every street corner. The only limitation to their use was the wire to the jack in the wall. Sure, the wires got longer—the TV or movie image of someone wandering around a room with a phone on a long lead is clear in my mind. But they were still tethered and needed to be set free. Wireless (radio) technology was the answer. First there were domestic cordless phones, then the first cellular phones. Many of these were in cars—hand-held units were somewhat cumbersome. The initial analog technology (first generation—1G) gave way to digital (2G). Then, more advanced digital (2.5G) became common, and now the next generation (3G) is beginning to dominate. In several countries, cell phones already outnumber fixed phones. In developing countries, they are installing cellular infrastructure from day one.

It is a shame that the language has not evolved so well. "Mobile phone" and "cell phone" are very awkward terms, and "cell" is just plain wrong. In Germany, "hand-held phone" is abbreviated to "handy" (yes, an English word). Maybe this terminology could be used more widely.

8.2 Who Needs a Web Server?

It is becoming quite common for web servers—or, more correctly, HTTP servers—to be incorporated into routers, gateways, and other networking equipment. There is every reason to consider doing this for a multitude of other types of embedded systems. In a piece written for the "Nucleus Reactor" newsletter in 2000, Neil Henderson (CEO of Accelerated Technology) gave a very good outline of what was possible and how to get started. This provided the basis for this article. It may be interesting to compare the use of HTTP with SNMP because in this context, their objectives are quite similar. Take a look at the article "Introduction to SNMP" later in this chapter.

CW

8.2.1 Introduction

You may view web servers in the same manner I did before I understood what they could do in an embedded system. In my mind, web servers were located on machines with huge disk drive capacity and served up pages to web browsers. Well, huge disk drives are not necessary, and web servers can do much more than just serve up web pages!

With an embedded web server you can, of course, serve up pages. But did you know you can use a web server to provide an interactive user interface for your embedded system? Did you realize that you could program that interface once and be able to use it independent of the type of machine your user has? Did you further know that you could monitor and control your embedded system from any web browser with very little programming? All of these things are made possible by this very small, very efficient, and very powerful piece of software.

In this paper I will provide information that you will most likely be able to use on your embedded system. All you need is a TCP/IP networking stack, an embedded HTTP server (i.e., web server), and a little imagination. So, let's get started.

8.2.2 Three Primary Capabilities

Web servers are capable of performing three basic functions:

- Serve web pages to a web browser
- Monitor the device within which they are embedded
- Control the device within which they are embedded.

We will be examining these functions in more detail in the remainder of this chapter. Here, I will give you a brief introduction to each of these functions so that you can better understand the sections that follow.

Serve Web Pages to a Web Browser

This is the most fundamental capability of a web server. The web server waits on the network for a web browser to connect. Once connected, the web browser provides a filename to the web server, and the web server downloads that page to the web browser.

In the most basic case, the web server can download simple HTML files (simple because there are no inherent capabilities other than to show information) from within its file system to the web browser. This feature is ideal for downloading user documentation from the embedded system so that it can be used in a web browser.

A more sophisticated and extremely powerful capability is for the web server to download Java programs or applets (encapsulated in an HTML file) to the web browser. Once loaded in the web browser, the Java program or applet executes and can communicate with the target (that contains the web server) using the TCP/IP protocol. The power of this capability lies in the ability to:

- Support legacy applications (existing TCP/IP applications that presently communicate with a Java application that runs in a browser rather than writing proprietary applications for different desktop operating systems).
- Write sophisticated TCP/IP-based applications between a host and server where you control both sides regardless of where the host is running.

Monitor a Device

Often there is a need to retrieve (i.e., monitor) information about how an embedded system is functioning. Monitoring can range from determining the current pixel resolution of a digital camera to receiving vital signs from a medical device.

By embedding certain commands within an HTML page, dynamic information can be inserted into the HTML stream that is sent to the web browser. As the web server retrieves the file from the file system, it scans the text for special comments. These comments indicate functions to be performed on the target. These functions then format dynamic information into HTML text and include the text into the HTML stream being sent to the web browser.

Control a Device

HTML has the capability to maintain "forms." If you have ever browsed the web and tried to download something, you probably have seen a form. A form is a collection of "widgets" such as text entry fields, radio buttons, and single-action buttons that can be assembled to collect virtually any type of data.

By constructing an HTML page with a group of widgets, information can be collected from the user in a web browser. That information can then be transmitted to the target and used to adjust or alter its behavior. For example, an HTML page could be constructed to configure a robot arm to move in certain sequences to perform some necessary function (e.g., to bend a piece of sheet metal). This could be done by placing specific text entry boxes in the HTML page that instruct the user to enter a number of specific data points. After being sent to the web server, the data points can then be analyzed by the embedded system's application, validated, and then executed (or, if the data is invalid, to have the user reenter the data) to move the robot arm in the proper directions.

8.2.3 Web Servers at Work

Once again we will explore the use of the web server based on the three primary capabilities. We will look at the processing that is done on the web server and how information is supplied both from and to the web browser. I will discuss the ability to serve pages to a web browser. Then, we will progress into the more complex tasks that can be achieved by implementing an embedded web server—namely, using the web server to provide dynamic information to a web browser and using a web server to control your embedded system.

Communication between the web server and the web browser is controlled by HTTP (HyperText Transfer Protocol). HTTP supplies the rules for coordinating the requests for pages to the web server from the web browser and vice versa. The pages are transferred in HTML (HyperText Markup Language) format.

Serving Pages
As discussed previously, the simplest use of the web server is providing HTML pages from the web server to the web browser. This is a straightforward operation where the server maintains a directory structure containing a series of files. The user, from the web browser, specifies the URL (Uniform Resource Locator) that includes the IP address of the web server and the name of the file to be retrieved. The web browser transmits an HTTP packet to the web server with the requested filename. The web server locates the file and sends it to the browser via the HTTP protocol. Finally, the web browser displays the page to the user.

This feature can be used to supply information such as the device's user manual from the embedded system to the user on a web browser. In most web server implementations, the ability to serve pages up to a web browser can be included in an embedded system with little or no coding effort.

Using the Web Server to Provide Dynamic Information to a Web Browser
By manipulating the HTML page that is sent to the web browser, the embedded system employing the web server can supply dynamic information to the user. The web server on the embedded device scans every HTML file that is sent to the web browser. If a certain string is encountered during the scanning process, the web server knows to call a function within the embedded system. The called function then knows how to format the dynamic information in HTML and append it to the buffer being sent to the web browser.

Let's assume, for example, that our embedded system is a router. Let's further assume that we want to display the router's IP address. The complete HTML file to display this information may look something like this:

```
<BODY>The IP Address of the Router is: <!-# IPADDR> </BODY>
```

As the web server scans this HTML, it encounters the `<!-#` symbol, performs a lookup on the string IPADDR, and determines that a function `display_IP_addr(Token *env, Request *req)` is to be called.

`display_IP_addr()` may look something like this:

```
/* Create a temporary buffer. */
char ubuf[600];
void display_IP_addr(Token *env, Request *req)
{
    unsigned char *p;
    /* Get the IP address. */
    p = req->ip;
    /* Convert the IP addr to a string and place in ubuf. */
    sprintf(ubuf, "%d.%d.%d.%d", p[0], p[1], p[2], p[3]);
    /* Include the IP string in the buffer on its way to
    the browser. */
    ps_net_write(req, ubuf, (strlen(ubuf)), PLUGINDATA);
}
```

Let's quickly review what we have just done. In the HTML file, we indicate that we want to display the string "The IP Address of the Router is." Additionally, there is a command to display the value of IPADDR. It is not evident in what we see here, but the IPADDR reference is actually in a table on the target. In the table, IPADDR has a corresponding element named `display_IP_addr` that is a pointer to the function call of the same name.

In the code, we assume that the web server has already found the <!-#> string and has located the IPADDR element in the table. This has resulted in the call to `display_IP_addr()`.

`display_IP_addr()` simply fetches the IP address from the `req` structure, formats it into the easily recognizable four-part IP number and then places the resultant string into the buffer that is on its way to the web browser.

From this simple example we can begin to see the power the web server possesses to transmit dynamic information to a web browser. By using more sophisticated HTML information, elaborate user displays can be created that are exciting and informative.

Using the Web Server to Control an Embedded System

For years, developers of network-enabled products (e.g., printers, routers, bridges) have had to develop multiple programs to remotely configure these devices. Since, in many cases, the products can be used on the Windows, Mac OS, and Linux operating systems, developers of these types of systems are forced to write applications for all three platforms. Using a web server can reduce this programming effort to developing one or more HTML pages and writing some code for the target. Using this paradigm, the users of the printers, routers, bridges, and so forth simply connect to the device using a web browser. I recently bought a SOHO router that had this capability. An IP address was specified in the literature that came with the router. I used that IP address in my web browser to communicate to the router's web server. It supplied a full screen of options to configure the router for my particular circumstances. Let's take a minute to look at a simple example of how something similar to this might be accomplished with an HTML file and a little code.

The HTML file will look as follows:

```
<BODY> Use DHCP to acquire IP Address? </BODY>
<br>
<br>
<INPUT TYPE="RADIO" NAME="RADIOB" VALUE="YES" CHECKED>YES
<br>
<INPUT TYPE="RADIO" NAME="RADIOB" VALUE="NO">NO
<br>
<br>
<INPUT TYPE="SUBMIT" VALUE="SUBMIT">
```

The code that will be used to process this request may be as follows:

```
int use_DHCP_flag;
int use_DHCP(Token *env, Request *req)
{
    /* Verify that we are looking at the right "command" */
    if(strcmp(req->pg_args->arg.name,"RADIOB")== 0)
        /* Should we use DHCP? */
        if(strncmp(req->pg_args->arg.value,"YES",3) == 0)
            /* Yes, use DHCP */
            use_DHCP_flag = TRUE;
}
```

Once again, we will review the elements just illustrated. First of all, let's look at the HTML. We have three sections in this file separated by two line breaks. The first section is simply a prompt for the user. The second section is the code necessary to display the radio buttons as shown in the browser display shown previously. The third section serves two functions. First of all, it dictates the drawing of the Submit button. Second, it triggers the browser to send the information from this screen to the web server once it has been clicked on. For our discussion, the format of the data in the packet sent from the web browser to the web server is unnecessary. However, as you can see in the preceding function use_DHCP(), the information is easily provided to a function that is capable of executing upon the user's request—in this case, for the router to use DHCP to acquire its IP address.

8.2.4 Brief Summary of the Web Server's Capabilities

We have looked at three distinct capabilities of the web server: transmitting HTML pages to a web browser, providing HTML files with dynamic information in them to a web browser, and using a web browser to command or control an embedded system. The examples and explanation of these features are simple. However, their use is limitless!

I have given presentations to hundreds of people on the benefits of an embedded web server. In those presentations, I always emphasize the importance of imagination in the use of this software. For about 20 K of code and a little effort, you can build systems that have sophisticated user interfaces allowing your users to understand, utilize, and control your embedded system.

What has been discussed thus far in this paper are the basic capabilities of the web server. In the section that follows, we will look at some additional capabilities that a specific implementation of a web server may or may not have.

8.2.5 What Else Should You Consider?

As you continue to pursue information about and use embedded web servers, you will find that vendors of commercial packages will vary in their offerings. Some things that you should look out for are:

- Authentication (security)
- Utilities for embedding HTML files
- File compression
- File upload capabilities.

HTTP 1.0 provides for basic network authentication. If you have ever tried to access a web page and received a dialog box that asks you to enter your network ID and password, you have seen the use of this capability. You should verify that the package provides the ability to add and delete users to the username/password database in the web server. In some cases, you will be required to add the code to do this.

In general, most embedded web servers will be using a very simple file system that resides in memory. Vendors that provide support for this should also provide support for building that file system on your desktop so that it can be included in ROM or Flash on your target. Furthermore, the vendor should also supply an ability to use a more full-featured file system that is capable of handling the myriad of offline storage capabilities available for embedded systems.

If a vendor supports the building of an in-memory file system (files included in ROM or Flash) as just discussed, they should also provide a file compression capability. HTML files can become large and consume a lot of space in an in-memory file system. The compression capability should be able to compress the files while building the in-memory file system and uncompress the file when requested from the web server.

HTML 3.2 provides for the uploading of files from the web browser's host machine to the web server. A vendor supplying a reasonable implementation of a web server should also provide the ability to support this feature.

8.2.6 Conclusion

Web servers will continue to proliferate in embedded systems. The capabilities afforded by this technology are as broad as the imagination of embedded developers like you. This is a technology that can be harnessed to build sophisticated user interfaces to embedded systems, maintain a local repository for user documentation, allow users of the embedded system to control it, and much more.

As the ubiquity of the web continues, embedded devices connected to it will increase. We know of no better method to monitor and control an embedded system right now than a web server. I hope you have the same excitement for this technology as I do. Most of all, I hope that you will be able to employ it in your system to its maximum advantage.

This paper was written primarily to introduce you to this technology and give you some idea how it might be beneficial. Hopefully it has encouraged you to get those creative juices flowing and find a way to use this great technology.

8.3 Introduction to SNMP

This article is based on "Getting Started with Enterprise MIB Design" by Richard Vlamynck, which appeared in NewBits in 1996. SNMP is widely used, but it is instructive to compare the approach with the use of an embedded web server—see "Who Needs a Web Server?" earlier in this chapter.

CW

8.3.1 Why SNMP?

SNMP (Simple Network Management Protocol) is a buzzword in the networking world. The key to understanding and using SNMP is the Enterprise MIB. An Enterprise MIB is a Management Information Base that you design and implement in an application-specific manner and customize for your real-time embedded system.

Why in the world would you want to manage a network? There are a number of situations in which it might be useful. For example, you may be building an embedded real-time system that is a network router, and you would like to have the ability to manage the router or a group of routers from a single or distributed control station. As another example, you may be building an embedded real-time system that is a networked color laser printer, and you would like to manage that resource using an off-the-shelf graphical user interface.

In considering why you may want to have SNMP embedded in your real-time system, ask yourself these questions:

- How do I currently manage networks in general?
- How do I currently manage the nodes on networks?
- Would I like to be able to allow for control of routing?
- Would I like to locate the nodes that violate protocol standards?
- Would I like to have some fancy off-the-shelf software (perhaps a GUI) for managing all the traffic that is on the network?

If you are building a networked resource, you may need to have a standard protocol for communicating information from one node to another node on the network.

8.3.2 The Role of Network Manager

Managers need to manage heterogeneous networks. The managed machines might not even use a common link-level protocol. Computers and embedded systems that are controlled by SNMP might be anywhere in the network, and end-to-end transport connection is required.

8.3.3 Architectural Model

The design architecture of SNMP is the client-server model. The client-server model can also be called the *manager-agent model*. It consists of:

- A program manager (client) on one machine that manages several other machines.
- A program agent (server) that runs on each machine that might be managed (host or gateway). See Figure 8.1.

Note that there is one client for many servers (hosts or gateways). Several different clients might manage the same servers (some only get the information, some change selected information on the gateway, etc.).

8.3.4 A Common Misperception

A common misunderstanding in SNMP implementation is the belief that all the user needs to do is get the agent on the hardware, and SNMP is ready to go. This is incorrect.

The standard MIB cannot cover everything that a real-time embedded system designer might require. Therefore, the system designer needs to create a MIB for the proprietary equipment and protocols.

Your RTOS vendor provides a kernel, but you must write the embedded application yourself. That is self-evident. You can purchase SNMP, but you must write the MIB yourself. SNMP, like a kernel, is a building block you can use to add value to your embedded system.

8.3.5 Application-Level Manager-Agents

There are distinct advantages to running the SNMP managers and agents at the application level:

- One set of protocols is used for the networks.
- There are no limitations in gateway hardware.
- A uniform approach is established by using the same protocols.
- The manager is using IP for communication without knowing which physical attachments are needed for the gateway or the network.

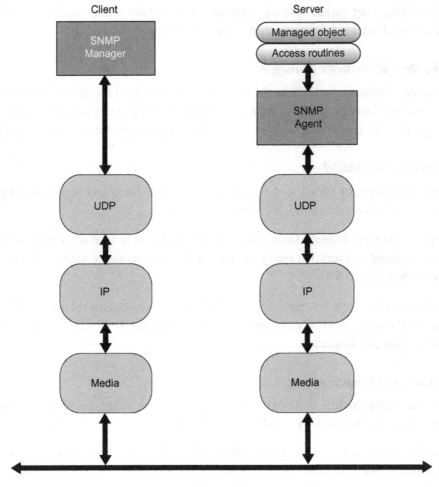

Figure 8.1
SNMP architectural model

There is one drawback to this scheme. In the case of a damaged data structure, table, etc., it is impossible to reach that machine and run the server application (agent) again without intervention.

8.3.6 How to Write Your MIB

Most SNMP implementations support MIB-II, which is used for monitoring generic configuration variables. It is only a starting point for building your embedded system. The embedded system designer must create and debug an applications-specific Enterprise MIB.

There is a general approach that applies to building any Enterprise MIB with SNMP:

1.   Write the ASN.1 specification of your Enterprise MIB.

2. Import that specification file into the GUI (usually via the "file import" button).
3. Use the MOSY (Managed Object SYntax) compiler (or similar) to prepare the MIB for use on the target.

8.3.7 Terminology

There is some minimal terminology that you must know to be conversant with other engineers about SNMP and Enterprise MIBs. The following list is a good starting point for the terms and definitions:

TCP/IP (Transmission Control Protocol/Internet Protocol): Set of communications protocols for networking dissimilar systems.

SNMP (Simple Network Management Protocol): A standard protocol that provides monitoring of IP gateways and the networks to which they attach. This standard defines a set of variables that the gateway must keep and specifies that all operations on the gateway are a side effect of fetching or storing to the data variables.

Gateway: A dedicated computer that attaches two or more networks to route packets from one network to the other. Gateways route packets to other gateways until the packets can be delivered to the final destination. For example, an IP gateway routes IP datagram packets among the networks to which the gateway connects.

Router: A machine configured to make decisions about which of several paths network traffic will follow. Within the TCP/IP environment, an IP gateway routes datagram packets using IP destination addresses.

Client-server model: An interaction concept in one type of distributed system. At the application level, network nodes are termed either "client" or "server," depending on what they are doing at any particular instance. A program at one site sends a request to a program at another site and awaits a response. The client is the requesting program; the server is the program that replies. Client software is traditionally easier to build than server software.

Manager-agent model: An interaction concept in one type of distributed system. At the application level, network nodes are termed either "manager" or "agent." The manager sends a request PDU (Protocol Data Unit), and the agent replies with a response PDU. To translate this to the client-server model, the SNMP manager is a client, and the SNMP agent is a server.

Transport layer: The level of operation that connects heterogeneous networks through gateways. Internet manager software needs to control and examine IP gateways to provide end-to-end transport connections. This concept will work only if everything functions correctly. For example, if the operating system crashes, the gateway's routine table gets damaged, etc., and it would be impossible for the application-level manager program to work properly.

CMOT (CMIP/CMIS over TCP): ISO CMIP/CMIS network management protocols that manage gateways in a TCP/IP Internet. This standard is recommended along with SNMP.

MIB-II (Management Information Base): The set of variables (database) that a gateway running CMOT or SNMP maintains. Managers can fetch and store data into these variables. MIB-II is an extended management database that contains variables used exclusively by SNMP and not shared with CMOT.

MIB-II-OIM (Management Information Base): The set of variables used exclusively by CMOT and not shared with SNMP.

SMI (Structure of Management Information): A specification of the rules used to define and identify MIB variables.

ASN.1 (Abstract Syntax Notation 1): A formal language to provide notation is used in documents that humans read and an encoded representation of the same information is used in communication protocols. In both cases, this formal notation removes any possible ambiguities.

8.3.8 Conclusions

SNMP is a very useful tool for keeping track of node activity on a network. The key to using SNMP successfully is to design and implement your application-specific Enterprise MIB.

Basic Overview of SNMP

The key points to understand about SNMP:

1. SNMP agent is a server for managed objects called MIBs.
2. SNMP manager is a client.
3. Typically, a developer purchases a network management GUI from one vendor and SNMP agent software from another (usually the RTOS supplier).
4. The managed object, the MIB, is application specific. It must be created, designed, built, and debugged by the developer.
5. The bulk of the work falls to the embedded systems developer to create and build the managed object, the application-specific MIB.
6. The developer does not change MIB-II; it is simply extended by writing an application-specific Enterprise MIB. This extension is straightforward because, lexically, the Enterprise MIB follows the generic MIB-II. When the users are at the end of MIB-II and do "get next," they "fall into" an Enterprise MIB.

8.4 IPv6—The Next Generation Internet Protocol

Other articles in this chapter discuss strategies for dealing with the limitations of Internet addressing: DHCP and NAT. Here we take a look at the real solution: IPv6. As expected, the shortage of IP addresses in the world is becoming critical, making IPv6 adoption urgent. This article is based upon an Accelerated Technology white paper written by Glen Johnson and Tammy Leino.

CW

8.4.1 Limitations of the Internet Protocol

The Internet Protocol (IP), developed during the mid-1970s, is the backbone of a family of protocols that includes TCP, UDP, RIP, and virtually every other protocol used for Internet communications. The current version of the IP, version 4 (IPv4), has been in use for more than 20 years. IPv4 has proven to be amazingly adaptable over time. However, the demands placed upon the protocol at its inception pale in comparison to the demands of the millions of hosts that are now connected to the Internet, and IPv4 is finally beginning to show some chinks in its armor. IPv6 deals with many of the shortcomings of IPv4 and introduces some new features. This paper discusses three of the major problems addressed by IPv6.

Depleted Address Space

The main motivation for replacing IPv4 with something better is that the IPv4 address space will ultimately be exhausted. Estimates for the total depletion of the IPv4 address space vary quite widely from 2005 until 2018. Most estimates put the date around 2008 to 2010. Despite the disagreement on when the address space will be depleted, most agree that it will definitely happen unless something better is put in place. Compounding the problem is the uneven distribution of the IPv4 address space. The shortage of addresses will be felt first in Asia, followed by Europe, as those areas received a much smaller number of addresses—9% and 17% of the address space, respectively. Compare that figure to North America's 74% of the address space. The transition to IPv6 is well under way in Japan and the rest of Asia, and it is beginning in Europe as well.

Flawed Addressing Architecture

IPv4 addresses do not provide an efficient and scalable hierarchical address space; that is, it is impossible for a single high-level address to represent many lower level addresses or networks. To picture what a hierarchical address space looks like, think of the telephone numbering system. Just by looking at the area code, one can immediately determine what city or region to route the call to. It is not possible to look at a portion of an IPv4 address and make such a judgment. Therefore, routing becomes increasingly complicated and expensive as the size of the Internet grows.

High Cost

Another criticism of IPv4 is the high cost and maintenance requirements of networks. A significant percentage of the cost of administering an IPv4 network is incurred in the initial configuration of network hosts. IPv4's limitations also aggravate the task of renumbering network devices, which is cumbersome to network administrators.

8.4.2 Introduction to IP Version 6

IP version 6 (IPv6) is a new version of the Internet Protocol, designed as a successor to IPv4. One of the myths associated with IPv6 is that the only reason to adopt IPv6 is the impending depletion of the IPv4 address space. The expanded address space of IPv6 is not the only improvement made in the protocol, however. IPv6 does solve not only the IPv4 address problem, but it also improves upon the current Internet Protocol in other areas including improved addressing architecture, a stateless address autoconfiguration mechanism, a less-expensive address resolution protocol, header format simplification, the ability to detect and recover from a failed forward route, and an improved method to join and leave multicast groups.

8.4.3 Dual Stack Eases Transition

The Internet will consist of a combination of IPv4 and IPv6 nodes for an indefinite period. Therefore, compatibility between IPv4 and IPv6 nodes is critical for a successful transition to IPv6. Because IPv6 is not backwards compatible with IPv4, a dual-stack approach is needed to enable nodes to communicate over both IPv4 and IPv6 simultaneously. This approach paves the way for transition mechanisms that will enable the Internet to move to IPv6.

Although the dual-stack approach is the recommended transition mechanism for networks, IPv6 implementations can also be used in IPv6-only mode for isolated IPv6 networks. This removes the additional overhead of the IPv4 stack for those devices that do not require IPv4 tunneling.

8.4.4 How IPv6 Works

Neighbor Discovery

Neighbor Discovery solves a set of problems related to the interaction between nodes attached to the same link. It defines mechanisms for solving each of the following problems:

- Stateless address autoconfiguration
- Router discovery
- Prefix discovery
- Parameter discovery
- Address resolution
- Neighbor unreachability detection
- Duplicate address detection
- Redirect.

Neighbor Discovery defines five different ICMPv6 packet types. The messages serve the following purpose:

- **Router solicitation:** Hosts send out messages that request routers to generate router advertisements.
- **Router advertisement:** Routers advertise their presence together with various link and Internet parameters either periodically or in response to a router solicitation message. Router advertisements contain prefixes that are used for on-link determination and/or address configuration, a suggested hop limit value, and so forth.

- **Neighbor solicitation:** Sent by a node to determine the link-layer address of a neighbor or to verify that a neighbor is still reachable via a cached link-layer address.
- **Neighbor advertisement:** A response to a neighbor solicitation message. A node may also send unsolicited neighbor advertisements to announce a link-layer address change.
- **Redirect:** Used by routers to inform hosts of a better first hop for a destination.

Stateless Address Autoconfiguration

Stateless address autoconfiguration is a new feature of IPv6 beneficial to network administrators, because it requires no manual configuration of hosts, minimal (if any) configuration of routers, and no additional servers. The stateless mechanism allows a host to generate its own addresses using a combination of locally available information and information advertised by routers and verifies that each generated address is unique on the link.

Stateless address autoconfiguration should greatly decrease the costs of administering an enterprise network. Also, the task of renumbering networks will be simplified since IPv6 can assign new addresses and gracefully time out existing addresses without manual reconfiguration or DHCP.

Duplicate Address Detection

To insure that all configured addresses are unique on a given link, nodes perform duplicate address detection on addresses before assigning them to an interface.

Router Discovery

Router discovery is used to locate neighboring routers as well as to learn prefixes and configuration parameters related to stateless address autoconfiguration.

Router advertisements allow routers to inform hosts how to perform address autoconfiguration and contain Internet parameters such as the hop limit that hosts should use in outgoing packets and, optionally, link parameters such as the link MTU. This facilitates centralized administration of critical parameters that can be set on routers and automatically propagated to all attached hosts.

Prefix Discovery

Router advertisements contain a list of prefixes used for on-link determination and/or stateless address autoconfiguration. Flags associated with the prefixes specify the intended uses of a particular prefix. Hosts use the advertised on-link prefixes to build and maintain a list that is used in deciding when a packet's destination is on-link or beyond a router.

Address Expiration

IPv6 addresses are leased to an interface for a fixed (possibly infinite) length of time. Each address has an associated lifetime that indicates how long the address is bound to an interface. When a lifetime expires, the binding (and address) become invalid, and the address may be reassigned to another interface elsewhere in the Internet. To handle the expiration of address bindings gracefully, an address goes through two distinct phases while assigned to an interface. Initially, an address is "preferred," meaning that its use in arbitrary communication is unrestricted. Later, an address becomes "deprecated" in anticipation that its current interface binding will become invalid. While in a deprecated state, the use of an address is discouraged but not strictly forbidden.

Address Resolution

Address resolution is the process through which a node determines the link-layer address (e.g., Ethernet MAC address) of a neighbor given only its IP address. Address resolution is redefined for IPv6 and does not use ARP (Address Resolution Protocol) packets, as is the case for IPv4.

Nodes accomplish address resolution of IPv6 neighbors by multicasting a request for the target node to return its link-layer address. The target returns its link-layer address in a unicast response. By using multicast and unicast addresses instead of the broadcast address, there are fewer needless interruptions of other nodes on the network.

Header Format Simplification

To simplify and optimize processing of IP packets, a few changes were made to the format of the IP header for IPv6. The length of the IPv6 header is fixed as opposed to the variable length IPv4 header. This helps to simplify processing of IPv6 packets as certain assumptions in the IP processing code can be made. Also, some IPv4 header fields have been dropped or made optional. Most notable is the lack of a checksum field for the IPv6 header. This greatly improves performance in routers. When an IPv4 packet is forwarded by a router, the Time-to-Live (TTL) field must be decremented, which forces the IPv4 header checksum to be recomputed; this is a CPU-intensive operation. Since this field is not present in the IPv6 header, routers simply decrement the hop limit—TTL in IPv6—and forward the packet.

Neighbor Unreachability Detection

Neighbor unreachability detection detects the failure of a neighbor or the failure of the forward path to the neighbor. Once failure has been detected, an alternate route can be found without interrupting the flow of data from the application's point of view.

Multicast Listener Discovery

The purpose of multicast listener discovery is to enable each IPv6 router to discover the presence of multicast listeners (i.e., nodes wishing to receive multicast packets) on its directly attached links, and to discover specifically which multicast addresses are of interest to those neighboring nodes. This information is then provided to whichever multicast routing protocol is being used by the router, to ensure that multicast packets are delivered to all links with interested receivers.

Tunneling

In most deployment scenarios, the IPv6 routing infrastructure will be built up over time. While the IPv6 routing infrastructure is being deployed, the existing IPv4 routing infrastructure can remain functional and can be used to carry IPv6 traffic. Tunneling provides a way to utilize the existing IPv4 routing infrastructure to carry IPv6 traffic.

IPv6/IPv4 hosts and routers can tunnel IPv6 datagrams over regions of IPv4 routing topology by encapsulating them within IPv4 packets.

Tunneling operates as follows:

1. The entry node of the tunnel (the encapsulating node) creates an encapsulating IPv4 header and transmits the encapsulated packet.
2. The exit node of the tunnel (the decapsulating node) receives the encapsulated packet, reassembles the packet if needed, removes the IPv4 header, updates the IPv6 header, and processes the received IPv6 packet as usual.

IPv6 defines numerous techniques to accomplish tunneling. Two of the most common tunneling techniques are *configured tunneling* and *6to4 tunneling*.

Configured Tunneling

In configured tunneling, the tunnel endpoint address is determined from configuration information in the encapsulating node. For each tunnel, the encapsulating node must store the tunnel endpoint address. When an IPv6 packet is transmitted over a tunnel, the tunnel endpoint configured for that tunnel is used as the destination address for the encapsulating IPv4 header. Configured tunneling uses IPv6 native addresses as the source and destination addresses of the IPv6 packet.

6to4 Tunneling

The IANA (Internet Assigned Numbers Authority) has permanently assigned the prefix `2002::/16` for the 6to4 scheme. The subscriber site is then deemed to have the address prefix `2002:V4ADDR::/48`, where `V4ADDR` is the globally unique 32-bit IPv4 address. Within the subscriber site, this prefix is used exactly like any other IPv6 prefix. The 6to4 address is used as the source address of all communications via the 6to4 tunnel.

IPv6 packets from a 6to4 site are encapsulated in IPv4 packets when they leave the site via its external IPv4 connection. `V4ADDR` must be configured on the IPv4 device.

DNS for IPv6

To support the storage of IPv6 addresses, the following extensions have been defined:

- A new resource record type, `AAAA`, is defined to map a domain name to an IPv6 address.
- A new domain, `ip6.int`, is defined to support lookups based on address.

IPv6 Extension Headers

Unlike in IPv4, the IPv6 header is a fixed length. Any additional information that needs to be provided to the IP layer is contained in extension headers appended to the basic IPv6 header. The following extension headers are commonly supported in IPv6 implementations: fragmentation, routing, destination options, and hop-by-hop options.

Ancillary Data

Ancillary data is used to transfer IPv6 extension headers and additional control information between the application and the network stack via socket options and the "send message" and "receive message" routines. This additional data is used by the local IPv6 stack, intermediate IPv6 stacks responsible for packet routing, and the destination IPv6 stack to properly process the IPv6 packet as is required by the sending application.

Ancillary data can be used to send/receive the following control information to the stack:

- Hop-by-hop options
- Destination options
- Routing header
- The interface index of the outgoing/incoming packet
- The source address of the outgoing/incoming packet
- The next-hop address to use for the outgoing/incoming packet
- The traffic class of the outgoing/incoming packet.

8.4.5 RFC Support

The following RFCs are key to IPv6:

- 1886: DNS Extensions to support IP version 6
- 1981: Path MTU Discovery for IP version 6
- 2080: RIPing for IPv6
- 2373: IP Version 6 Addressing Architecture
- 2452: IP Version 6 MIB for the Transmission Control Protocol
- 2454: IP Version 6 MIB for the User Datagram Protocol
- 2460: Internet Protocol, Version 6 (IPv6) Specification
- 2461: Neighbor Discovery for IP Version 6
- 2462: IPv6 Stateless Address Autoconfiguration
- 2463: Internet Control Message Protocol (ICMPv6) for the IPv6 Specification
- 2465: MIB for IP Version 6: Textual Conventions and General Group
- 2466: Management Information Base for IP Version 6: ICMPv6 Group
- 2710: Multicast Listener Discovery (MLD) for IPv6
- 2711: IPv6 Router Alert Option
- 2893: Transition Mechanisms for IPv6 Hosts and Routers
- 3019: IPv6 MIB for the Multicast Listener Discovery Protocol
- 3056: Connection of IPv6 Domains via IPv4 Clouds
- 3493: Basic Socket Interface Extensions for IPv6
- 3542: Advanced Sockets Application Program Interface (API) for IPv6
- 3810: Multicast Listener Discovery Version 2 for IPv6

8.5 The Basics of DHCP

DHCP is another important technology that has grown largely from the need to conserve scarce IP addresses on networks connected to the Internet. This article is based upon an Accelerated Technology white paper written by Steven Lewis.

CW

The Dynamic Host Configuration Protocol (DHCP) provides a means for automating the configuration of Internet hosts. DHCP can be used to automatically assign IP addresses, to deliver TCP/IP stack configuration parameters such as the subnet mask and default router, and to provide other configuration information such as the IP address of a print, time, or news server.

DHCP is built on the client-server model. Upon receiving a request from a DHCP client, a DHCP server is responsible for the allocation of an IP address and other configuration parameters the client may require. The server allocates the IP address from a pool of addresses that it is responsible for managing. The client is responsible for requesting each of the parameters that are required for proper operation on each connected network.

The primary goal of DHCP is to reduce the administrative costs of managing a network. By utilizing DHCP, the administrative costs can be limited to the configuration of the DHCP server. The DHCP server will need to be manually configured with one or more pools of IP addresses that can be used to satisfy requests. The DHCP server will also need to be manually configured with the other parameters that network hosts will require. However, there is no need to configure any of the DHCP clients. New network hosts can be plugged into the network and function with no manual configuration. Parameters such as default gateway address, DNS server addresses, and network domain name can be changed at the server for every client on a network that uses DHCP rather than requiring a manual change of each host.

8.5.1 A DHCP Server

Flexible Address Assignment

Typically, when a client requests an IP address, a DHCP server will return one of many IP addresses from a pool. For most network hosts this is fine. However, it is desirable for a known IP address to be given to servers. A DHCP server may perform both dynamic and static address assignment, and both methods can be utilized simultaneously. It is also possible to configure a DHCP server with multiple noncontiguous blocks of IP addresses. This is useful when multiple blocks of addresses are used for dynamic address assignment, which are separated by one or more addresses that are reserved for static address assignment. Finally, the lease times can be configured to match the dynamics of the local network. On a very stable network, the lease times can be very long. On networks that are very dynamic, it is generally better to shorten the lease times.

Support for Multiple Subnets

Multiple subnets can be serviced with a single DHCP server. Through the use of multiple network interfaces and/or use of relay agents, the DHCP server can be configured to handle DHCP requests from clients on multiple network subnets. Also, each subnet can be fully configured with its own set of network parameters or can use certain parameters from a global configuration, which can be used for parameters that are common across all subnets.

8.5.2 Theory of Operation

There are four transactions that occur during a successful DHCP operation. All DHCP operations begin with a client broadcast of a DISCOVER message. The DISCOVER message is formatted with client characteristics, such as the client's hardware address and client ID, that will enable a DHCP server to differentiate between clients. By default, DHCP DISCOVER messages are sent to port 67, the well-known DHCP server port. The server will reply to DISCOVER requests from clients on any network serviced by the server. If the server decides that the client is one that it should service, the server will choose an IP address to offer to the client. After selecting an IP address for the client, the server will broadcast an OFFER message. The OFFER message will contain the selected IP address, lease times, and very basic network parameters. By default, OFFER messages are sent to port 68, the well-known DHCP client port.

Upon accepting the OFFER, the client will broadcast a REQUEST message. The REQUEST message will contain the offered IP address, as well as requests for each of the network parameters that it desires. The REQUEST message will be sent to the server port 67. If the client receives more than one OFFER message in response to its DISCOVER, only one will be chosen.

When the server receives the REQUEST message, it will determine whether the requested IP address is still available and whether the requested network parameters are valid. Then it will reply to the client with an ACK message. (Note that some clients, such as certain Microsoft DHCP clients, when restarted, will not send out a DISCOVER message; rather, the client will jump right to a REQUEST. Therefore, if the lease had expired while the client was restarting, the server may have provided this address to another client. The server will send a NAK to the requesting client to let it know that address is not available, and the client will send out a DISCOVER message.) The ACK message will contain the requested IP address and the data corresponding to the requested network parameters. The ACK will be broadcast to client port 68. Once the client receives the ACK message, it is free to use the IP address and network parameters for the duration of the lease. Figure 8.2 illustrates the process of acquiring an IP address via DHCP.

Although the DHCP client can use the IP address only for the duration of the lease, the client will request that the lease be extended as necessary. The expiration of the lease provides a means for IP addresses to be returned to the pool of available IP addresses in the event that the client is removed from the network. The DHCP protocol defines a mechanism for lease renewal and milestone times for clients to adhere to when attempting to renew an IP lease.

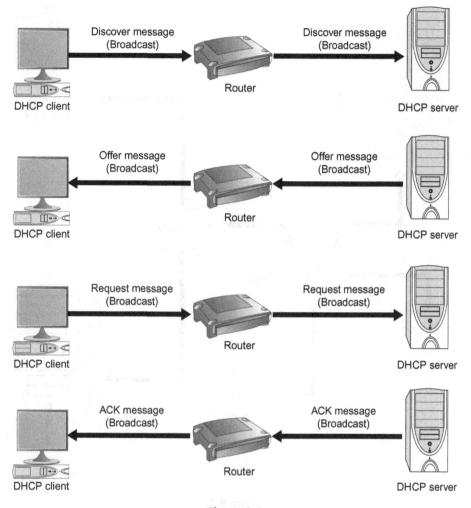

Figure 8.2
Acquiring an IP address via DHCP

When the client was presented with the OFFER and ACK messages, the total lease time for the IP address was provided. This lease time is determined by the administrator of the DHCP server and can vary from infinite to less than a day. In addition to the total lease time, two other lease parameters were provided to the client: a renew time and a rebind time. The renew time is the time at which the client should attempt to renew the lease with the server that currently holds the IP lease. To renew the lease, the client would unicast a REQUEST message to the server that provided the IP address that it is using. The renewal time is usually 50% of the total lease time. This is done so network traffic, due to DHCP messages, will be minimized and to provide the client with a standard. The client will periodically attempt lease renewal until success or until the rebind time is reached.

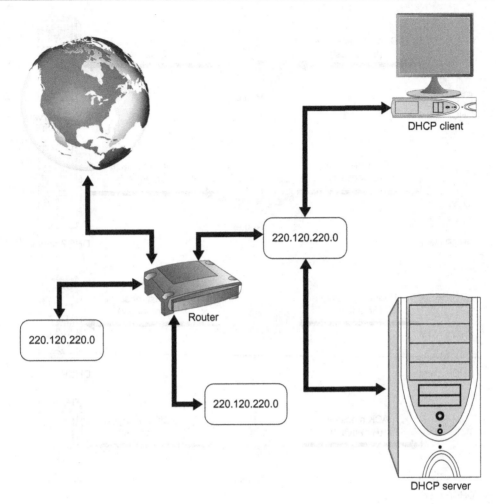

Figure 8.3
DHCP server servicing a single local subnet

If all attempts to renew the lease fail, the rebind time will eventually be reached. At this point the client will attempt to extend the lease with any server. During the rebinding time, the client will broadcast a REQUEST message to server port 67. The REQUEST message will contain the IP address that the client is currently using, in hopes that some other server will be able to extend the lease for the client's current address. The rebinding time is usually 87.5% of the total lease time. If the client is unable to extend the lease before the total lease time has expired, the client must discontinue its use of the IP address.

The simplest case for the use of a DHCP server is the servicing of a single local subnet. This is illustrated in Figure 8.3.

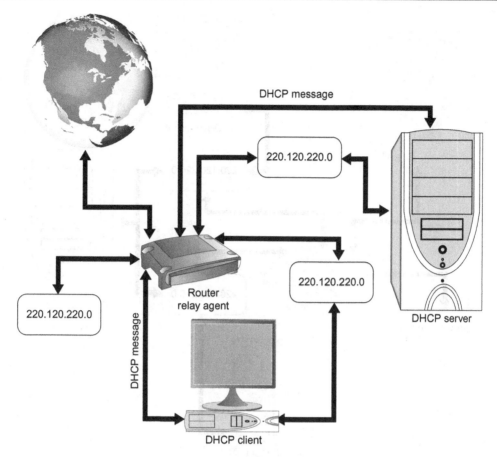

DHCP message

220.120.220.0

Router
relay agent

220.120.220.0

220.120.220.0

DHCP server

DHCP message

DHCP client

Figure 8.4
Using a DHCP relay agent

There are several options for the management of multiple subnets with DHCP. The obvious solution is to place a DHCP server on each subnet. This may not be possible for technical or financial reasons.

A second option is to use a DHCP relay agent, which is illustrated in Figure 8.4.

If a relay agent is implemented, it must be configured to forward all DHCP messages to the DHCP server. The relay agent does not keep track of the DHCP messages that it forwards. The relay agent only modifies the DHCP message by inserting the IP address of the interface that received the message into the gateway IP field of the DHCP message and incrementing the hops field. The server will send its replies to the interface of the relay agent that initially received the DHCP message. Since the network segment that is accessing the server through the relay agent is completely isolated from the server, a separate configuration must be used for the IP addresses that will be offered to the clients on that segment.

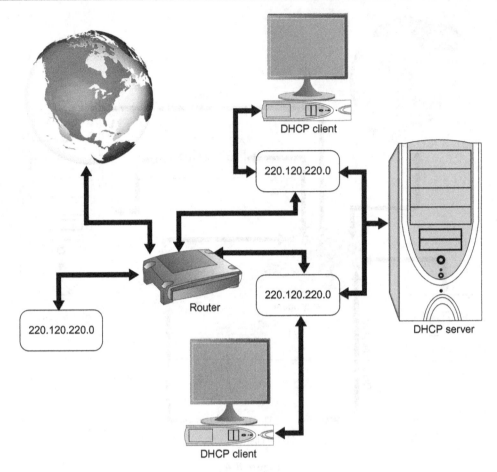

Figure 8.5
A DHCP with multiple network interfaces

A final option, for those DHCP servers that have more than one network interface, is to connect the server to each subnet that must be serviced, as illustrated in Figure 8.5.

In this case, the DHCP server must be configured for multiple subnets. When the server is configured in this manner, each DHCP network parameter can differ from one network segment to the next. Also, if it is desired to have a set of network parameters to be constant across all network segments, the network manager may configure a global configuration from which all segments may draw all or some of their parameters.

8.5.3 RFC Support

The requirements for a DHCP server are outlined in RFC 2131. DHCP options are outlined in RFC 2132. Not all options listed in RFC 2132 are necessarily supported by a given DHCP server.

8.6 NAT Explained

Network address translation is a critical technology, as the Internet becomes ever more widely used, with more devices becoming connected, and the availability of IP addresses is becoming a serious problem. NAT is essentially a "kluge" that deals with the issue extremely well in some situations. Longer term, the next generation of IP protocol, IPv6, is likely to be the solution—this is described in "IPv6—The Next Generation Internet Protocol," earlier in this chapter. This article introduces NAT and is based upon an Accelerated Technology white paper written by Glen Johnson and Tammy Leino.

CW

The IP Network Address Translator (NAT) protocol is a router protocol that allows nodes on a private network to transparently communicate with nodes on an external network and vice versa. Nodes on a private network have not been assigned a globally unique IP address; therefore, communication with the external network would otherwise be impossible. This transparent communication is accomplished by modifying the IP and protocol-specific headers of packets flowing to and from the private network. NAT solves three common problems with growing networks: shortage of globally unique IP addresses, firewall-like protection for the private network, and flexibility of network administration.

8.6.1 NAT Explained

There are a variety of flavors of NAT. Basic NAT maps an IP address on the private network to a globally unique IP address. Basic NAT performs translation on only the IP address and requires the NAT router to have a pool of globally unique IP addresses, which can be mapped. Basic NAT also limits the number of nodes on the private network that can communicate with the external network to the number of globally unique IP addresses that are available. This means that in order for five nodes on the private network to communicate with the external network at the same time, there must be five globally unique IP addresses available for translation. While these five addresses are in use, no other nodes on the private network can communicate with the external network.

NAPT (Network Address Port Translator) solves some of the problems with basic NAT and does a much better job of solving the problem of a shortage of globally unique IP addresses by allowing all nodes on a private network to communicate with the external network by sharing a single globally unique external IP address. This is advantageous for homes and businesses with limited globally unique IP addresses, because all users can access the external network simultaneously. NAPT accomplishes this by replacing, within the protocol headers, the IP address and TCP/UDP port number of the private node with the globally unique external IP address and TCP/UDP port number. In other words, NAPT performs translation on the UDP/TCP port numbers as well as on the IP address. With NAPT, the theoretical limit is up to 64,000 simultaneous sessions (address/port combinations) at a time. NAPT is also known as *IP masquerading*.

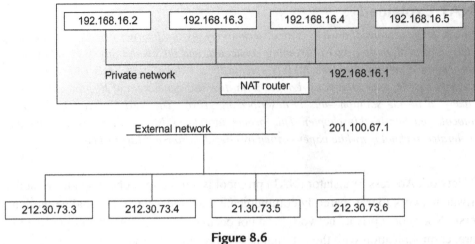

Figure 8.6
NAT theory of operation

Bidirectional NAT enables connections to be initiated from hosts on the external network as well as the private network. Specific ports on the NAT router are mapped to services on a private node or server via a portmap service (see the section "The Portmap Service" later in this article). The NAT router relays all matching requests from the external network to the specific private server. This enables servers on the private network to be accessible to nodes on the external network. For example, an FTP client on the external network could establish a connection with an FTP server on the private network. Without bidirectional NAT support, all connections have to be initiated from nodes on the private network.

Since all connections must be initiated from the private network or registered with the portmap service, NAT provides firewall-like protection for the private network. An intruder would have to first gain access to the NAT router to infiltrate the private network. Also, the size and topology of the private network are hidden behind the NAT router. Note that NAT does not necessarily preclude the need for a real firewall.

No modifications need to be made to the NAT router when a new node is added or existing nodes are removed or reconfigured. This provides for flexibility of network administration.

The theory of operation for NAT is illustrated in Figure 8.6.

The private network 192.168.16.x is hidden from the external network behind a NAT router. The NAT router has one external interface (201.100.67.1) used to communicate with the external network and to protect the anonymity of the private nodes. The NAT router has one private interface (192.168.16.1) used to communicate with the private network.

When a private node sends a packet to the external network, the NAT router intercepts the packet and replaces all instances of the private source IP address (192.168.16.xxx) and

TCP/UDP source port with the external IP address (201.100.67.1) and an assigned external TCP/UDP source port. NAT assigns the port number. No user intervention or configuration is necessary for private nodes to initiate communication with external nodes. However, if a server on the private network needs to service clients located on the external network, then the server's port must be registered with NAT via the portmap service.

When an external node responds to a private node or initiates an acceptable connection with a private node, the NAT router intercepts the packet and replaces all instances of the external destination IP address (201.100.67.1) and assigned external destination TCP/UDP port with the private IP address (192.168.16.xxx) and destination TCP/UDP port.

The Portmap Service

As mentioned previously, NAT may be bidirectional. This means that servers can be supported on the private network. NAT achieves this via a portmap service, which is used to register services (servers) on the private network as accessible to the external network.

Multiple nodes on the private network may be registered on the same port using the same protocol. For example, multiple nodes may be registered as FTP servers. As requests for connections come in through the NAT router from the external network, NAT will forward these requests in a round-robin manner to the respective servers on the private network. This is done to distribute the work evenly across multiple servers.

Note that since the external network sees the NAT router as the one and only final destination, there is no way to specify to which of the multiple private servers the packet may have been intended. For example, if a certain file is stored on one of three private servers, and an external user FTPs to retrieve that file, it is not guaranteed that the request will be sent to the proper server. If multiple servers of the same type are to be used effectively on the private network they must be mirrored.

8.6.2 RFC Support

The requirements for a NAT router are outlined in RFC 1631 and clarified in RFC 2663. Since the implementation of a NAT router is so closely related to the private network it is hiding, the RFCs are more informational overviews than stringent requirements documents.

8.6.3 Protocol Support

NAT may support a wide variety of networking protocols. Note that *support* in this case means that NAT can forward data sent by these protocols from the private network to the external network. Any networking protocol can be executed on the NAT router itself. For example, if TFTP client is not listed as supported by a particular NAT implementation, this means that NAT does not support a TFTP client on the private network communicating with a TFTP server on the external network. However, a TFTP client could execute on the NAT router. This

TFTP client could communicate with TFTP servers on both the private and external network. Examples of protocols that are likely to be supported include IP, TCP, UDP, DNS, ICMP, HTTP client-server, Telnet client-server, TFTP client-server, and FTP client-server.

8.6.4 Application Level Gateways

An Application Level Gateway (ALG) is an extension to NAT, which modifies the payload of a packet aside from the IP and/or protocol headers. Note that only those applications that embed IP addresses and/or port numbers within the application payload require an ALG.

Commonly implemented ALGs are:

- ICMP: Provides functionality for ICMP error codes.
- FTP: Provides functionality for FTP PORT and PASV commands.

8.6.5 The Private Network Address Assignment

The Internet Assigned Numbers Authority (IANA) has reserved the following three blocks of the IP address space for private Internets:

- `10.0.0.0-10.255.255.255`
- `172.16.0.0-172.31.255.255`
- `192.168.0.0-192.168.255.255`

An organization that decides to use IP addresses in this address space can do so without coordination with any Internet registry. This address space information is taken from RFC 1918.

8.7 PPP—Point-to-Point Protocol

Although PPP has been around for many years, it has recently become even more common with the adoption of broadband (xDSL) Internet connections. This article is based upon an Accelerated Technology white paper jointly written by Glen Johnson, Kevin George, Fakhir Ansari, and Uriah Pollock.

CW

8.7.1 Introduction

PPP (Point-to-Point Protocol) provides a standard method for transporting multiprotocol datagrams over point-to-point links. In the context of a network application, PPP allows IP datagrams to be exchanged with a node at the other end of a point-to-point link. Typically, a client will initiate a PPP connection by using a modem to dial into a foreign server through the public telephone system. However, PPP is also used in environments where the physical medium is not always point-to-point. One such example is Ethernet. The PPPoE and L2TP protocols enable support for transmission of PPP packets over Ethernet.

A PPP implementation may include support for a PPP client and a PPP server, perhaps even being utilized as both at the same time. Applications only have to be aware that PPP is being used as the underlying link-layer driver when establishing and breaking the physical link—that is, during dial-up and hang up. In all other respects, the application is not aware that PPP is the low-level driver being used.

Abstracted Link-Layer Interface

Because PPP is now being adapted for use over various types of physical mediums, including ATM and broadcast mediums such as Ethernet, it is necessary to recognize PPP as providing for communications over logical point-to-point links as well as physical point-to-point links. To provide for flexibility in supporting multiple link layers, the interface to the link layer is commonly abstracted. The interface to each link layer is thus a self-contained module. Serial (HDLC) and Ethernet (PPPoE, L2TP) link layers are examples. This modularization of PPP results in greater system flexibility, efficient code reuse, and hardware transparency for easier application development. It also makes it straightforward for users to plug in support for new link layers; for example, PPPoA (PPP over ATM).

HDLC and Modem Support

PPP originated as a protocol for sending datagrams over serial point-to-point links. These links were usually dial-up links. Today this is still by far the primary use for PPP. As a result, PPP usually includes support for HDLC framing, as well as basic support for driving a Hayes-compatible modem.

8.7.2 How PPP Works

Theory of Operation

The process of establishing a PPP connection begins with establishing a physical link with a foreign peer. This can be done in one of two ways. When using serial communication with HDLC, a client will use a modem to connect with a server over standard telephone lines. Once the physical connection has been established, the client and server can exchange packets. When using the optional PPPoE interface, the physical connection is always up. Instead of using a modem to dial into a server, a PPPoE client performs "discovery" to find the PPPoE server. To the application, the connection process looks the same for both PPPoE and HDLC. The PPPoE and HDLC modules take care of the details. However, when HDLC is used over a dial-up link, it is likely that the modem will require some initialization.

Once the circuit between the peers has been created, via either dial-up (HDLC) or other network mediums (PPPoE, L2TP), PPP will begin Link Control (LCP) negotiation with the foreign peer. During LCP negotiation, each peer tells the other what its capabilities are, so that they can make adjustments for each other's limitations. For example, one peer may be configured to send packets no larger than 700 bytes in size, while the other peer may send a full-size packet (1500 bytes).

Authentication is performed during the LCP negotiations. Authentication is optional, and some PPP servers may not authenticate clients. PPP supports a number of user-authentication protocols: Password Authentication Protocol (PAP), Challenge Handshake Authentication Protocol (CHAP), Microsoft PPP CHAP Extensions (MS-CHAP v1), and Microsoft PPP CHAP Extensions version 2 (MS-CHAP v2), one of which is selected during LCP negotiation.

Once the client is authenticated, LCP negotiation will be complete, and the next phase—Network Control (NCP)—will begin. Each higher layer network protocol will have its own NCP specification for use over PPP. PPP is likely to support two NCPs: the Internet Protocol Control Protocol (IPCP) and the Internet Protocol version 6 Control Protocol (IPV6CP). Both are used for configuring the client and server peers with IP address information. This information consists of the client and server's IP addresses, and optionally a primary and secondary DNS address, all of which are typically provided by the PPP server.

All of these stages make up the PPP negotiation, and upon completion, the link is ready for normal IP and/or IPv6 network communication. There are no restrictions on the types and number of applications that can utilize the PPP link. Obviously, available bandwidth will impose some limitation on the number of applications that can effectively utilize the link simultaneously.

Once communications are complete, PPP provides an API for closing the connection (hang up). This step is the same for HDLC, PPPoE, and L2TP, and it is taken care of by the LCP layer of PPP.

Application Development

Developing a network application that uses PPP is not very different than developing one for any other network medium. There are two minor differences. First, if PPP will be used over dial-up, the modem will have to be initialized. Second, a dial-up step is necessary to establish the link. Note that dial-up is necessary even when PPPoE or L2TP is used. However, in this case, the software virtually performs this step because a modem is not actually dialed.

The responsibilities of the application developer, when creating a PPP link, are limited to initialization of PPP and the modem, assigning minimal properties of the link, and initiating the connection with a peer. The negotiation parameters are set to the most common defaults recognized by most PPP clients and servers. In most cases, there is no need to worry about these details.

The following steps are taken to establish a PPP link within an application:

1. **Device initialization:** This step is not unique to PPP. Device initialization is similar to that of other networking applications. Parameters, such as serial port number, baud rate, and data bits, that are specific to the serial device, need to be set up and registered.
2. **Modem initialization:** If a modem is being used, it must be initialized. This is done using standard AT command strings.
3. **Establish the connection:** Next the link must be established. This step varies for PPP clients and PPP servers. A PPP client must perform dial-up, while a server will wait for a client to dial in.

4. **Disconnect:** When communications on the link are complete, hang up is performed to break the link.

8.7.3 PPP Details

NCP/IPCP/IPV6CP

The PPP RFCs define multiple NCPs (Network Control Protocols). IPCP (Internet Protocol Control Protocol) and IPV6CP (Internet Protocol version 6 Control Protocol) are the NCPs commonly supported by PPP implementations. Both are responsible for configuring, enabling, and disabling the IP/IPv6 protocol for use on the point-to-point link.

Authentication

The following protocols are generally supported by PPP to authenticate PPP clients:

- None (no authentication)
- PAP (Password Authentication Protocol)
- CHAP (Challenge Handshake Authentication Protocol)
- MS-CHAP v1 (Microsoft PPP CHAP Extensions)
- MS-CHAP v2 (Microsoft PPP CHAP Extensions, Version 2).

LCP

The Link Control Protocol is responsible for establishing, configuring, and testing the data-link connection. Options include the following:

- Maximum Receive Unit
- Asynchronous Control Character Map
- Protocol Field Compression
- Address and Control Field Compression
- Magic Number
- Authentication Protocol.

HDLC

HDLC provides support for sending PPP packets over serial links. This is HDLC-like framing as hardware-based HDLC is not used. Instead, a software implementation of the HDLC framer is used as specified by the PPP protocol. Typically, when using HDLC, the target system will use a serial modem to establish a physical connection with another peer over a telephone line. The services necessary to drive a modem are normally supported by PPP. These include dialing, waiting for incoming calls, hanging up the modem, and Caller Line Identification (CLI), if allowed by the underlying hardware.

Alternatively, the HDLC may also support the Microsoft Windows Direct Cable Connection (DCC) protocol. This allows the target system to establish a connection to another peer directly over a NULL modem cable. Using DCC, a phone line and modem are not required to perform network communication over PPP. This method helps to simplify the development environment by removing those unnecessary parts. The DCC protocol can also be used in applications that do not require modem dial-up functionality.

PPPoE

PPPoE combines the best points of Ethernet and traditional PPP, allowing a virtual point-to-point connection to be established over Ethernet. Utilizing PPPoE, it is possible to leverage Ethernet for its costs and PPP for the ability to manage user access. Also, solutions that utilize PPPoE present a familiar interface to an ISP's existing customer base, who are used to the dial-up paradigm.

Because of these advantages, virtually all of the ISPs that provide xDSL access require their customers to "dial-up" using PPPoE. As a result, residential gateways will need to include support for PPPoE.

Support for Both Hosts and Access Concentrators

Although residential gateways will generally only have to act as a host (client), PPPoE implementations include support for both hosts and access concentrators (servers).

Support for MSS Replacement

PPPoE implementations may have optional support for MSS (Maximum Segment Size) replacement. The PPPoE header adds 8 bytes of data to each Ethernet packet. As a result, the effective MTU becomes 1492 instead of then normal Ethernet MTU of 1500.

When PPPoE is used in a gateway, clients on the network will have no knowledge of this fact. When establishing TCP connections, hosts will advertise a MSS of 1500 rather than the correct value of 1492. This can result in the oversized segments getting dropped by a gateway enabled with PPPoE. This is solved by dynamically replacing the MSS in TCP packets with the correct value.

L2TP

L2TP is a Virtual Private Network (VPN) protocol. It has been designed to carry PPP packets over any routed protocol like IP, IPX, and Apple Talk. It also supports any WAN technology including ATM, Frame Relay, and X.25. L2TP can be used as a tunneling protocol over public and private networks.

L2TP may be used in an L2TP Access Concentrator (LAC) and L2TP Network Server (LNS). A LAC tunnels PPP packets over a public network to the LNS, which acts as a gateway to the private network. The LAC may act as a PPP server to which remote clients connect, as shown in Figure 8.7.

Once the remote client connects to LAC, the remote client's PPP packets are then tunneled by the LAC to the LNS, in effect connecting the remote client to the private network. The LAC may also act as the remote client, as shown in Figure 8.8.

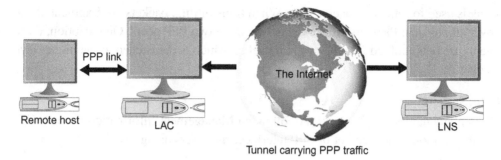

Figure 8.7
Remote host dials into the LAC to connect to the private network

Figure 8.8
Remote host acting as the LAC to connect to the private network

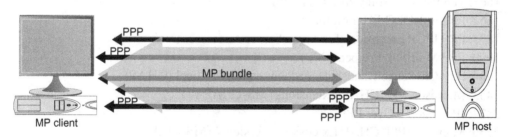

Figure 8.9
Multiple links form one virtual link

Multilink Protocol

Rising demands for speed, combined with the desire to utilize the existing infrastructure, makes Multilink Protocol (MP) an ideal choice. Multilink Protocol provides greater bandwidth by providing multiple independent PPP links between two network nodes. The multiple links form one virtual link, which is the only link visible to the higher-layer protocols (see Figure 8.9).

This link is used to send and receive data. When transmitting, packets are fragmented and sent over all physical PPP links that exist between the two PPP peers. On reception, these fragments are reassembled into the original packet, which is dispatched to the higher layer protocols.

PPP MIB Support

A PPP implementation may support the following Management Information Bases (MIBs). These allow remote management of PPP interfaces and users using SNMP:

- LCP MIB: This MIB includes link statistics counters, status variables, and LCP option configuration.
- IPCP MIB: This MIB includes IP information and configuration for each PPP device.
- Security Protocols MIB: This MIB manages authentication for each link and for each PPP user account.

8.7.4 RFC Support

Following is the list of RFCs that pertain to PPP:

1332: PPP Internet Protocol Control Protocol (IPCP)
1334: PPP Authentication Protocols (PAP)
1471: MIB for Link Control Protocol of the PPP
1472: MIB for Security Protocols of the PPP
1473: MIB for IP Control Protocol of the PPP
1661: The Point-to-Point Protocol (PPP)
1662: PPP in HDLC-like Framing
1877: PPP IPCP Extensions for Name Server Addresses (DNS only)
1990: PPP Multilink Protocol (MP)
1994: PPP Challenge Handshake Authentication Protocol (CHAP)
2433: Microsoft PPP CHAP Extensions (MS-CHAP v1)
2472: IP version 6 Over PPP
2759: Microsoft PPP CHAP Extensions, Version 2 (MS-CHAP v2).

The RFC that pertains to PPPoE is 2516: Method for Transmitting PPP over Ethernet (PPPoE).

8.8 Introduction to SSL

Security and confidentiality of data communications are high priorities in modern society. So it is unsurprising that embedded systems have an increasing need to support the appropriate protocols and SSL is a key component. This article, which is based upon an Accelerated Technology white paper by Doug Phillips, provides an introduction to the topic.

CW

8.8.1 Introduction

Secure Sockets Layer (SSL) and Transport Layer Security (TLS) provide a secure protocol by which two networked peers may perform encrypted communications. SSL is most commonly used for sending private data from a web browser to a web server. This private data may include credit card numbers or other personal information. Virtually everyone who has browsed the web has at some point used SSL but may not have realized it. When browsing the Internet, if the browser states that it is entering a secure site, then SSL is being used to authenticate the web server and to encrypt the data exchanged.

Netscape created SSL 1.0 in 1994. However, version 1.0 was never shipped. The first working version of SSL was 2.0. This was packaged in Netscape Navigator in late 1994. Netscape continued to work on SSL with the help of outside engineers. A year later, SSL 3.0 was released. In 1996, the Internet Engineering Task Force (IETF) took over responsibility for the protocol from Netscape. The IETF chose to rename the protocol to TLS (Transport Layer Security)—perhaps thinking a name not linked to Netscape would be more acceptable. This newly named protocol can be found in RFC 2246. Since that time, only one version of TLS was released in early 1999. Current widely used standards are SSL 2.0, SSL 3.0, and TLS 1.0. The name SSL is still more commonly used to refer to all versions of the protocol. As a result, within this article, SSL will be used to mean either SSL or TLS.

The web, or HTTP (HyperText Transfer Protocol), is the most common, but not the only, use for SSL. Other protocols such as FTP, SMTP, and POP3 (e-mail) can be modified to utilize SSL, or SSL can be used with basic TCP or UDP communications. Once the data to be sent has been created, SSL encapsulates the data in much the same way TCP would. There is one difference: SSL modifies the data by encrypting it.

Many SSL implementations are based upon OpenSSL. OpenSSL is a free, open source SSL solution. OpenSSL was not developed with the embedded industry in mind. So care is needed for it to be used efficiently in this context.

8.8.2 How SSL Works

Certificates

The SSL standard specifies not only secure negotiations, but also the ability to authenticate who is on the other side of the connection. This is especially important when sending private data. Authentication is performed through the use of certificates. A certificate is much like a driver's license. It holds the name and address of the certificate holder, along with some encryption information and a signature. The signature, however, is not the holder's, but is placed there by a third party known as the Certificate Authority (CA). Verisign is an example of a CA.

The CA places their signature on the certificate to verify that the entity holding the certificate is who they say they are. This signature is really a hash of the certificate itself, then encrypted by the CA. This ensures that no two certificates will be alike, and therefore, a signature cannot be copied from a trusted certificate to a fake certificate.

To obtain a certificate from a CA, it is necessary to create a Certificate Signing Request (CSR). The CSR is simply a certificate without a signature. The certificate has many fields where specific information can be placed to describe the owner of the certificate. When a CSR is created, it can be given to a CA to process. Once the CA signs the certificate, it will be returned and can then be uploaded by the client-server for use as its primary certificate.

Initial Handshake

An SSL session begins with an SSL client contacting an SSL server. This session will be composed of two separate parts. First is the initial handshake. The initial handshake determines if any data will be exchanged, and how that data will be encrypted. The second part will be the actual data exchange.

During the initial handshake, the version of SSL is determined. The valid versions are SSL 2.0, SSL 3.0, and SSL 3.1 (TLS). At this time, certificates may be exchanged. These certificates, as discussed previously, help to verify the client and/or the server. A certificate is required only if the opposing application specifically requests it.

At this point, the client and server must decide upon a public key encryption method. This encryption method will be used to finish the initial handshake. A public key encryption is one where the key to encrypt a message is announced publicly, while the key to decrypt the message is kept private. An example of a popular public encryption is RSA. If certificates are exchanged, the public key will appear on the certificate itself.

Once the public key encryption has been resolved, all further communications will be encrypted. For the public key encryption to be successful, however, a very large key is required. This large key makes encrypting and decrypting very demanding on the processor. To reduce processor usage, a private key encryption will be used to encrypt and decrypt the data to be sent. A private key encryption is one where the key to encrypt and decrypt are the same, and so must not be announced publicly. Since all communications at this point are encrypted with the public key encryption, it is safe to share the key that will be used for the private key encryption. This key can be much shorter than the public key, and thus is less demanding on the processor.

Once this encryption is verified, negotiations can end, and encrypted data communication can begin. Either the server or client may begin sending data. The server and client titles are only designated for the negotiation process.

The initial handshake can fail at any time. This failure can be based on invalid certificates or failure to agree upon an SSL version, public key encryption, or private key encryption.

SSL API

Many aspects of programming SSL are implementation dependent, but some general guidance may be given here. Though there are many steps in creating a secure communication, the SSL library will handle most of them. The library is first initialized with the application's preferences. These preferences include SSL versions, encryption algorithms allowed, the setting up of certificates, and the handling of buffers among other things. These preferences remain in place until the application modifies them, which is generally unnecessary after the application is running.

For the application to utilize SSL, it must first establish a connection with a peer. This connection is then passed to the SSL library. The library then begins the initial handshake. The library does not return control to the application until an SSL session has been established, or until it is determined that the two peers cannot establish a secure connection.

Once a valid SSL session has been established, the application will need to send all outgoing data through the SSL library. This outgoing data will be encrypted before it is sent to the network. All incoming data must also be received through the SSL library. All incoming data the application receives will be decrypted. Once the SSL session is terminated, and before the connection is closed, the application must notify the SSL library, which will then free up all structures pertaining to the session.

8.8.3 Some SSL Details

Client and Server

An implementation of SSL may include functionality to be the client, the server, or both.

Encryption

It is common to use a separate library for encryption. This library is only accessed by the main SSL library and not by the application itself.

SSL implementations commonly include the following symmetric encryption algorithms:

Algorithm	Key length (bits)	Mode
DES	40, 56	CBC (cipher block changing)
3DES	168	CBC
RC2	40, 128	CBC
ARC4	40, 128	Stream cipher

and the following asymmetric encryption algorithms:

Algorithm	Modulus (bits)
RSA	512, 1024
DH	512, 768

Several encryption algorithms that are included in OpenSSL are commonly left out of SSL implementations. A complete list of such algorithms is IDEA, RC5, and DSA. These are encumbered by patent issues. If the user were to acquire licensed software for these algorithms, they could be plugged into the SSL library.

Certificate Request

The ability of a product to create a certificate is fundamental to the idea behind SSL. Once the CSR has been created, it can either be downloaded from the application for signing by a CA, or the application can create a self-signed certificate.

SSL and Web Servers

The most obvious context in which an embedded system can take advantage of the secure socket layer is when it employs a web server. HTTP connections use TCP port 80, as set by the HTTP standard. SSL through HTTP is termed HTTPS and uses port 443. Embedded web servers are normally set up to watch this port for any connections. Once one is made, the web server begins the secure negotiations through the SSL API.

Export Restraints

SSL implementations tend to include encryption above the level allowed by export with approval. When the end product is produced, a strict following of the laws concerning the export of encryption technology is required.

8.9 DHCP Debugging Tips

From his experience in technical support, Dan Schiro did a piece on DHCP in NewBits in early 2004, which was the basis for this article.

CW

DHCP (Dynamic Host Configuration Protocol) has become a common means of network administration. DHCP is to networking what plug and play is to PCI devices. It allows network administrators to control configuration options remotely and allows end users to move their network devices around subnets without needing to know the intricacies of the network topology. Developers of network devices wishing to utilize a DHCP client need to be aware of a few key features of DHCP to avoid possible troubles in the field.

Before discussing possible issues, an understanding of how DHCP works is needed. DHCP works using a small caveat in RFC 1122: the TCP/IP software *should* accept and forward to the IP layer any IP packets delivered to the client's hardware address before the IP address is configured. This means that as long as the packet has the right MAC address, it will be passed up the stack, regardless of the destination IP, as long as an IP address has not been attached to the stack. For the purposes of the following discussion, only the procedure for obtaining an IP lease for the first time will be discussed. It will also be assumed that the DHCP server is on the same subnet as the client.

DHCP relies on UDP for data exchange. The DHCP client normally uses port 68 and the server uses port 67. These ports must be available for use. These port allocations should never be changed unless required due to special circumstances.

While negotiating an IP lease, the client and server send a packet between them with each of them filling in bits of information as it is passed back and forth. The DHCP packet structure shares the BOOTP packet structure as per RFC 2131 (see Figure 8.10).

The client will initiate the communication by sending a DHCPDISCOVER message to the hardware broadcast address/IP broadcast address/port 67. It will fill in the op, htype, hlen, xid, flags, chaddr, and options fields. The op field will be set to 1. The flags field will normally be set to 0 by default—this becomes important later. The options field will be filled in to specify a message type of DHCPDISCOVER. The DHCP server, listening on port 67, will send a DHCPOFFER message containing the configuration parameters to port 68. The configuration parameters include yiaddr and options that typically include the message type of DHCPOFFER, the server IP address, the lease length, the renewal timer value, the rebind timer value, and the subnet mask. The op field will be set to 2 for a BOOTREPLY. The server's message can be sent either to the broadcast address or the address it has filled into yiaddr, depending on the value of the flags field specified by the client. The server has the MAC address of the client and also knows the IP address it is willing to assign the client. Therefore, it can send a unicast reply to the client using the offered IP address and the client's MAC address. This greatly reduces disruption to other nodes on the network by eliminating broadcast traffic. Problems arise from improper implementation of this concept. Some servers will set the client MAC address and the IP broadcast address as the destination.

This can cause problems with the client depending on how the Ethernet driver is tied into the stack. The Ethernet driver will think it has a unicast packet, but the TCP/IP stack will think it has a broadcast packet. The packet could end up being rejected. To fix this, the networking stack should be made to specify a complete broadcast reply.

Once the client receives the DHCPOFFER message, it will send a DHCPREQUEST message to the server including the options it accepts. The client need not accept all options specified by the server. The server will send a DHCPACK or DHCPNACK message to the client either validating the configuration settings the client has chosen or rejecting them. If the client receives a DHCPNACK message, it must start the negotiation process over again.

Since DHCP relies on broadcast traffic, it is restricted to communication within its broadcast domain. Routers typically set the broadcast domains and would not forward broadcast traffic from one subnet to another. This limitation would require each subnet to have its own DHCP server. This is not practicable in today's large enterprise LANs. Often one DHCP server handles multiple subnets requiring the use of DHCP relay agents. Relay agents, while useful, can add confusion while debugging a negotiation issue. Relay agents inspect broadcast traffic

op [1]	htype [1]	hlen [1]	hops [1]
xid [4]			
secs [2]		flags [2]	
ciaddr [4]			
yiaddr [4]			
siaddr [4]			
giaddr [4]			
chaddr [16]			
sname [64]			
file [128]			
options [variable]			

Field	Octets	Description
op	**1**	**Message op code / message type.**
		1 = BOOTREQUEST, 2 = BOOTREPLY
htype	1	Hardware address type, see ARP section in "Assigned Numbers" RFC; e.g., "1" = 10 M ethernet.
hlen	1	Hardware address length (e.g.6' for 10 M ethernet).
hops	1	Client sets to zero, optionally used by relay agents when booting via a relay agent.
xid	4	Transaction ID, a random number chosen by the client, used by the client and server to associate messages and responses between a client and a server.
secs	2	Filled in by client, seconds elapsed since client began address acquisition or renewal process.
flags	2	Flags.
ciaddr	4	Client IP address; only filled in if client is in bound, RENEW OR REBINDING state and can respond to ARP requests.
yiaddr	4	"Your" (client) IP address.
siaddr	4	IP address of next server to use in bootstrap; returned in DHCPOFFER, DHCPACK by server.
giaddr	4	Relay agent IP address, used in booting via a relay agent.
chaddr	16	Client hardware address.
sname	64	Optional server host name, null terminated string.
file	128	Boot filename, null terminated string; "generic" name or null in DHCPDISCOVER, fully qualified directory path name in DHCPOFFER.
options	var	Optional parameters field.

Figure 8.10
DHCP packet structure

for BOOTP/DHCP messages and forward them to a DHCP server from a preconfigured list. When forwarding the BOOTP/DHCP message, the relay agent fills in the giaddr with the IP address of the interface it received the packet on and increments the hop count. It then forwards the packet on to the server. The forwarding process may require the packet to go through multiple relay agents. If a relay agent receives a packet that already has the giaddr field filled in, it simply forwards the packet without making any changes to the giaddr field.

When the DHCP server receives a DHCPDISCOVER that has the giaddr field filled in, it sends the DHCPOFFER message unicast to the IP address specified in giaddr. The relay agent will then receive the DHCPOFFER message and broadcast it on the interface that has the IP address specified in giaddr. The relay agent may also unicast the message to the client depending on the flags value.

Knowing how relay agents work will help debug issues. If the DHCP server is on the same subnet, the giaddr field should not be filled in the DHCPOFFER/DHCPREQUEST/DHCPACK messages. If it is, then the lease is coming from another server. If the DHCP server is not on the same subnet, make sure that a relay agent is present, and it is properly configured. Taking simultaneous packet traces on the client subnet and the relay agent configured forwarding subnet will show if the giaddr is properly set in the packet. It will also allow for observation of the traffic coming from the DHCP server to the relay agent.

If the packet traces show the server responding to a DHCPDISCOVER message with a DHCPOFFER message but the stack does not respond, the stack may be silently discarding the packet. This can be verified by placing a breakpoint at the beginning of the code in the stack that interprets an IP address. If this is not called, then the Ethernet driver is not passing the packet to the stack.

Hopefully the preceding information will be helpful in tracking down issues.

8.10 IP Multicasting

Networking is very complex, but anyone involved in embedded software needs to know something about it. Fortunately guys like Dan Schiro, who wrote the 2003 NewBits piece upon which this article is based, deal with technical support. Those guys are used to explaining things.

CW

In a world that demands information distribution in a quick, seamless manner, IP multicasting has stood out as a method of providing data to a large number of hosts without generating proportionally large traffic loads on network infrastructures. IP multicasting utilizes Internet Group Management Protocol (IGMP) to send information to selected hosts only. The highest level of support for IGMPv1 is level 2. Setting up an application to use multicasting can seem daunting, but it becomes a simple task when broken down into initialization, receiving, and transmitting. Note that many of the details of the implementation of multicasting will be dependent upon the specific networking stack. So, some of the guidance provided in the following sections should be taken as an example.

8.10.1 Initializing Multicasting

IP multicasting provides a means for a network application to send a single IP datagram to multiple hosts. Multicasting differs from broadcasting in that every host on a network segment receives a broadcast packet. This can lead to unnecessary interruptions for those hosts that are not interested in the broadcast. With multicasting, only those hosts that have explicitly joined an IP multicasting group will receive a multicast packet to that address/group. The class D IP address space defines the multicasting IP addresses. These addresses do not define individual interfaces, but instead define groups of interfaces. Hence class D addresses are referred to as *groups*. The class D IP addresses are those in the range 224.0.0.0 to 239.255.255.255.

Membership in a multicast group is dynamic. An application can join and leave a group on an interface at any time. Applications join or leave a multicast group by utilizing a networking stack service call. Many IP multicasting groups are reserved for specific applications. RFC 1700 provides a current list of registered groups.

All level 2 conforming hosts are required to join the 224.0.0.1 group at initialization. The 224.0.0.1 or "all hosts group" is the group of all hosts on the local subnet. At initialization, a networking stack, which supports IP multicasting, will join this group on interfaces that support multicasting (this would normally be a build option). Multicasting can only be enabled on UDP sockets. This is because UDP is a connectionless protocol. TCP on the other hand establishes a connection with a specific host. Communication over a TCP socket is possible only with that one specific host.

IGMP (Internet Group Management Protocol) is the means by which IP hosts report their host group memberships to any immediately neighboring multicast routers. A networking stack may also have support for IGMP. IGMP is transparent to the user in that no API service calls directly invoke IGMP. Rather, each time an application joins a multicast group, the IGMP services will be invoked by the IP layer to report the new group membership. Also, if there are any multicast routers on the local network, they will periodically send requests for updated group membership information. At this time, IGMP will report all group memberships to the router. These requests are sent by routers to the group address of 224.0.0.1, hence the necessity of joining this group during boot up on all interfaces that support multicasting.

It is important to keep definitions straight. In this article, references to multicast packets will mean packets that contain application data and are sent to a multicast group address. IGMP packets refer to packets that contain instructions for multicast group maintenance.

8.10.2 IGMP Protocol

The Internet Group Management Protocol is a layer 4 protocol and lies at the heart of multicasting. IGMP packets contain the instructions to add and/or remove an interface to or from a multicast group and perform group maintenance functions. IGMP packets do not contain any application data. The IGMP header replaces the TCP or UDP header in the packet structure.

IGMP is an asymmetric protocol and is specified here from the point of view of a host, rather than a multicast router. (IGMP may also be used, symmetrically or asymmetrically, between multicast routers. Such use is not covered here.)

Like ICMP, IGMP is an integral part of IP. It is required to be implemented by all hosts conforming to level 2 of the IP multicasting specification. IGMP messages are encapsulated in IP datagrams, with an IP protocol number of 2. All IGMP messages of concern to hosts include the following fields:

- **Version:** This memo specifies version 1 of IGMP. Version 0 is specified in RFC 988 and is now obsolete.
- **Type:** Two types of IGMP messages are of concern to hosts: 1 = Host Membership Query; 2 = Host Membership Report.
- **Unused:** Unused field, zeroed when sent, ignored when received.
- **Checksum:** The checksum is the 16-bit one's complement of the one's complement sum of the 8-octet IGMP message. For computing the checksum, the checksum field is zeroed.
- **Group address:** In a Host Membership Query message, the group address field is zeroed when sent, ignored when received. In a Host Membership Report message, the group address field holds the IP host group address of the group being reported.

8.10.3 Implementing Multicasting

Multicasting requires the interaction of the Ethernet driver and the networking stack. How to receive multicast and IGMP packets will be discussed first.

Receiving

To receive a multicast packet or IGMP packet, the Ethernet driver must be initialized to utilize multicast. Driver initialization varies with the hardware, but one common component is the initialization of filters for handling multicast and IGMP packets. These packets will have a multicast group MAC rather than the individual MAC. The driver needs to initialize the hardware to tell it to filter for packets that have either the individual address or the group addresses that the device belongs to.

Since the driver works at the second layer, it only has access to the Ethernet header. That header contains a type field along with a source and destination MAC field. This means that the driver needs to be able to determine by the destination MAC address if the packet is for itself or another device. The Ethernet driver needs to parse out the individual/group bit in the destination address field. If the destination MAC address is marked as individual, then it is compared against the Ethernet device's MAC address and the packet is accepted if they match or rejected if they do not.

Before discussing destination MAC addresses with the group bit set, it is necessary to discuss how the Ethernet driver maintains the list of multicast groups to which it belongs. Whenever a multicast group is joined, the Ethernet device will generate a multicast MAC for a particular group using a defined algorithm to assure that other devices on the network utilizing multicast will be using the same MAC for the same group. The Ethernet device will then store this multicast MAC for filtering incoming packets. This may be done in a number of ways, such as a hash table.

If the packet's destination MAC address is set as group, then the destination address is filtered against the stored multicast MAC addresses. If a match is found, then the packet is passed into the networking stack, which will determine if the packet is an IP or ARP packet. Since it is either a multicast or IGMP packet, it will be determined to be an IP packet. Then it will determine if the packet is TCP, UDP, IGMP, or ICMP. It is at this point that the processing of multicast and IGMP packets differ. If the packet is a multicast packet—meaning it contains application data—then the packet will be determined to be UDP and will be processed accordingly. If the packet is an IGMP packet—meaning it contains group maintenance information—it will be determined if the IGMP packet is a host membership report or host membership query and be processed accordingly. If the IGMP packet is neither of those two options, it will be discarded.

For an application to receive multicast messages, it must first join a particular multicast group. This is done through a network stack API call. This is likely to include options to allow for adding or dropping multicast group membership, changing the Time To Live on multicast packets, and invoking the necessary functions to update the Ethernet driver's multicast MAC filter. At this point, the stack is able to receive multicast packets for the groups that have been joined.

Transmitting
The transmission side of multicasting is also dependent on the type of packet. If the packet is IGMP, the transmission is transparent to the developer.

To transmit a multicast packet (application data), the developer will use the standard UDP transmission processes with a few modifications. If the device will be a client, a server address structure for the multicast group will need to be created prior to the "send" call. However, instead of placing the server's IP address in the address structure, the multicast group address will be placed in it instead. This will result in the packet being received by all members of the multicast group. The "send" call will be processed as normal. The IP address in the server address structure will be used to resolve the multicast destination MAC address for the header creation—hence the need for each multicast group to use the same multicast MAC. However, the individual host information will still be placed in the source fields of all headers. Following this procedure, it is not necessary for an application that will only transmit to a multicast group to join the group. The only time it is required to join a multicast group is if the application needs to receive messages from the group.

8.10.4 Bringing it All Together

If the device is to be a server, and depending on the application, then the developer may need to create a client address, in addition to the local address structure, prior to the "receive" call for a particular multicast group. If the application is to respond to the entire multicast group after receiving a message sent to the group, then the local and client address structures must be created prior to the call. In this case, the address structure fields filled in by the "receive" call cannot be used in the following "send" call because they will contain the information for the specific host that sent the multicast packet; they will not have the multicast group information in them. The client address structure will need to be recreated prior to the "send" call so that it will contain all the correct multicast information. If the application only needs to respond to the individual host that sent the multicast message, the standard methods of using "receive" and "send" will suffice.

CHAPTER 9

Open Source, Embedded Linux, and Android

Chapter Outline

For many years, open source in the embedded software world was all about tools—the GNU compiler, etc. But, with the increasing interest in embedded Linux, the scope of open source has become broader. For some time, open source implied a "home brew" approach to operating system or tools, but in recent years it has matured and numerous companies provide various levels of support, enabling embedded software engineers to safely and economically take advantage of the opportunities that open source offers.

CW

9.1 GNU Toolchain for Embedded Development: Build or Buy

For many engineers, their first taste of open source software has been the GNU compiler, which is often touted as a free alternative to expensive commercial toolchains. The reality is that there are costs of deploying an open source toolchain for real embedded projects. I have based this article on a white paper by Mark Mitchell (a founder of CodeSourcery, now part of Mentor Graphics) in which the challenges entailed in the deployment of open source tools is considered in some detail.

CW

9.1.1 Introduction

Increasingly, embedded software developers are choosing the GNU toolchain for open source development. The GNU toolchain contains an optimizing compiler targeting most embedded processors. The toolchain also supports programming in C, C++, assembly language, and compiler and linker extensions which are specifically designed to assist embedded programmers. Further, support for multiple target platforms makes porting code between processors simpler since developers can use the same tools on multiple platforms.

Of course, the core components of the toolchain (an IDE, C compiler, C++ compiler, assembler, linker, debugger, and other tools) are available as open source software, so

Embedded Software: The Works. DOI: 10.1016/B978-0-12-415822-1.00009-X developers have the option of building the toolchain themselves. But deciding to use the GNU toolchain, one has to make a "build vs. buy" decision: should you build all the components yourself, or should you buy a pre-built toolchain?

This article presents some of the technical issues involved in building and validating the toolchain. By considering these issues, developers will be better able to decide whether they can commit the resources required to "do it yourself."

Note: A question, printed in bold, **like this**, indicates an important decision. If you're seriously considering building your own toolchain, you'll need to answer these critical questions.

9.1.2 Toolchain Components

Which tools and libraries will you include in your GNU toolchain?

The GNU toolchain is more than just a compiler. For embedded software developers, a complete toolchain generally includes:

- C/C++ Compilers
- Assembler
- Linker
- Runtime Libraries
- Debugger
- Debug Stub(s)
- Integrated Development Environment.

This section discusses each of the components listed above. Other options include profiling tools, a flash programmer, or a simulator (which are not described in this section).

GNU C/C++ Compilers
The GNU C and GNU C++ compilers (GCC and G++) are the heart of the GNU toolchain. The compilers transform source code written in C or C++ into assembly code for your target processor. The GNU C and C++ compilers support most popular CPUs and operating systems.

Do you develop both C and C++ code?

If you are confident that you will not need to use C++, you can save yourself time and effort by not building the C++ compiler. Although G++ works on virtually all systems supported by GCC, it can be difficult to build the GNU C++ runtime library for some target operating systems. Some C++ features, like exceptions or automatic initialization of global objects, require additional configuration and need to be validated carefully.

GNU Binary Utilities
The GNU binary utilities include the GNU assembler, GNU linker, and tools that can display the contents of object files and convert object files from one format to another. Some of these

tools are invoked automatically by the compiler driver in the process of building an application; you will use others directly in creating, debugging, and deploying your application.

Runtime Libraries
Several runtime libraries are required to build complete applications.

The compiler support library is called libgcc and is provided by the compiler itself to supply low-level compiler support routines such as software floating-point emulation and support for exception-handling. This library is used by virtually all programs compiled with GCC.

Which C runtime library is appropriate for your target system?

The C runtime library contains ISO C functions, like `printf()` and `memcpy()`. The three most popular C libraries used with the GNU toolchain are the GNU C Library (GLIBC), uClibc, and Newlib:

> GLIBC is a full-featured, POSIX-compliant C library designed for use with the Linux kernel. In addition to those functions required by ISO C, it includes functions required by POSIX, functions provided in other UNIX C libraries, and GNU extensions.
> uClibc is a smaller footprint library (also designed for use with the Linux® kernel) that contains a subset of the GLIBC functionality. You can use uClibc on either GNU/Linux or uClinux.
> Newlib, generally used on bare-metal targets, has an even smaller footprint, but provides significantly reduced functionality.

If your system already has a C library, you may not need to build a C library at all—but you may need to modify GCC or the C++ runtime library to work with your C library.

The C++ runtime library includes classes required by the ISO C++ standard, including `std::ostream` and `std::vector`. There is only one C++ runtime library in widespread use with the GNU toolchain. The GNU C++ Library (also called `libstdc11`) is provided in the same source package as the GNU C and C++ compilers.

GNU Debugger
The GNU Debugger (GDB) provides both source- and assembly-level debugging. GDB allows you to view registers on your target, connect to a running system, set both software and hardware breakpoints and watchpoints, step through your application, disassemble code, and modify data on the target system. You can use GDB directly (from the command line) or as the "back end" for a graphical environment, such as Eclipse.

Debug Stub
How will the GNU debugger communicate with your target system? Does your ICE or JTAG unit work with GDB?

How will you program Flash memory?

In an embedded environment, it's generally not practical to run the debugger directly on the target system. Instead, the debugger runs on the host system and communicates with the target via a "stub" over an ICE or JTAG unit, TCP/IP, or serial connection. The stub acts as an intermediary between GDB and the target, allowing GDB to start and stop programs, look at the contents of registers, and perform other necessary functions on the target system.

GDB, as distributed by the Free Software Foundation, does not contain any stubs that can be used with ICE or JTAG units, but it does provide a TCP/IP for communicating with third-party stubs. Stubs are available from some JTAG and board vendors, either in firmware or as part of their software kits, and from toolchain distributors. You may also write your own stub using the documentation in the GDB manual.

Without a full-featured stub, you cannot take advantage of all of GDB's features for debugging applications on your actual target hardware. You should look for a stub with support for all of the following features:

- Burning Flash memory
- Semihosting (i.e., permitting target applications to read and write files on the host system)
- Displaying extended registers (such as control registers, memory-mapped I/O registers, etc.)
- Setting both hardware and software breakpoints and watchpoints
- Operation on both Microsoft Windows and GNU/Linux.

Eclipse IDE

Are you willing to work exclusively from the command line? Or would you prefer a graphical development environment?

Eclipse is the integrated development environment of choice for use with the GNU toolchain. You should include the Eclipse IDE in your toolchain unless you are comfortable developing entirely from the command line.

Eclipse contains editors for writing code, a class hierarchy viewer, and support for debugging. If Eclipse, GDB, and a GDB stub are properly integrated, developers can edit, compile, link, and debug applications running from Flash memory on an embedded target board without ever leaving Eclipse. In addition, a large number of third-party Eclipse plug-ins are available that go beyond the functionality provided by the base IDE.

9.1.3 Building the Toolchain

Building a complete GNU toolchain is a complex process. You must:

- Create your build environment
- Obtain the source code

- Make decisions about configuration options
- Understand and perform the build process
- Package the toolchain.

The remainder of this section examines each of these steps in more detail.

Build Environment
Do you have the right tools to build your toolchain?

In order to build your GNU toolchain, you'll need tools, of course. And these "build tools" (i.e., the compilers, libraries, and related tools that you have installed on your system) can influence the quality of the resulting toolchain. For example, it is most convenient to build the toolchain using another version of GCC—but if the version of GCC you are using to build your toolchain has defects, then it may incorrectly compile the toolchain.

Because the libraries on your build system will be linked into the toolchain, your toolchain will only run on machines compatible with yours. So, you must be careful to build on a system that is the "lowest common denominator" of the set of systems you wish to support.

Do you want to use "Canadian cross" compilers to build toolchains for multiple host systems?

If you need to support multiple host operating systems (i.e., if your toolchain will be used on both Microsoft Windows and GNU/Linux), then you must either (a) set up multiple build environments or (b) build "Canadian cross" toolchains.

A Canadian cross toolchain is a toolchain that is built on one system, runs on a second system, and generates code for a third system. For example, you might use a GNU/Linux build system to build a toolchain that will run on Microsoft Windows and generate code for an embedded target. The advantage to building a Canadian cross compiler is that you can then do all of your builds on a single build system. However, in order to build a Canadian cross toolchain, you must first build an ordinary cross toolchain (e.g., from GNU/Linux to Microsoft Windows) and then use that cross toolchain to build the Canadian cross toolchain.

If you need to apply a patch and rebuild the toolchain, how will you ensure that you have not changed the build environment?

If, after deploying your toolchain, you find a defect that requires a rebuild, you will need to perform the entire build process again. Therefore, you should archive all of the source code for the toolchain, the build tools used to build the toolchain, and the commands you used to build the toolchain.

On Windows, will you use Cygwin or the Win32 API?

If your toolchain will run on a Microsoft Windows host system, you must decide whether you want to use the Win32 API or the Cygwin UNIX-emulation environment. A Cygwin toolchain will behave somewhat differently from a Win32 toolchain. In particular, a Cygwin toolchain will use UNIX-style path names, while a Win32 toolchain will use Windows path names.

Because the toolchain was originally developed to run on UNIX-like systems, it is easier to build the toolchain to depend on Cygwin. In addition, a Cygwin toolchain is convenient if you plan to build GNU/Linux software for your target system.

On the other hand, many Windows tools do not understand UNIX-style paths, so it may be challenging to integrate your toolchain into the rest of your environment. In addition, Cygwin imposes some performance overhead relative to the Win32 API. Finally, all Cygwin binaries require a copy of the Cygwin DLL. If you provide this DLL yourself, it may conflict with other copies of the DLL that users of your toolchain already have. On the other hand, if you do not provide this DLL, users of your toolchain will have to download it themselves.

Source Code and Patches
What versions of the various toolchain components will you use?

Before you can build the toolchain, you must obtain all the relevant source code. Your first step should be to determine which version of each component will suit your needs. You should ensure that the version you select contains support for your target hardware and that it is compatible (at the source and binary level) with any software that you will be incorporating into your project.

For example, some source code may rely on language features that were not supported in older versions of GCC. Other source code may contain errors that prevent compilation by the very newest (and most ISO-compliant) versions of GCC. If you are not sure what version to use, you should generally use the most recent release.

You should remember that in the free- and open source software communities, version numbers do not mean the same thing that they do in other environments. For a proprietary software package, a particular version number or build number indicates the binary in use. In contrast, a version of GCC (such as 4.1.1) only denotes a version of the source code. The binary code generated from the source code depends on the configuration options selected and the build environment used.

What patches should be applied to the toolchain?

While there is a single FSF source release of GCC (or any other toolchain component) with a given version number, there are myriad patches to that release available from various locations.

Between FSF releases, the open source community makes substantial enhancements to add support for more hardware architectures, improve code generation, and eliminate defects.

Some of these changes may be useful for your toolchain, and some may not. Some patches may introduce new problems of conflict with other patches. You must evaluate which of these patches you wish to apply to the base source version.

Unfortunately, there is no way to know whether you have all the patches that you need, and some important patches may not be easy to find. It often takes months for a patch to make its way into the FSF source repositories. Some patches are available from public mailing list archives or defect-tracking systems. In other cases, the patches may only be available to direct customers of a toolchain distributor.

Configuration Options
How should you configure the various toolchain components?

Most toolchain components contain a "configure script" which you must use to select the target platform and to specify which of the available features will be included in your toolchain. Taken together, these configure scripts have thousands of options, covering everything from the locations in which components should be installed to whether or not the `_cxa_atexit` routine should be used to register destructors for C++ objects with static storage duration.

Some configuration options are purely a matter of preference, some affect performance but not correctness, and still others have a definitive right answer on particular systems. After building your GNU toolchain, you should validate it to ensure that you have picked options that work correctly in your target environment.

What CPUs will be used on your target systems? Will both big-endian and little-endian code be required?

The default runtime libraries provided for a given architecture may not be appropriate for your target environment. For example, the default runtime libraries for a "bare-metal" ARM environment are optimized for a little-endian ARM V5 processor. If you have an ARM V4 processor, or a big-endian XScale processor, these libraries will not work on your system. On the other hand, if you have an ARM V6 processor, these runtime libraries will not take full advantage of your hardware. To adjust the set of runtime libraries present in your toolchain, you will often need to make modifications to the source code for GCC. You must also build the C library multiple times, once for each of the configurations you require.

Build Process
The build process is itself complex, especially for a GNU/Linux toolchain. While the GNU coding standards suggest that all packages should be built with a uniform "configure, make, make install" sequence, there are variations on that theme in the various toolchain components. In general, it is necessary to provide additional options to each of the three phases in order to obtain optimal results.

There is an interdependency between the compiler and the C library. In particular, GCC examines the C library to determine certain configuration parameters, but, of course, you cannot build the C library until you have a compiler.

The Eclipse IDE (which is written in Java) uses a build process entirely different from that used by most of the other toolchain components. The Eclipse web site contains additional information about configuring and building the IDE.

Packaging

Are there standard package formats for your host operating system(s)?

Would a graphical installation process make it easier to install your toolchain?

Once you have built your toolchain, you will want to package it for easy installation throughout your organization. On Microsoft Windows, a graphical installer makes it easier for users to install the toolchain, place the tools in the PATH, and to manage uninstallation. On GNU/Linux, you may want support for Red Hat Package Manager (RPM) packages, which are used on Red Hat Enterprise Linux, SuSE Enterprise Linux, and other popular GNU/Linux distributions.

In designing your packaging scheme, you should consider the possibility that users will install multiple toolchains. It's useful to share certain components (such as Eclipse) so that users do not have to set up their preferences for each toolchain. But, if files are shared between installations, then you must consider the impact of upgrading one toolchain independently of the others or uninstalling a single toolchain.

Once you have selected your packaging software and layout scheme, you must build the installers. You should automate that process so that you can easily rebuild your installation packages whenever you rebuild your toolchains.

9.1.4 Validating the Toolchain

Will your toolchain work?

Having built the toolchain, you should validate it before deploying it in a production environment. In general, the following techniques are used to test a compiler (and related tools):

- **Compiling programs and running those programs in the target environment.** This technique verifies that the generated code behaves as intended. You can also use this approach to test the performance of the code generated by the toolchain.
- **Compiling programs and inspecting the generated object file or executable image, without running the generated code.** This technique can be used to check that the

generated code contains expected machine instructions, correct debugging information, conforms to a specified Application Binary Interface, etc.

- **Compiling fragments of a program with multiple compilers and linking the fragments together.** This technique checks that the compiler you wish to validate interoperates with a known-good compiler for the target platform.
- **Compiling invalid program fragments and checking for appropriate error messages.** This technique (called "negative testing") checks that the compiler is correctly enforcing constraints.

There are a number of additional techniques that you should apply to other components of the toolchain, including:

- **Interactive testing of the debugger.** You should compile and debug programs using the debugger with your target system to ensure that all debugger functionality works as expected.
- **Interactive testing of the IDE.** Like debuggers, IDEs are, by their nature, interactive. You should check that all desired IDE functionality (including, in particular, integration with GDB, the GDB stub, and the target system) works as intended.

A variety of test suites are available to perform the various validation steps described above. The following sections provide a brief overview of some of the most useful test suites.

GNU Regression Test Suites
Does your toolchain meet basic correctness requirements? Do GNU source extensions work correctly?

Many of the GNU toolchain components (including the GNU Compiler Collection, GNU Binary Utilities, and GNU Debugger) include test suites based on the DejaGNU framework. Every GNU toolchain build should be validated using these test suites because only the DejaGNU test suites provide test coverage for GNU extensions to the C and C++ programming languages and the wide variety of features available in other tools. The DejaGNU GDB test suite performs live tests of the debugger on a running target system to ensure appropriate interactive behavior. Taken together, the DejaGNU test suites contain tens of thousands of tests, and are usually expanded to include new tests whenever a defect is corrected or a new feature is added.

To use each of these test suites, you must first develop a DejaGNU board configuration file for your target system. The board configuration file will contain code to perform a variety of basic operations, including, most critically:

- Running programs on your target system. This code must upload the program to the target system (or make it available via a network file system), execute the program, and report the results.

- Rebooting your target system. Many embedded systems lack memory protection. Therefore, when a test fails, the target system is likely corrupted, and the system must be rebooted. You may require specialized hardware to manage automatic rebooting, such as managed power strips.

The board configuration is written in the Expect programming language, which is an extension to the Tcl programming language. Unfortunately, documentation for DejaGNU is very sparse, so you will likely need to make use of existing board configurations as a starting point.

It is advisable to perform testing on installed toolchains. The DejaGNU test suites have customarily been run from the build directories in which the toolchain was built. However, testing in this manner requires invoking the tools in a substantially different way from that in which users will actually invoke them. More accurate results can be obtained by installing the toolchain first and then running the tests. Testing installed components ensures that the tools tested are the same binaries, invoked in the same way, as they will be by users.

However, testing installed toolchains is more complex than testing from the build directory. In particular, you must create a DejaGNU "site file" to describe your installation. You may also find that some of the DejaGNU test suites require modification to support testing installed toolchains. If you wish to test installed toolchains, you may find it helpful to automate both the installation process and the generation of appropriate DejaGNU site files so that you can easily reproduce your testing.

The GNU C library should also be tested. The GLIBC test suite does not yet make use of the DejaGNU framework. However, it does contain a set of tests that you should run. These tests cover much of the functionality provided in GLIBC, including, in particular tests for the Native POSIX Threads Library (NPTL), which provides high-performance threading support on GNU/Linux systems. These tests provide a powerful mechanism for testing the compiler, C library, and kernel, all of which must cooperate to provide support for threads. Because the GNU C Library test suite does not support cross-testing (i.e., compiling the tests on one system and testing on another), you will have to modify the test suite to support testing on embedded systems.

Conformance Test Suites
Does your toolchain meet the requirements of relevant standards and specifications? Will standard-compliant source code be processed correctly?

You should use conformance test suites to validate the behavior of your toolchain relative to published specifications. You may wish to seek out additional conformance test suites to validate functionality specific to your intended use of the compiler.

For example, the Plum-Hall C and C++ Validation Suites are comprehensive tests for conformance to the C and C++ programming language specifications. The Plum-Hall test suites contain tests for nearly every sentence of the published specifications, including both the programming languages proper and the associated runtime libraries. In addition, the Plum-Hall test suite can automatically generate a number of "expression tests" which contain complex arithmetic expressions. These expression tests have proven useful in identifying instances of incorrect code generation.

The OpenPOSIX test suite checks conformance of a compiler, C library, and operating system to the POSIX specification. This test suite runs C programs that make heavy use of the C library and checks that the results returned by the library routines are correct. It is common to discover problems in the C library and/or operating system kernel through the use of this test suite.

Performance Test Suites

Does your toolchain meet performance requirements? Will generated code run as quickly as possible?

Performance test suites are designed to provide data about the speed at which the code generated by the toolchain executes. These test suites can help to identify misconfigurations of the toolchain, such as situations in which the generated code is using software floating point on a system that supports hardware floating point. They are also useful in comparing newer versions of the toolchain with older versions. For example, before deploying an upgraded version of the toolchain for production use, you might wish to ensure that the code generated is in fact better (or, at least, no worse) than that generated by the current toolchain.

The EEMBC benchmarks are widely used as measurements of embedded system performance. These benchmarks are divided by application areas; for example, there are EEMBC benchmarks for networking applications, automotive applications, and for office automation.

The SPEC CPU benchmarks are highly regarded benchmarks for C, C++, and Fortran applications. Some of these benchmarks are "scientific" code, while others focus on general-purpose computing.

9.1.5 Testing Multiple Options

Has your toolchain been tested in all the ways it will be used?

Once you have validated the compiler using a single set of options, you should expand your "validation matrix." There are three important dimensions to the validation matrix:

- Target System
- Level of Optimization
- Host System.

Target system options include the choice of target CPU and operating system, whether to compile big-endian or little-endian code, and other related options which specify the system on which the code generated by the toolchain will execute. It is not at all uncommon for the GNU toolchain to behave correctly for one target system, but not for a seemingly related system. For example, even though little-endian code works, big-endian code may not. These problems are especially likely if you have incorporated patches from hardware manufacturers designed to support a particular CPU or CPU family, as these vendors may well not have tested other configurations. If you intend to use your version of GCC with multiple targets, you should validate each target independently.

Optimization options include whether to generate code best-suited to debugging, code designed to run quickly, or code optimized for size. Different optimization options exercise different code paths in the compiler and can therefore have a substantial impact on test results. The three compilation modes used most often in development are (a) debug (-g), (b) optimized for time (-O2), and (c) optimized for space (-Os). You should ensure that your testing exercises all of these operating modes.

Finally, you should validate the toolchain on all host systems (Microsoft Windows, GNU/Linux, etc.) on which you will be using the toolchain. In addition to checking that the generated code is correct, you should verify that the code generated is in fact identical on all host systems. Because most GNU toolchain developers use IA32 GNU/Linux systems for their own development, host support for Microsoft Windows has tended to be particularly problematic. For example, some versions of GCC have made incorrect assumptions about the behavior of the Windows C library that resulted in the generation of incorrect assembly code. Similarly, reliance on pointer values as hash-table keys has resulted in different code generated on different hosts.

Analyzing and Correcting Failures
Do failing tests indicate serious problems?

Having gathered data about which tests pass and which fail, you must now evaluate the results. The GNU toolchain (like all toolchains) has defects. Therefore, in evaluating the toolchain you have built, you should attempt to determine whether or not the failures that you observe are unique to your toolchain. If the failures you encounter are not present in other builds of similar toolchains, then you may have built the toolchain incorrectly. However, even if the failures are not specific to your toolchain, you must evaluate whether or not they are sufficiently severe as to impede use of your toolchain.

Failing tests can be divided into the following categories:

• **Defects in the tests themselves.** This category includes tests that make incorrect assumptions about the target environment, such as tests that assume the target is a little-endian machine, has 32-bit pointers, or treats plain char as a signed data type. These tests should be corrected. If the tests cannot be readily corrected, these failures should be ignored.

- **Defects in the hardware platform.** In some cases, testing the toolchain reveals microprocessor defects (such as incorrect handling of instructions in delay slots). In other cases, the target board may contain faulty parts or may use faulty software. It may be necessary to implement toolchain work-arounds for some of these problems, such as the insertion of NOP instructions to avoid CPU defects. In other cases, it may be sufficient to ignore the failures.
- **Resource limitations.** For example, some tests may require more memory than is available on the target system, or may require that operations complete faster than can reasonably be achieved on the target system. These failures should be ignored.
- **Defects in the toolchain.** These defects must be further analyzed to determine (a) the significance of the failure (i.e., its likely impact on software developers using the toolchain) and (b) the difficulty of fixing the defect. You should categorize each of the failures you observe so that you can fully evaluate the quality of your toolchain.

How will you correct defects in your toolchain?

If you have found problems through the validation process, you will need to fix them before deploying your toolchain. Even if all of your validation has been successful, you may encounter problems in the course of using the toolchain. In either event, you will need to determine the cause of the problem and then develop a solution.

Since the toolchain contains several million lines of code, the first step is to identify the component that is causing the problem. Then, you will have to debug that component. You may have to spend some time becoming familiar with the source code for the toolchain before you can correct the defect. You may wish to post a description of your problem to the public mailing list for the affected component asking for help. Some components also have publicly accessible defect-tracking systems that can be used to report problems. When you have fixed the problem, you should consider contributing the change that you have made to the public source repository for the affected component so that others can benefit from your improvement, just as you have benefited from theirs.

9.1.6 Conclusions

This article has looked in detail at all the issues involved in deploying an embedded software development toolchain based upon GNU technology. The developer, who has opted for an open source approach, has broadly two options: build a toolchain for themselves, using this article as a guide or obtain a pre-built, configured, optimized, augmented, and tested toolchain from a reputable supplier. A classic "build vs. buy" decision.

9.2 Introduction to Linux for Embedded Systems

Linux has rapidly increased in popularity for desktop computers over the past decade and I guess it was a natural progression for engineers to consider its use for embedded. Christopher Hallinan is an expert in embedded Linux and I based this article on a white paper that he authored.

CW

9.2.1 Introduction

Few doubt that Linux now dominates the embedded systems marketplace. Linux has found its way into nearly every product segment, from consumer electronics to the "big iron" boxes used to power global wireless and landline networks. It is not a stretch of one's imagination to assume that many modern households have one or more Linux-powered appliances.

Linux® can be found in nearly every part of our lives, from diagnostic medical devices, the navigation, entertainment, and engine control systems of our automobiles, to the wireless access points powering our homes. Linux has even found its way into demanding hard real-time applications as well as mission-critical commercial, military, and aerospace applications.

9.2.2 Challenges Using Open Source

One of the major reasons why Linux has become so successful in embedded systems also contributes to the challenges inherent in developing and deploying embedded Linux-based devices. Open source software in general and Linux, in particular, move at a dizzying pace.

The Linux kernel is but one small component of an embedded Linux system. A complete Linux system consists of a bootloader, Linux kernel, and many dozens, often hundreds of application packages. These packages contain the programs, libraries, and configuration files that make up a Linux system.

The embedded Linux developer also needs to have a functional cross-toolchain and other development tools to help build and customize his embedded Linux system. The cross-toolchain and other tools, such as analysis and profiling tools need to run correctly on his choice of development host.

Embedded Presents Unique Challenges

It is not so difficult today to download a Linux kernel from kernel.org and compile it to run on a modern Intel-based x86 machine. However, when the underlying machine is of a different architecture, such as Power Architecture or MIPS, it is a different story altogether. It is no secret that most of the mainline kernel developers focus on traditional desktop and server applications that are dominated by Intel and AMD x86 machines.

Embedded Linux developers are far smaller in number. It is not uncommon that patches and fixes for non-x86 architectures lag significantly behind those for x86. In fact, many patches and board ports never reach mainline kernel.org. This is especially true for hardware platforms that are not in the popular market, such as the majority of reference platforms produced by the leading semiconductor manufacturers.

Building an embedded Linux distribution for non-x86 architectures is much more challenging than for x86. For starters, you need a toolchain of high quality, and these are usually only available from commercial sources. You will also need a bulletproof kernel ported not only to your architecture, but also to your particular board. On top of that, you will need the runtime packages that make up the system, including the standard C library family and other runtime libraries and system configuration files that make up an embedded system. All of these components must play nice together, and that alone is a nontrivial challenge. Package version dependency has always been challenging in the Linux world.

One Size Does Not Fit All

Traditional embedded Linux distributions have been an all-or-nothing proposition. They were created to be generic—to apply to a large variety of embedded applications. This often meant that you got many packages and components you did not need. Worse, it often meant there were critical components that your system requires that were not present.

Large Scope of Expertise Required

To build your own embedded Linux distribution, you need a broad range of specialized expertise. At a minimum, you will need developers with experience in the following disciplines:

- Hardware/BSP—This includes detailed knowledge of the complex System-on-Chip (SOC) being offered today, including a solid understanding of the underlying hardware architecture.
- Compilers/toolchain—Compilers are bug-free, right? Open source compilers are subject to the same rapid development and therefore instability as other open source components, and they can be particularly difficult to diagnose and fix.
- Linux kernel—Back to revision 0.9, the Linux kernel was reasonably easy to understand, modify, and fix. Today's Linux kernel has something north of 20 million lines of text, not including documentation, and has become highly complex requiring specialized skills even for simple bug fixes.
- Userland—The challenge with userland packages is simply the vast number and variety of packages. Many are not cross-compiler friendly, and some are quite specialized.
- Development tools—There are many options for development tools, from sparse (`printf()` debugging and occasional use of gdb) to full-blown IDEs. Assembling a capable suite of tools is no small task.

Despite the challenges outlined above, a Linux-based operating system offers the richest set of applications and widest choice of hardware platforms and devices than any other operating system available today.

The challenge becomes how to leverage the available open source components and technology in a way that is both cost effective and timely. This entails the integration of open source components, together with other intellectual property such as in-house developed code, software from a semiconductor vendor, and perhaps third-party solutions from ISVs.

9.2.3 OpenEmbedded

OpenEmbedded is a successful open source build system derived from the OpenZaurus project *circa* 2003. OpenEmbedded was created to overcome the limitations in the existing build systems of the era. Since its inception, OpenEmbedded has found its way into a significant number of commercial and open source projects. The build engine—bitbake—is based on concepts from Gentoo/portage.

OE Project Goals
The OpenEmbedded development team had specific goals for the project. These goals included the following:

- Handle cross-compilation even for complex packages, including gcc and glibc.
- Manage and resolve package dependencies.
- Emit packages in any possible type, the default being the lightweight .ipk package type.
- Create images and feeds from packages.
- The system must be highly configurable, with support for multiple platforms, architectures, machines, and distributions. In this context, the term machines refers to an embedded board.
- The metadata language used to describe projects must be easy to use and reuse.

OE Components
OpenEmbedded is composed of two primary components—bitbake and metadata. Bitbake is the build engine and is the heart of the build system. Bitbake reads instructions (metadata) and acts on them in order to build an embedded Linux distribution.

Metadata is a structured collection of instructions that tell bitbake what to build. It's important to keep in mind what metadata is, and what it is not. Metadata exists in four general categories: recipes, tasks, classes, and configuration metadata. Metadata is not source code. A recipe tells bitbake where and how to download source code, but metadata itself contains none.

Attributes of Open Embedded
OpenEmbedded is capable of building for multiple target devices from a single code base. The bitbake tool automatically builds any required dependencies. This applies to runtime

dependencies as well as to build dependencies. Multiple types of flashable root file system images can be created for booting directly on the target device.

OpenEmbedded supports a wide variety of package formats, including .rpm, .deb, .ipk., and more. OpenEmbedded supports automatic building of cross-tools if required. Support for using external toolchains is also possible with OpenEmbedded. OpenEmbedded can build host-compatible (native) utilities required during the build process, which helps provide robust support across a wide variety of host systems.

9.2.4 Understanding Metadata

Metadata exists in four general categories. These are recipes, tasks, classes, and configuration. Recipes usually describe build instructions for a single package. Tasks are often used to group packages together for root file system creation. Classes, analogous to classes in object-oriented languages, are a mechanism to encapsulate common functionality that can be inherited by recipes and tasks. Configuration metadata is used to drive the overall attributes of the embedded Linux distribution that System Builder creates.

Recipe Basics

Each recipe has a set of default tasks, provided by a base class in OpenEmbedded called base. bbclass. The default tasks provided for every recipe are:

- `do_setscene`—initial housekeeping, stamp checks
- `do_fetch`—locate and download source code
- `do_unpack`—unpack source into working directory
- `do_patch`—apply any patches
- `do_configure`—perform any necessary pre-build configuration
- `do_compile`—compile the source code
- `do_install`—installation of resulting build artifacts
- `do_populate_staging`—stage any required files for subsequent builds
- `do_build`—make packages.

These default tasks can be overridden by a given recipe to accomplish a particular task. Let's look at a simple minimal recipe to see how this works.

A simple hello world recipe with a file name like `hello-1.0.0.bb` might look like this:

```
DESCRIPTION = "Hello demo project"
PR = "r0"
SRC_URI = "http://localhost/sources/hello-1.0.0.tar.gz"
```

The description field is informational and can contain any text. It is meant as a short explanation for what the recipe builds. PR is a specific OpenEmbedded variable and refers to the recipe version. If you make changes to a recipe, prudence dictates that you bump the PR.

The SRC_URI tells bitbake where to find the source code for the package being built. In this example, the source is contained in a tarball on the local machine, served up via the HTTP. The source code can be a simple C source file accompanied by a one-line makefile, or it can be a large software project, with many subdirectories and its own custom-built system. In the former case, this simple recipe is all that is needed to build it. In the latter, you might have to override some of the default recipe tasks in order to issue the proper build instructions to your project.

In our example hello recipe, the source tarball is trivial—it contains the usual hello world C program, and a one-line makefile containing the single binary target and its dependency:

```
hello: hello.o
```

Bitbake takes care of the rest. When this recipe is built, using a simple command such as "bitbake hello," the default tasks are applied to the recipe to cause it to be built. do_fetch downloads the tarball to the working directory, do_unpack unpacks the tarball into the working source directory, and do_compile calls make on the resulting source directory.

Other recipe tasks are responsible for placing any build artifacts into a staging area, where they are made available in case other packages depend on them for building, and so that the artifacts can eventually be packaged into a .ipk named after the original (hello-1.0.0) recipe.

Finally, a special recipe called a task (Note that the term "task" is overloaded in OpenEmbedded terminology. In the case of a recipe, a task is a method or function specific to a given step in the build, such as do_build or do_install. When speaking of metadata, a task is a special type of recipe that is often used to group packages together.) takes the output package and places its contents into a root file system.

Tasks
A task is a special kind of recipe that groups packages together for inclusion in a file system image. A task can define an image or a collection of packages for use in an image. A simple task may specify the base functionality for booting a particular target, or the packages required for a specific functionality, such as task-perl-module-all.bb. A more complex task might use python code to pull packages into an image based on features defined with a particular machine. A good example of this type of task is task-base.bb.

Classes
Classes are used in a similar manner to those used in traditional object-oriented programming languages such as C++ and Java. Classes are used to abstract common functionality that can then be inherited by any recipe.

A good example of a class is autotools.bbclass. This class implements the methods typically used to build autotools-based packages. Among these methods is the familiar ./configure step familiar to anyone who has compiled packages on a Linux host. The default configure method provided by `autotools.bbclass` knows how to specify the proper parameters for cross-compiling in your given environment.

9.2.5 Project Workflow

After a project has been created, bitbake is invoked on an image recipe, similar to the following:

```
$ bitbake devel-image
```

Once bitbake finishes building an image recipe, your output directory typically contains all the images necessary to boot your target board. The output images include the following:

- Bootloader image for your platform
- Device Tree Binary (if required for your platform)
- Kernel and device drivers (loadable modules)
- Root file system images (jffs2, ext3, etc.)
- Packages for all root file system contents
- Bill of Materials (BOM) if desired

9.2.6 Summary

Linux has been deployed as the *de facto* standard for a wide variety of embedded applications across virtually every market segment. Developers today are facing huge development challenges brought on by insanely short development cycles and ever-increasing hardware and software complexity driven by customer demand and competitive pressures. One of the distinguishing characteristics of embedded systems is that each is a unique combination of hardware and software components.

9.3 Android Architecture and Deployment

As Android has become increasingly popular on smart phones and tablets, there has been increasing interest in is deployment elsewhere. I have made numerous presentations at conferences and seminars and lots of articles about Android. This article is an introduction to the technology, based upon my various writings.

CW

9.3.1 What Is Android?

It is easy to think of Android as being yet another operating system for high-end mobile phones. It's really a software platform rather than just an OS—with the potential to be utilized in a much wider range of devices. In practical terms, Android is an application framework on top of Linux, which facilitates its rapid deployment in many domains. A key to its likely success is licensing. Android is open source and a majority of the source is licensed under Apache2, allowing adopters to add additional proprietary value in the Android source without source distribution requirements.

Another way to appreciate the significance of Android is to take a historical perspective. In the early days of PCs, the operating system was DOS. This presented some interesting challenges to application developers, as DOS provided a minimal number of services. The result was that every application needed a complete framework to provide the full functionality that was required. For example, a word processing program would need to have a driver for every imaginable printer. This was a major headache for developers and a serious ongoing maintenance problem. The solution came in the early 1990s with the release of Windows, or rather the development of Windows 3.0. Although we think of Windows as being primarily a GUI, it's really much more than that. Nowadays, a word processor just talks to a logical printer. The manufacturer of the printer hardware simply needs to provide a Windows driver and everything works together properly.

In some respects, a similar situation exists today when developers want to deploy Linux for embedded applications. Android is the enabler for a broad application developer base, a complete stack on top of the Linux kernel.

9.3.2 Android Architecture

An Android system is a stack of software components. At the bottom of the stack is Linux—Linux 2.6 with approximately 115 patches. This provides basic system functionality like process and memory management and security. Also, the kernel handles all the things that Linux is really good at such as networking and a vast array of device drivers, which take the pain out of interfacing to peripheral hardware.

Libraries
On top of Linux is a set of libraries including `bionic` (the Google `libc`), media support for audio, video, and graphics along with a lightweight database that serves as a useful repository for storage and sharing of application data.

Android Runtime
A key component of an Android system is the runtime—the Dalvik Virtual Machine (VM). This is not strictly a Java virtual machine. It was designed specifically for Android and is optimized in two key ways. First, it is designed to be instantiated multiple times—each application has its own private copy running in a Linux process. And second, it was designed to be very memory efficient, being register based (instead of being stack based like most Java VMs) and using its own byte code implementation. The Dalvik VM makes full use of Linux for memory management and multi-threading, which is intrinsic in the Java language.

It is important to appreciate that Android is not a Java virtual machine, but does use the Java language.

Application Framework

The Application Framework provides many higher-level services to applications in the form of Java classes. This will vary in its facilities from one implementation to another. A key Android capability is the sharing of functionality. Every application can export functionality for use by other applications in the system, thus promoting straightforward software reuse and a consistent user experience.

Applications

At the top of the Android software stack is the Applications layer. There are a number of supplied standard applications. As mentioned, each application may also expose some of its functionality for use by another application. For example, the message sending capability of the SMS application can be used by another application to send text messages.

The supplied applications are not particularly "special"—all Android applications have the same status in a given system.

Although there are other options, Android applications are commonly implemented in Java utilizing the Dalvik VM. Not only is Dalvik highly efficient, but it also accommodates interoperability which results in application portability. While all of these attributes are attractive, many developers will also want their C/C++ applications to run on an Android-based device.

9.3.3 Application Development

Development Environment

The standard Android development environment from Google is, as you might expect, Eclipse-based, using a plug-in to provide the necessary facilities. You need to define your target configuration by specifying an Android Virtual Device. You can then execute code on either the host-based emulator or a real device, which is normally connected via USB.

This environment only supports Android development on ARM-based target devices.

Programming Model

An Android application consists of a number of resources, which are bundled into an archive called an Android package. Programs are generally written in Java, built using the standard Java tools, and then the output file is processed to generate specific code for the Dalvik VM. Each application runs in its own Linux process—an instantiation of the Dalvik VM—which protects its code and data from other applications. Of course, there are mechanisms for applications to transfer, exchange, and share data.

An application is a set of components which are instantiated and run as required.

There are four types of application components: activities, services, broadcast receivers, and content providers:

- An **Activity** is a functional unit of the application, which may be invoked by another activity or application. It has a user interface of some form. An application may incorporate a number of activities. One activity may be nominated as the default which means it may be directly executed by the user.
- A **Service** is similar to an activity, except it runs in the background without a UI. An example of a service might be a media player that plays music while the user performs other tasks.
- **Broadcast Receivers** simply respond to broadcast messages from other applications or from the system. For example, it may be useful for the application to know when a picture has been taken. This is the kind of event that may result in a broadcast message.
- A **Content Provider** supplies data from one application to other applications on request. Such requests are handled by the methods of the `ContentResolver` class. The data may be stored in the file system, the database, or somewhere else entirely.

When you develop an Android application, you'll need to describe it to the system and this is achieved by means of a manifest file. This is an XML file called `AndroidManifest.xml`, which is stored in the root folder of the application's file system.

This outline example of a manifest file includes the definition of a single activity called `MyActivity`:

```
<?xml version="1.0" encoding="utf-8"?>
<manifest ...>
<application ...>
    <activity android:name="com.example.project.MyActivity"...>
        </activity>

    ...

</application>
</manifest>
```

When an Android application wishes to obtain some functionality from another application or from the system, it can issue an *Intent*. This is an asynchronous message that is used to activate an activity, service, or broadcast receiver. For an activity or service, the specific action and location of data is included.

Although an Intent may include the specific activity required, it can be more generalized and the request resolved by the system. This mechanism is governed by Intent Filters. These filters specify what kind of Intents the activities and services of an application can handle. They are described in the manifest file, thus:

```
<application ...>
    <activity android:name="com.example.project.MyActivity"...>
<intent-filter ...>
<action android:name="android.intent.action.MAIN" />
<category android:name="android.intent.category.LAUNCHER" />
</intent-filter>
    </activity>
    ...
</application>
```

9.3.4 The Android UI

With so many Android-based devices under development, one question needs to be asked: How can one product be properly differentiated from another Android-based device? A highly effective way of achieving this is through user interface (UI) differentiation. One method is to customize the home screen as this allows standard third-party Android applications to run without modification. It is possible to implement these UI changes through hard coding; however, the UI effects that can be achieved will be limited (without detailed knowledge and a lot of hard work).

Fortunately, there are a small number of tools available that allow changes to be made to the look and feel of the UI without touching the code. This means that product variants or new products can be achieved with very little engineering effort.

9.3.5 Extending Android Beyond Mobile

Until very recently, Android deployment has been focused on mobile handsets. This was Google's target market and the available software IP and development tools are designed and configured with this in mind. The potential for Android is enormous in other market areas—anywhere that sophisticated software, including connectivity and a user interface, encapsulates the functionality of a device. Consumer, telecom, automotive, medical, and home applications are all attractive candidates for the deployment of Android. However, there are challenges in moving away from mobile handsets.

To expand into other markets, investment is required in order to optimize and tune the Dalvik VM and the libraries to the selected SoC, develop or integrate drivers and libraries for industry-specific peripheral devices, and to customize the UI for the required market. In addition, it is not uncommon for a large amount of legacy native C/C++ code to exist which needs to be ported to Android. This code needs to be integrated within the Android environment and potentially interface to the Dalvik VM so that developers can make use of this functionality in their Java applications. At the same time it is imperative that modules and their licenses are tracked and managed and that open source compliance is achieved.

This is best achieved through using an experienced Android partner who has the required experience, knowledge, and toolset to ensure that the Android development is successful and that the considerable benefits of using Android are fully realized.

9.3.6 Conclusion

Android is a disruptive technology. It was introduced initially on mobile handsets but has much wider potential today as it moves to handheld video, digital, medical, automotive, and industrial applications. There are challenges in customizing Android both within the traditional handset market and other markets. In addition, there are challenges in using Android in other market.

9.4 Android, MeeGo, and Embedded Linux in Vertical Markets

Although Android is very high profile, there are other open source operating system options that may be employed in a variety of embedded applications. Vlad Buzov and John Lehmann wrote a white paper to consider the options and I have based this article on that piece.

CW

9.4.1 Introduction

Linux®-based systems have been dominant in enterprise networking for more than a decade. They have been the basis of consumer electronics devices dating back to the original TiVo DVR in 1999, and wireless handsets beginning with the NEC phones for DoCoMo in 2003. However, many vertical markets are only now seeing new devices built on open source software. The reasons lie both in the nature of these markets and in the specialized devices themselves, as well as the emergence of advanced pre-configured software stacks like Android and MeeGo.

9.4.2 How Vertical Markets Are Different

Although devices in vertical specialized markets (like medical or transportation) can work side-by-side, share many of the same features, or even compete with consumer products, they usually have more stringent requirements, as well as market-specific functions and features that may be proprietary to the device maker.

By their very nature, many of these devices cannot fail. While the failure of a consumer device, like a dropped call on a cell phone can cause annoyance, and occasionally business or personal issues, it is rarely life-critical. However, if a medical device malfunctions, or the infotainment system in a car causes a sudden distraction to the driver, the device has the potential to trigger a life-threatening event. Safety, security, and reliability are therefore paramount. Security and high availability have always been a priority for Linux-based network systems, but safety can be a challenge with some open source software that may not be designed with these requirements in mind.

Code that meets these stringent requirements, as well as market-specific applications and unique differentiated features, often are found in similar or previous versions of a device. This software needs to be repurposed and reused whenever possible on new and next generation products. The open source development philosophy is evolutionary by design and moves on quickly to work on the "latest and greatest" rather than legacy software no matter how valuable it may be for a particular device.

All of the above requirements require extra cycles and extensive testing before a device can come to market. Longer development cycles are common, although they grow shorter with every new product, but development will always take longer than consumer devices. This is balanced by typically long product lives, both in the length of time a device is sold and in the time it is in use and supported by the manufacturer. Open source software changes dramatically multiple times during these long life cycles, even during the development phase, and is often beyond the control of the device maker.

Finally, vertical markets adhere to standards and certification that can be several orders of magnitude more complex, costly, and time-consuming than on consumer devices. Certification can differ from country to country and pose a particular set of challenges for devices to be sold globally. The development process itself is often part of the certification, which can be a problem when building a solution by integrating and testing existing open source software, rather than starting from scratch or using pre-certified commercial components. In fact, the certification process may take longer than the life cycle of a typical electronics product.

Likewise, mixing proprietary and open source licensing requires knowledge and skill to design the system correctly, choose the appropriate models and terms, and apply proper management to ensure the proprietary code stays proprietary while respecting open source requirements.

9.4.3 The Appeal of Android

In just a couple of years, wireless handsets using Google's Android™ have captured the majority share of the smart-phone market. This market dominance, combined with its royalty-free, open source model and Google's commitment of resources, has made Android an appealing candidate for use in vertical market devices. Companies like Embedded Alley (acquired by Mentor Graphics in 2009) noticed early the potential for Android beyond mobile phones and have been pioneers in helping bring Android to vertical market devices.

Chief among Android's appeal is its fairly mature full-stack integration, updated as often as twice a year by Google. This frequent updating is both an advantage and a problem, since vertical market development cycles often overlap one or more of these releases, and unless the device maker is a member of Google's hand-picked Open Handset Alliance (OHA), it's hard to know what is coming next. Although Android uses open source licensing, it is controlled and developed by Google, not the open source community, and Google alone decides what features will be added or enhanced and when.

In fact, Android is more than an integrated software stack. It is complete development environment with its own ecosystem, including a unique build system, configuration management tools, and a well-defined development process. While these tools are very effective for building an Android-based device, integrating them into a device maker's development flow can be difficult.

With the current release of Android, Google has finally made it easier to integrate legacy C/C++ applications, so it's important to the reuse model in vertical market devices but other challenges still remain. In spite of its frequent releases and feature updating, Android still retains some "hardwired" phone assumptions. For example, Android requires "real" Home and Back buttons, just fine for an Android-based phone, but a real design constraint for in-flight entertainment systems or back-seat in-vehicle infotainment (IVI) consoles where buttons may or may not be needed for other functions, and space is at a premium.

Getting support for a non-phone development project can also be a challenge because Google has thus far only shown interest in handsets (and more recently general-purpose tablets and Google TV). Other companies provide commercial support on several key non-handset Android platforms: ARM, MIPS, and Power Architectures.

9.4.4 The Promise of MeeGo

MeeGo is a much purer open source play. It is developed in the community, building releases by integrating "upstream" open source projects. Device makers can participate in consensus-driven roadmaps and new releases provide few surprises, which can be a very good thing for product development in vertical markets.

MeeGo provides an excellent set of application development tools (QT). The platform development process, however, is less exposed to developers.

Well resourced by the Linux Foundation (which includes Intel who contributed Moblin, its open source environment for Atom processors—and until very recently Nokia, who contributed Maemo and QT), MeeGo is a true open source model, not controlled by a single company.

MeeGo is of particular interest in the automotive IVI market since the GENIVI Alliance, an active group of over a hundred of the top automakers and suppliers in the world, has chosen it as the basis of its open source IVI specifications.

Few MeeGo-based devices have yet to come to market. However, if its sponsors and the community as a whole continue to embrace MeeGo, it has some key advantages over Android for vertical market devices.

9.4.5 The Versatility of Embedded Linux

Of course, a device maker could chose to build the entire open source stack themselves. (This had been the only choice before Android and MeeGo appeared, and was an important reason why both were developed.) The unlimited choice and lack of commercial control is a double-edged sword, since the development team must find, evaluate and test the options, then do all the integration and customization themselves, which can add significant time and cost to product development. Further, open source "in the wild" rarely has a clear roadmap or timetable.

9.4.6 Conclusion

While vertical market devices are a good fit for the advantages of open source software in general, their exacting requirements, and long development and product life cycles can gain a significant advantage by using the integrated software stacks of Android or MeeGo. While technical and business challenges may need to be overcome, both can save significant time-to-market and help build better devices more efficiently.

Multicore Embedded Systems

Chapter Outline

It is becoming apparent that, before long, using multiple CPUs—multiple cores on a chip, multiple chips on a board, multiple boards in a system or any combination of these—will soon be the norm, and single core design will become less common. With that in mind, a chapter gathering together some articles that address this topic from various angles seemed logical.

CW

10.1 Introduction to Multicore

To start with, a short article setting the scene with the basics of multicore, including some comments on a key motivation for such design: power consumption. This article draws content from a white paper that I jointly authored with Russell Klein and Shabtay Matalon in 2011.

CW

10.1.1 System Architecture

An increasing number of embedded systems are being implemented with multiple CPUs. This may be multiple CPU cores on a chip, or multiple chips on a board, or multiple boards in a system, or any combination of these. Although the term "multicore" is usually applied to the multiple cores on a chip situation, the other architectures exhibit similar characteristics from the software developer's perspective.

A multicore device or system may consist of a number of identical cores (homogeneous multicore) or cores of multiple architectures (heterogeneous multicore). In either case, new challenges are presented to the software developer.

In a homogeneous multicore system, there may be the possibility to use an operating system that supports SMP (Symmetrical Multi-Processing), where a single instance of the OS runs

Embedded Software: The Works. DOI: 10.1016/B978-0-12-415822-1.00010-6
365

across all the cores and manages distribution of work between them. A well-designed multi-threaded application ports readily to an SMP environment. Operating systems that support SMP include Linux and a number of real-time operating systems.

Otherwise, as system is AMP (Asymmetrical Multi-Processing) and a separate OS instance runs on each processor (or a CPU may have no OS). Each OS may be selected to be optimal for the function of a particular core, depending on its application. A standardized inter-core communication mechanism needs to be adopted and MCAPI (Multicore Communications API) provides a viable option for many applications.

10.1.2 Power

There may be a number of motivations for pursuing a multicore design strategy, but a key one is optimization of power consumption.

Multicore system offers the potential of a significantly reduced power consumption for a processor-based system. It may not be obvious, but smaller simpler processors are far more power efficient than larger more complex ones. This is true even if the complex processor have sophisticated power management facilities which are used effectively.

As an example, consider the ARM Cortex A9 when compared with the ARM Cortex R4. When implemented on a 65 nm process, the Cortex A9 delivers 2075 DMIPS and has a power efficiency of 5.2 DMIPS/mW and runs at 830 MHz (see http://www.arm.com/products/processors/cortex-a/cortex-a9.php performance tab—Single core 65 nm, optimized for performance). While the ARM Cortex R4 implemented on the same process delivers 1030 DMIPS and has a power efficiency of 13.8 DMIPS/mW and runs at 620 MHz (see http://www.arm.com/products/processors/cortex-r/cortex-r4.php performance tab, 65 nm process, optimized for performance).

Recall that Watts are Joules/Second and MIPS is Millions of Instructions/Second. Dimensional analysis allows us to multiply the numerator and denominator by Seconds and get "Millions of Instructions/Joule". Inverting this we get Joules/Instruction—or the amount of energy needed to accomplish a given workload on the core. For the A9 we get 1/5.2 or 0.192 MilliJoules/Million Instructions, and for the R4 we get 1/13.8 or 0.072 MilliJoules/Million Instructions—or a difference of a factor of 2.65. Thus you will burn almost 3 times the energy implementing an algorithm on an A9 as compared with the R4. This is energy and does not include any time aspects. If we look at power (Watts/Second) you need to include the speed of the processor. Looking at the power differential between these processors, the A9 burns 400 mW, while the R4 burns a mere 75 mW. That is 5.3 times more power, but just twice the compute throughput. Two R4s will deliver the same compute capability while consuming only 37% of the power. And the story gets more compelling as the processors get smaller and simpler.

10.1.3 Challenges

The challenge, of course, is to enable algorithms to take advantage of the two cores. But the benefit of going to multiple smaller cores, in terms of power consumption, is significant.

10.2 Multiple Cores: Multiple Operating Systems

For many multicore designs, it is attractive to use an AMP architecture, which opens up the opportunity to select different operating systems for different cores. This raises a number of interesting issues. This article is based on a 2009 white paper by my colleagues Dan Driscoll and Stephen Olsen.

CW

Multicore systems have been around for some time. Today many of these systems are expected to deliver exceptional performance and power efficiency while managing the ever-increasing functionality demanded by consumers across the broad range of market segments such as mobile phones, portable medical devices, and consumer electronics.

Multicore processors and the software that runs on these cores can be categorized in a variety of ways. In one type of configuration, the cores (or processing units) in a device can be homogeneous, which means each core has the same instruction-set architecture and view of the system. In another configuration, the heterogeneous design, the cores have different instruction-set architectures or views of the system. The software running across these multiple cores can be categorized as either symmetric (a single instance of the software) or asymmetric (different software or multiple instances of software running on each core).

Symmetric Multi-Processing (SMP) is used for load distribution across homogenous cores with shared resources. SMP consideration often includes compatible scheduling mechanisms, thread/interrupt affinity spinlocks, and virtualization. Unlike SMP, Asymmetric Multi-Processing (AMP) systems often use load partitioning with multiple instantiation of the same or different operating systems running on either homogenous or heterogeneous cores. With the use of an optimized Inter-Processor Communication (IPC) framework, AMP can implement common middleware and Graphical User Interface (GUI) services while providing separation between deterministic real-time components from the rest of the system. There are several examples of such tightly coupled heterogeneous AMP architectures such as Texas Instrument's OMAP and DaVinci Platforms or the MIPS1004K processor, where a specialized companion chip has been integrated with a general-purpose processor in an SoC.

In fact, TI's OMAP is a good example of the multicore evolution—whereas the number of cores or units of execution have increased in every new iteration to the point where today, Texas Instruments is adding more general-purpose cores as evidenced by the OMAP 4 (2 X Cortex A9 MPCores).

10.2.1 SMP Hardware for AMP

SMP hardware allows for a single instance of an OS to manage the resources across all SMP cores, the SMP hardware can also be divided to allow several operating systems to run on different cores. Even though the memory is shared, it can be divided between the OSes running in the system—allowing each OS to run independently. This yields an AMP environment on SMP hardware. The advantages can be realized even with the same GPOS or RTOS on different cores, to provide either to support legacy applications or a layer of security between OS domains.

For example, a two core SMP system can be divided between two OS instances with one core assigned to each environment. Since they both share the same memory, each core is dedicated to use a predefined portion of the memory as well as peripherals divided between the two operating systems. One core/OS can be dedicated specifically to security applications, where the only method to communicate between the two OSes is via an Inter-Processor Communication (IPC) mechanism. Further, the MMU prevents the one core/OS from accessing the memory owned by the other core.

While the concept of AMP has not changed over the years, what has changed—and rather quickly—is the complexity of hardware and drivers that employ AMP. An AMP design, for example, presents unique challenges to operating systems which traditionally assume homogeneous hardware. Generally, the Real-Time Operating System (RTOS) is responsible for delivering real-time aspects of an AMP system such as low-latency performance, determinism, and minimal footprint; however, it must also coexist with a General-Purpose OS (GPOS) which operates within a different set of parameters to manage many of the user applications.

10.2.2 AMP Hardware Architecture

Modern day homogeneous multicore SoCs are more complex and have many more capabilities than most older multicore designs. In the past, multicore systems were implemented on customized ASICs or FPGAs using Intellectual Property (IP) better suited to unicore environments. This IP did not have many capabilities built specifically for multicore environments, and the additional logic to support these environments was provided by the ASIC or FPGA developer, resulting in a custom hardware solution for each design.

New multicore IP such as the ARM Cortex-A9 MPCore has specific logic and capabilities to support both symmetric and asymmetric environments with vastly improved efficiency. These modern designs have multiple levels of cache that can be unified via hardware cache controllers. They also include advanced power management capabilities to maximize power and performance across multiple execution units, interrupt controllers that are focused on interrupt routing to multiple cores, and specific configuration settings to support both AMP and SMP environments.

The complexity of these designs extends beyond the cores themselves—the surrounding environment continues to evolve as well. Mesh fabrics and interconnects are replacing standard bus topologies, hardware acceleration units continue to be added for off-loading video, audio, etc, and the number of on-chip peripherals continues to grow as well. It is not uncommon to see SoC designs containing IP blocks for networking, USB, encryption/decryption, multiple busses (SPI, I2C, PCI-e), LCDs/touch panels, SD/MMC, UARTs, timers, etc. on a single chip with multiple cores.

All of these improvements have a compounding effect. New multicore IP coupled with new architecture improvements (ARMv7 architecture) and significant SoC enhancements result in complex multicore hardware designs that are well suited for both AMP and SMP environments.

10.2.3 AMP Software Architecture

As mentioned earlier, embedded devices using AMP designs are no secret, but there are some prevalent design patterns for such systems that are worth discussing. Figure 10.1 shows a typical, high-level software architecture of an AMP system.

This design is found in many types of embedded devices, crossing numerous vertical markets (handsets, consumer electronics, medical, industrial control, etc.). It includes both a GPOS and an RTOS in which each serves a different purpose. The work of the system is partitioned to allow the GPOS to handle control plane activities (user interface, application management, complex protocols, etc.), while the RTOS handles data plane activities that require hard real-time tasks (motor control, sensor input, data computation, time-sensitive processing, encryption/decryption, etc.). For these reasons, Linux and more modern incarnates of Linux (such as Android) are increasingly becoming the OS of choice for the GPOS, while an RTOS is often used to do time-sensitive, deterministic, or computationally intensive operations.

Additionally, it should be noted that an RTOS can be a better candidate to run certain middleware requiring certification. The main reason for this is that with the typical RTOS

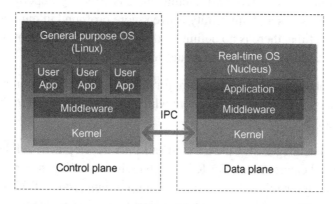

Figure 10.1
A typical, high-level software architecture of an AMP system

there is often much less code to certify as compared to a GPOS. Considering the extremely high price per line-of-code to certify software, this decision can greatly reduce the overall system cost.

The AMP software design pattern described above will continue to be used in AMP designs in the foreseeable future. The main areas changing in these designs are in the complexity of operations on both the GPOS and RTOS—both sides of the system have more advanced applications, middleware, etc. running on them. The increase in intricate software is directly correlated to the increase in complexity of newer SoCs and the ever-increasing number of features these modern embedded devices support.

10.2.4 The Importance of Inter-Process Communication

Operating systems communicate via some method of Inter-Process Communication (IPC) that, in the past, varied from design to design. There are different open standards for IPC today, with the more prevalent standards being Transparent Inter-Process Communication (TIPC) and Multicore Communication API (MCAPI). TIPC is a completely transparent and heavyweight IPC protocol supported by Linux and other operating systems, while MCAPI is a newer, lightweight message-passing API for closely distributed systems (SoCs). Adherence to such open standards is critical to scaling AMP systems and allowing easier porting of legacy software to these systems. Of note, having optimized IPC mechanism between the GPOS and the RTOS can be a critical factor to the AMP system's viability. Therefore, an open standards based, highly optimized IPC mechanism is very important to AMP designs.

10.2.5 AMP Development Tools

A complex software design using a single operating system running on a single core can often fail to be completed on time due to lack of good development tools. An integrated development environment suited to AMP architectures is critically important to solving complex system level issues and this reality is even more profound in multicore systems. Unfortunately, at this time, there is no definitive development environment, which makes the development process somewhat challenging at times.

When it comes to debugging an AMP system, for example, there are a number of challenges to consider. First, the tools and environment in which the tools run is often different for GPOS and the RTOS portions of the system.

GDB is the standard debugger for Linux and most Linux development will be performed within a Linux-hosted environment. For RTOS debugging, usually a proprietary debugger is utilized though many reputable software vendors support numerous RTOSes. Bottom line— the debugging environment will often contain at least two different debuggers potentially running in two different host environments (Windows and Linux).

To further complicate matters, there are different types of debugging for the different OS environments. For Linux, user applications are typically debugged using GDB and a remote-target connection to a GDB server. JTAG debugging is more common for RTOS debugging, though agent-based debugging can be used as well. In essence, to accomplish all of the types of debugging necessary for this environment, multiple connections and support for each will need to exist on the target and in the debugger.

Finally, the profiling tools used by a GPOS and a typical RTOS have no relationship between the two. Correlating data gathered from Linux profiling and an RTOS profiling session will be difficult and tools to facilitate this effort are hard to find, especially if debugging on different platforms and host operating systems.

Having a unified tools solution for both operating systems, in an AMP design, and an engineering team that can effectively use these tools will become more and more important as the device sophistication continues to increase. Embedded tools vendors are working to fill this gap.

10.2.6 Difficulties

While AMP designs have been around for a long time, the difficulty of integrating multiple operating systems onto a new SoC and platform continues to be a burdensome effort. Figure 10.2 illustrates a few of the design decisions involved when integrating the GPOS and RTOS into a system.

While Figure 10.2 shows only the high-level partitioning of the resources between the GPOS and the RTOS, there are many other areas that must be considered in this integration effort. For instance, the booting strategy for the device must be determined—who boots first, the GPOS or RTOS? Is a boot loader used (none, u-boot, etc.)? How will memory be partitioned and protected between the different operating systems? How can the system be optimized or fine-tuned for the specific platform? These are just a few of the many hard questions that must be answered just to integrate the multiple operating systems on to a single AMP SoC.

Unfortunately, time spent integrating and optimizing this environment detracts from solving the real problem—developing the applications that need to run on the GPOS and RTOS to realize the features of the device being created. This integration effort continues to grow with the ever-increasing feature sets of these designs and with the fast-paced advances in capabilities of the SoCs they execute upon.

Having the ability to purchase such tightly integrated GPOS and RTOS solutions and the development services that can support and optimize these solutions is becoming increasingly important. Access to these solutions and services will allow development teams to focus solely on the problem space being addressed by their device—this problem space find its answer at the application level and consequently, the underlying operating systems and the

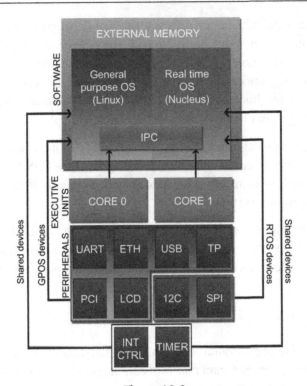

Figure 10.2
A few of the design decisions involved when integrating the GPOS and RTOS into a system

underlying communication strategy should not be a burden to the application development efforts.

10.2.7 AMP Use Cases

As streaming multimedia wireless communications and compelling user interface technologies become more prevalent in today's electronic devices, there is an ever-growing demand for system software to accommodate the above-mentioned functionality. Efficient system design in an AMP environment often relies on the system designer's ability to partition the system resources between the RTOS and the GPOS and optimize their mapping to the underlying hardware components. System requirements along with partitioning and mapping of resources usually vary from one system to another based on variables such as need for peripherals, subsystems and their timing requirements, connectivity I/Os, etc.

To illustrate this complexity, a patient monitoring system might incorporate services for data collection, data analysis, storage and display, and communication of data—along with alerts across a secure wireless network. Multiple biosensors are used for collecting data for a patient's vital signs such as respiration, blood pressure, body temperature, and heart rate. Timing requirements are an important consideration when designing such a system. Collection of ECG waveform data requires sampling synchronized with the patient's heart beat within a few millisecond window and thus necessitates a low-latency, highly deterministic RTOS. On the other hand, human directed functions, such as printing, linking to hospital database, and display controls can be off-loaded to a GPOS.

The mobile phone is another example where AMP offers a compelling value for embedded developers. Consider the smart phone, a multifunctional device with a myriad of applications ranging from multimedia streaming and gaming, to placing and receiving voice calls. Many of the user applications can be off-loaded to a General-Purpose Processor (GPP) running a GPOS, whereas the wireless baseband subsystem requires a low overhead, real-time operating system that is often integrated with a multiband protocol stack with power optimization to deliver QoS while maintaining an "always-on" air link.

Applications requiring multimedia processing often entail computationally intensive algorithms such as data compression and decompression, image processing, image stabilization, manipulation, and resolution enhancements. As referenced earlier in this chapter, many of today's AMP SoCs incorporate companion accelerators for performing intense data computation tasks and are often managed or controlled by the RTOS. Using a multicore processor architecture is one way to separate divergent system requirements with a design that is also power optimized for battery-operated handheld devices.

10.2.8 The Use of a Hypervisor

In addition to traditional SMP and AMP designs, the use of a hypervisor can allow the system resourced to be shared in the time domain. Such a design could share a set of cores in time between an RTOS and a GPOS. The RTOS, with its system latency restraints, will still need to be met, but by time slicing the available time between both operating systems, the design has access to all system resources. This can be achieved on either a single core or multiple cores. The hypervisor is part of the IPC mechanism and is involved in protecting access from one core to the other. This allows the system to be divided in space, as well as time. Each OS has the impression that it owns the entire system, but in fact, it does not. The hypervisor is in control of how much time is dedicated to each OS instance.

The hypervisor not only virtualizes the CPU cores, but it also can virtualize the devices below the OS. For instance, there could be only a single networking interface, but both cores need to share the resource. The hypervisor could either pass initial control to one OS instance and, when it is decided the data is destined for a different OS, pass it along using the IPC mechanism. Alternatively, the hypervisor could handle the initial networking packet and decide which OS it is destined for, and interrupt the OS associated with the packet thus minimizing system latency.

10.2.9 Conclusions

As multicore adoption continues to grow, so too will multi-OS systems and their applicability. Initially, multi-OS may carry a cost in added design and integration complexity arising from partitioning and distribution of system resources, and in dealing with the challenges posed by development and deployment of these devices. However, in a great many situations, the benefits will soon outweigh these initial hassles.

The use of a hypervisor, as discussed, may be a necessary addition to a multicore design to provide enough processing power to the right applications while meeting some real-time system constraints. The additional complexity of adding a hypervisor has some advantages that cannot be achieved otherwise. The system designer must weigh the needs of the applications and their processing capabilities against dedicating system resources to a particular OS.

There is no single design approach that delivers an optimal solution for every multicore multi-OS use case, so it is important for developers to realize that such systems are inherently different from the single core environments. Utilizing commercially available solutions and services ensures their application is optimized to fully leverage either the SMP or AMP environment. In many cases, developers may consider seeking value-added design services to leverage an outside source with system expertise to manage the design complexity and meet their time-to-market constraints. It further ensures that device manufacturers are focused on incorporating differentiated features in their device rather than spending valuable time on non-differentiated components.

10.3 Selecting Multiple Operating Systems for Multiple Cores

Continuing the theme of the previous article, this one focuses on the selection of operating systems. During 2010/11 I wrote a number of pieces and did a Web seminar on OS selection. This article is based on that material and is expanded to accommodate multicore.

CW

10.3.1 Introduction

The increasing deployment of multicore embedded system designs presents some interesting challenges to software engineers. The usual selection process for an operating system becomes considerably more complex. The selection process is somewhat influenced by the hardware architecture, but is primarily driven by the application of the device being implemented.

If the design is homogenous multicore—i.e., all the cores are identical—a symmetrical multiprocessing (SMP) approach may be appropriate. A single instance of an OS runs over all the cores in the system. This makes sense if the multicore architecture was chosen in order to offer raw processing power for a compute-intensive application. Choice of OS would then be from products that offer an SMP option, which includes Linux and a number of real-time operating systems.

If the multicore design approach was chosen in order to use specific CPUs for specific purposes, SMP would not be appropriate—an asymmetrical multiprocessing (AMP) approach, where each CPU has its own, local OS, is the best option. The OS for each CPU needs to be selected in a similar way to that used for single core systems.

10.3.2 Types of OS

Broadly, available operating systems for embedded applications may be categorized as real time or not—the latter category are mostly derived from desktop operating systems. The most popular option for non-real time is Linux (or Android, which actually incorporates Linux).

Real-Time Operating Systems
A real-time operating system (RTOS) is designed from the ground up for hard real-time applications, where a predictable response is the key requirement. Most RTOSes are compact and efficient, so make minimal impact on memory footprint and CPU utilization.

Linux
Linux is not designed to be real time. It has the benefit of a very wide user base and a large range of available middleware, which can make it attractive for many applications.

10.3.3 OS Selection

Selecting an OS for each core, like the OS selection for single core applications, is driven by a number of specific criteria, which may be identified by answering a series of questions:

Is the application real time?

A real-time system is not necessarily fast, but responds to external events in a predictable and reproducible way. An OS that exhibits a high degree of determinism is termed a real-time operating system. Determinism can vary. A "hard real-time" system is one where there are very tight time criteria; "soft real time" is more relaxed.

For a hard real-time application, a true RTOS is really the only option. For a soft real-time application, an RTOS may still be used, but Linux (running on a fast enough CPU) may be acceptable. There are real-time extensions to Linux available, but there are doubts about their usefulness, as they can render much middleware unusable.

Is the memory size constrained?

All embedded devices have some kind of limitations on memory size, but can still be quite large. However, unless there are many megabytes of RAM, OSes like Linux are not an option.

How much CPU power is available?

If the available microprocessor/microcontroller/core is only just fast enough for the application, the additional overhead of a heavyweight OS may be a problem. An RTOS is likely to make much more efficient use of the available CPU power.

Is device power consumption a concern?

Power consumption is almost always a design parameter in modern devices, particularly for portable, battery-driven equipment. Software has a significant bearing on the power efficiency of a device. Factors like memory and CPU usage efficiency are important, but also an OS may include power management facilities built in.

Does the design include unique or obscure peripherals?

Although many RTOS products offer very large ranges of middleware and drivers, Linux is likely to provide even more. So, for an obscure device, there may well be a Linux driver. For custom peripheral hardware, a driver would need to be written. This is a specialized activity. There is a plethora of Linux driver writing expertise available.

Care should be taken with the acquisition of middleware. In particular, networking protocols should be fully validated.

Does the design feature a memory management unit (MMU)?

If there is no MMU, the heavyweight OSes are unlikely to be an option; most RTOSes are fine without an MMU. If there is an MMU, Linux becomes a possibility. May RTOSes can take some advantage of an MMU, if one is available.

Does the application need high security?

Specifically, is it necessary for tasks to be protected from one another?

If so, a process model OS is ideal, as each task can be a process; the OS uses the MMU to provide protection. This has a significant CPU time overhead and is offered by the heavyweights and a few RTOSes.

Another option, supported by some RTOS products, is the implementation of "thread protected mode" using an MMU. This has a lower overhead than process model, but most of the security advantages.

Does the application require certification?

Some devices need to pass specific safety and quality standards and receive certification before they can be sold. Notably, this is the case for aerospace and medical applications. Such a certification process is expensive and requires access to all the source code (including the OS). The cost is significantly affected by the volume of source code to be analyzed, so a compact OS is advantageous.

In general, only a whole application can be certified. So, an OS cannot be certified by itself. However, selecting an OS which has a track record of successful certifications makes sense.

Does the device need interoperability with enterprise systems?

In this context, the Microsoft products tend to win out.

What is the end cost and shipping volume of the device?

Any OS has some costs associated with it and these have an influence on the development and manufacturing costs of a device. A royalty bearing OS has a modest upfront cost and a charge for each device shipped, which may change according to volume. This is clearly good for low-volume applications. Of course, support costs may be ongoing. A royalty-free OS may have a slightly larger initial cost, but no ongoing licence fees. Open source OSes sound as if they are "free," but, in reality, there are support costs which may be quite significant.

What past experience of OSes is available in the team?

Acquiring expertise is expensive, so leveraging existing knowledge and expertise is vital. An example is the API used by Linux: POSIX. There is a very large body of experience with POSIX, which is also supported by many of the available RTOS products on the market.

Naturally the availability, quality, and cost of documentation and support are critical factors to consider when selecting an OS.

10.3.4 Multicore Systems

In order to look at the selection of multiple operating systems in a multicore system, it is useful to introduce some terminology, which may be familiar to developers of networking devices, but can be applied more widely: the idea of the control plane and the data plane.

Essentially, the control plane/data plane separation is applicable to any device having both real-time requirements and non-real-time requirements. An RTOS is much better suited to handle deterministic, data-process-intensive operations than an OS like Linux. On the other hand, Linux is preferred for handling a user interface, control, and management functions.

A broad categorization can be defined of the functionality handled by one or more CPUs on the data plane compared with that performed by CPUs on the control plane.

Inter-Core Communications

In a system with multiple cores, it is inevitable that communication between the cores will be a necessity. If the cores are running different operating systems, this may be a challenge. The emerging MCAPI (MultiCore Applications Program Interface) offers a realistic solution, as implementations are available for numerous OSes. MCAPI is covered in more detail in a later article in this chapter.

Examples

There are numerous examples of embedded applications which may be implemented using multiple cores and where the use of multiple operating systems may be advantageous. Two possibilities:

- A high-performance network laser printer. Such a device has needs for real-time performance in handling paper, etc. which may be readily accommodated by a core running an RTOS. A user interface does not need real-time behavior, so Linux may be a good option to support a modern GUI. Support for a wide range of networking protocols is likely to be necessary; again Linux may be a good choice, as it has an unrivaled array of middleware available.
- A sophisticated medical instrument. Like a printer, such a device is likely to need a carefully designed graphical UI, which may be supported by Linux. It is likely to need networking, but its needs are modest, so either Linux or an RTOS may be suitable. For the core functionality of the instrument, an RTOS is likely to be a good choice, as, along with real-time capabilities, the software is likely to need certification. The modest size of an RTOS makes certification more economic.

10.3.5 Conclusions

With an increasing number of embedded systems being implemented using multiple cores, the software developer has the opportunity to select the optimal operating system for each core. The selection of the OSes is a challenge, but broadly some combination of Linux and RTOSes covers most situations. MCAPI provides a good approach to inter-core communications.

10.4 CPU to CPU Communication: MCAPI

A particular challenge in AMP multicore systems is communication between cores. An emerging standard, MCAPI, addresses this issue. In 2010, I wrote a white paper on the topic and this article is based on that piece.

CW

10.4.1 Introduction

Embedded designs incorporating several CPUs—either multiple cores on a chip, multiple chips on a board, or a combination of the two—are becoming increasingly common. It's no secret that software development for multicore systems has its challenges, but fortunately, the industry is working toward addressing these key issues and challenges.

The primary focus of this paper is on the emerging standard from the Multicore Association, the Multicore Communications API, or "MCAPI." MCAPI provides a rational way for software developers to implement communication between cores, even between multiple disparate operating systems.

10.4.2 Multicore

It is becoming increasingly common for designers to choose a multicore approach to embedded system design. Indeed, there are indications that this approach will be the norm in just a year or two. By the term "multicore," we generally mean multiple CPU cores on a chip, but the same ideas really apply to multiple chips on a board or any combination of the two.

AMP and SMP

Although it is a simplification, there are broadly two multicore architectures: symmetrical and asymmetrical multiprocessing, SMP and AMP, respectively. It is with AMP designs, where each CPU is running its own OS (or is running none at all), that inter-COPU communications is a challenge.

Multicore Processors

A wide variety of multicore processors exist right now. Many of these are relatively new or very new, but processors like the OMAP family from Texas Instruments has been around for a while. Actually, the OMAP is a good example of multicore evolution—the number of cores or units of execution has increased in every iteration and the increase was first in specialized cores (adding graphics processors or accelerators), and now TI is adding more general-purpose cores in the OMAP 4—2 × Cortex-A9 MPCores to be precise.

Freescale's QorIQ includes 2, 4, or 8 PowerPC cores, which is an example of a very up-to-date homogeneous multicore device with a clear evolutionary path for the designer.

10.4.3 The MCAPI

MCAPI is an applications program interface (API) for passing data between cores in an AMP system. It was first published by the Multicore Association (MCA) in 2008.

The key objective is to provide source code compatibility across multiple operating systems, while providing the flexibility to tune for performance or footprint in true embedded software fashion.

What MCAPI Is Not

It's important to note, however, that MCAPI is **not** a protocol specification. The protocol is, by definition, a type of implementation. Interoperability between different vendors' MCAPI implementations was never an intended part of this API. However, it is very much intended that all application source code would be totally portable between different MCAPI implementations.

The MCAPI Design

The design of MCAPI follows a clear set of philosophical principles. The plan was for a small API, which would be suitable for a number of multicore architectures, and would be straightforward to understand and apply. In good embedded software tradition, low-memory and high-performance implementations must be possible. Also, other inter-CPU communications methods are not excluded or locked out.

MCAPI is also an opportunity for silicon vendors to optimize hardware for inter-CPU communication and even implement the API in hardware. MCAPI can run on an OS, a hypervisor, or with no OS at all and it plays well with hardware acceleration. A design with multiple core architectures and different operating systems is fine and no assumptions are made about memory system design.

The key value of MCAPI results from this philosophical starting point.

A user does not need to know anything about the operating system or any details of the connection between the CPUs, as the API is quite generic. This results in readily portable application code. A typical MCAPI implementation exhibits low latency and high performance, supports multiple channels and facilitates prioritization of communications. Compared with a "home brew" solution, a key measurable benefit of MCAPI is faster time to market.

MCAPI Concepts

Some of the key MCAPI concepts may feel quite familiar, if you know about networking. The first is a **domain**, which may be likened to a subnet in networking terms. A system, which uses MCAPI, includes one or more domains and each domain includes a number of nodes. Each node can only belong to one domain, so there is a true hierarchy. The domain concept is used by other MCA APIs, so it should be applied consistently.

A **node** is an abstract concept, but may broadly be thought of as a stream of code execution. So it might be a process, thread, task, core, or one of a number of other possibilities. The exact nature of a node is specified for a given MCAPI implementation. A node is initialized by a call to the `mcapi_initialize()` API. This may only be called once for a given node. To reinitialize a node, you need to call `mcapi_finalize()` and then call `mcapi_initialize()` again.

An **endpoint** is a destination to which messages may be sent or to which a connection may be established—rather like a socket in networking. Each endpoint has a unique identifier, which is the tuple <domain ID, node ID, port ID>, and a node may have multiple endpoints. The creating node receives from an endpoint, but any node may send to one.

Data Transfer

MCAPI offers a lot flexibility for the transfer of data. Broadly speaking, there are three options:

- **Messages** are datagrams—chunks of data—sent from one endpoint to another. No connection needs to be established to send a message. This is the most flexible form of communication, like UDP in networking, where senders and receivers may be changing along with priorities.

- A **packet channel** is a first-in/first-out unidirectional stream of data packets of variable size, sent from one endpoint to another, after a connection has been established.
- A **scalar channel** is similar to a packet channel, but processes single words of data, where a word may be 8, 16, 32, or 64 bits of data.

Connection-based communication potentially removes the message header and route discovery overhead and is, thus, more efficient for larger volumes of data.

API Call Example

MCAPI is fairly small, with calls in five categories addressing the key functionality required for node-to-node communication. Although there are not a great many functions in MCAPI, it would be redundant to cover the whole API here, as that information is freely available in the MCAPI specification, which is available from the Multicore Association web site.

However, to give a flavor of the calls, here is one example.

```
void mcapi_msg_send(
    MCAPI_IN mcapi_endpoint_t send_endpoint,
    MCAPI_IN mcapi_endpoint_t receive_endpoint,
    MCAPI_IN void* buffer,
    MCAPI_IN size_t buffer_size,
    MCAPI_IN mcapi_priority_t priority,
    MCAPI_OUT mcapi_status_t* mcapi_status
);
```

This call sends a message from one endpoint to another. It is a blocking call and only returns when the buffer may be reused by the application. A non-blocking variant is also available.

The endpoints are identified by the parameters `send_endpoint` and `receive_endpoint`. The memory, in which the application has placed the message, is referenced by the pointer buffer, the size of which (in bytes) is specified by `buffer_size`. The priority of the message is indicated by the parameter priority. A value of zero commonly represents the highest priority level, but this is implementation specific. The pointer `mcapi_status` points to a variable where the result of the call will be placed.

On return from this API call, the variable referenced by `mcapi_status` will normally contain the value `MCAPI_SUCCESS`. But a number of error conditions are possible, which are described in the MCAPI specification.

MCAPI Implementation

Many aspects of MCAPI are somewhat implementation dependent. So, if you are considering a commercial MCAPI product, there will be questions that you will need to ask of the vendor.

To understand the issues, some insight into the details of the Mentor Embedded MCAPI implementation are included here. This implementation for Linux has been declared open source and called OpenMCAPI—it is available from OpenMCAPI.org for free download.

A number of high-level design decisions were made.

First, it was decided to avoid the use of dynamic memory allocation. This is commonly good practice for embedded applications. The implementation has a minimal dependence upon the underlying operating system, with which it communicates via an abstraction layer. This improves portability.

The implementation features static configuration—the entire topology is determined at compile time—as required by the MCAPI specification. A management service is included to register and find the endpoint associated with a service; this allows a more flexible topology for sending and receiving data.

Shared memory initialization must be performed by the system or by application code, and the master/slave relationship between nodes is established at runtime as a result of boot order.

As a number of API calls are blocking, a thread suspension mechanism is required. For Nucleus, an event flag was used; for Linux, a condition variable is employed.

The transport layer is implemented as an array of buffers in shared memory, which is managed by a ring queue for each node and an allocation bitmap. The platform-specific layer is designed for maximum portability.

There is an issue with address space. A thread model RTOS like Nucleus uses physical addresses, but Linux, being process oriented, remaps memory into virtual addresses. To avoid problems, the MCAPI implementation works in terms of offsets within the shared memory area.

10.4.4 Conclusion

Although the implementation of software for multicore designs remains challenging, the MCAPI specification offers a good solution to the key issue of inter-CPU communications and is an increasingly popular standard.

Afterword

Great Expectations

A little story to close with. It might make you think ...

CW

Unless it is now 2025 and you have just found this book in a remainders bookshop, it may be assumed that you are a child of the twentieth century—you were born in nineteen-something-or-other. As such, you have perceptions and expectations about the world around you in general and embedded systems in particular. But take a moment to consider what citizens of the twenty-first century might expect.

Meet Archie.

At the time this photo was taken, he had just had his third birthday, so he is very much a citizen of the twenty-first century. Archie is interested in embedded systems. For example, he loves his photo being taken. Just produce a camera and he is ready—posing for a snap. And, as you can see, he makes a good picture.

As a citizen of the twenty-first century however, he has expectations. Having had his picture taken, he wants to see results—*right now*. He expects to see the picture on the LCD on the back of the camera.

If you were to use an old-fashioned film camera and say "It's OK Archie, we'll get the pictures back tomorrow," he would be unimpressed. Unimpressed 3-year-olds are no fun to be with …

So, when you are designing your next embedded system, think about the expectations of your users.

And think about Archie.

Index

This index is designed to lead you to the article(s) that cover the topic you have looked up. If the page number is **bold**, it indicates that this is a key topic of the article. In most cases, the page references refer you to the first occurrence of a term in a given article

Printed in the United States
By Bookmasters